27
Topics in Heterocyclic Chemistry

Series Editor: Bert U. W. Maes

Topics in Heterocyclic Chemistry

Series Editor: Bert U.W. Maes

Recently Published and Forthcoming Volumes

Halogenated Heterocycles
Volume Editor: J. Iskra
Volume 27, 2012

Heterocyclic Scaffolds II: Reactions and Applications of Indoles
Volume Editor: G.W. Gribble
Volume 26, 2010

Synthesis of Heterocycles via Multicomponent Reactions II
Volume Editors: R.V.A. Orru, E. Ruijter
Volume 25, 2010

Anion Recognition in Supramolecular Chemistry
Volume Editors: P.A. Gale, W. Dehaen
Volume 24, 2010

Synthesis of Heterocycles via Multicomponent Reactions I
Volume Editors: R.V.A. Orru, E. Ruijter
Volume 23, 2010

Heterocyclic Scaffolds I: β-Lactams
Volume Editor: B. Banik
Volume 22, 2010

Phosphorous Heterocycles II
Volume Editor: R.K. Bansal
Volume 21, 2009

Phosphorous Heterocycles I
Volume Editor: R.K. Bansal
Volume 20, 2009

Aromaticity in Heterocyclic Compounds
Volume Editors: T. Krygowski, M. Cyrański
Volume 19, 2009

Heterocyclic Supramolecules I
Volume Editor: K. Matsumoto
Volume 17, 2008

Bioactive Heterocycles VI
Flavonoids and Anthocyanins in Plants, and Latest Bioactive Heterocycles I
Volume Editor: N. Motohashi
Volume 15, 2008

Heterocyclic Polymethine Dyes
Synthesis, Properties and Applications
Volume Editor: L. Strekowski
Volume 14, 2008

Synthesis of Heterocycles via Cycloadditions II
Volume Editor: A. Hassner
Volume 13, 2008

Synthesis of Heterocycles via Cycloadditions I
Volume Editor: A. Hassner
Volume 12, 2008

Bioactive Heterocycles V
Volume Editor: M.T.H. Khan
Volume 11, 2007

Bioactive Heterocycles IV
Volume Editor: M.T.H. Khan
Volume 10, 2007

Bioactive Heterocycles III
Volume Editor: M.T.H. Khan
Volume 9, 2007

Bioactive Heterocycles II
Volume Editor: S. Eguchi
Volume 8, 2007

Heterocycles from Carbohydrate Precursors
Volume Editor: E.S.H. ElAshry
Volume 7, 2007

Bioactive Heterocycles I
Volume Editor: S. Eguchi
Volume 6, 2006

Marine Natural Products
Volume Editor: H. Kiyota
Volume 5, 2006

Halogenated Heterocycles

Synthesis, Application and Environment

Volume Editor:
J. Iskra

With contributions by

A. Decker · R. Dembinski · A.A. Gakh · D. Gundapuneni ·
J. Hayakawa · E. Heath · J. Iskra · T. Kosjek · Y. Li ·
S. Minakata · S. Pazenok · G. Sandford · M. Schnürch ·
Y. Shermolovich · Y. Takeda

Springer

The series *Topics in Heterocyclic Chemistry* presents critical reviews on "Heterocyclic Compounds" within topic-related volumes dealing with all aspects such as synthesis, reaction mechanisms, structure complexity, properties, reactivity, stability, fundamental and theoretical studies, biology, biomedical studies, pharmacological aspects, applications in material sciences, etc. Metabolism will also be included which will provide information useful in designing pharmacologically active agents. Pathways involving destruction of heterocyclic rings will also be dealt with so that synthesis of specifically functionalized non-heterocyclic molecules can be designed.
The overall scope is to cover topics dealing with most of the areas of current trends in heterocyclic chemistry which will suit to a larger heterocyclic community.
As a rule, contributions are specially commissioned. The editors and publishers will, however, always be pleased to receive suggestions and supplementary information. Papers are accepted for *Topics in Heterocyclic Chemistry* in English.
In references, *Topics in Heterocyclic Chemistry* is abbreviated *Top Heterocycl Chem* and is cited as a journal.

Springer www home page: springer.com
Visit the THC content at springerlink.com

Topics in Heterocyclic Chemistry ISSN 1861-9282
ISBN 978-3-642-25102-3 e-ISBN 978-3-642-25103-0
DOI 10.1007/978-3-642-25103-0
Springer Heidelberg Dordrecht London New York

Library of Congress Control Number: 2012931923

Printed on acid-free paper

Springer is part of Springer Science+Business Media (www.springer.com)

Series Editor

Prof. Dr. Bert U.W. Maes

Organic Synthesis
Department of Chemistry
University of Antwerp
Groenenborgerlaan 171
B-2020 Antwerp
Belgium

Volume Editor

Dr. Jernej Iskra

"Jozef Stefan" Institute
Department of Physical and
Organic Chemistry
Jamova 39
1000 Ljubljana
Slovenia
jernej.iskra@ijs.si

Editorial Board

Topics in Heterocyclic Chemistry Also Available Electronically

Topics in Heterocyclic Chemistry is included in Springer's eBook package *Chemistry and Materials Science.* If a library does not opt for the whole package the book series may be bought on a subscription basis. Also, all back volumes are available electronically.

For all customers who have a standing order to the print version of *Topics in Heterocyclic Chemistry*, we offer free access to the electronic volumes of the Series published in the current year via SpringerLink.

If you do not have access, you can still view the table of contents of each volume and the abstract of each article by going to the SpringerLink homepage, clicking on "Chemistry and Materials Science," under Subject Collection, then "Book Series," under Content Type and finally by selecting *Topics in Heterocyclic Chemistry.*

You will find information about the

- Editorial Board
- Aims and Scope
- Instructions for Authors
- Sample Contribution

at springer.com using the search function by typing in *Topics in Heterocyclic Chemistry.*

Color figures are published in full color in the electronic version on SpringerLink.

Aims and Scope

The series *Topics in Heterocyclic Chemistry* presents critical reviews on "Heterocyclic Compounds" within topic related volumes dealing with all aspects such as synthesis, reaction mechanisms, structure complexity, properties, reactivity, stability, fundamental and theoretical studies, biology, biomedical studies, pharmacological aspects, applications in material sciences etc. Metabolism is also included which provides information useful in designing pharmacologically active agents. Pathways involving destruction of heterocyclic ring are also dealt with so that synthesis of specifically functionalized non-heterocyclic molecules can be designed.

Overall scope is to cover topics dealing with most of the areas of current trends in heterocyclic chemistry which suits a larger heterocyclic community.

The individual volumes of *Topics in Heterocyclic Chemistry* are thematic. Review articles are generally invited by the volume editors.

In references *Topics in Heterocyclic Chemistry* is abbreviated *Top Heterocycl Chem* and is cited as a journal.

Preface

The study of heterocyclic compounds covers a broad area of chemistry. Their structures contain at least two different atoms as members of a saturated, unsaturated or aromatic ring, which forms either a simple monocyclic compound or a part of a larger structure of fused rings. Equally, halogenation is a similarly broad field and both topics are interwoven within these chapters. Research on both fields is ongoing and broad, and a comprehensive overview is not possible but rather, this book is oriented towards current research involving representative transformations and strategies for the synthesis of halogenated heterocycles and their use.

The subject of Chap. 1 is the recent developments in nucleophilic aromatic substitutions, using the example of perfluoroheteroaromatic compounds as precursors for the synthesis of highly functionalized heterocycles, macrocycles, ring-fused systems and glycosyl donors. Chapter 2 presents the chemistry of a single fluorine atom in the heterocyclic molecule and describes the latest methods of synthesizing monofluoro-substituted heterocycles via nucleophilic and electrophilic fluorination. Chapter 3 goes further and describes halogenation and halocyclization reactions resulting in β-halofurans discussed according to the halogen atom, i.e. iodination, bromination, chlorination and fluorination, and their synthesis from the halo-substituted starting compounds. Chapter 4 explores the synthesis of halogenated five- and six-membered sulphur containing heterocycles. Electrophilic halogenation often requires reagents having a heterocyclic structure with an N-halo bond, and Chap. 5 focuses on representative reaction patterns in which N–X heterocyclic reagents are used as halogenating reagents and in other transformations such as oxidation, substitution, addition, cyclization and asymmetric reactions. Chapter 6 shows how "halogen dance" reactions became an important tool for synthesizing halogenated heterocycles substituted at specific positions. Chapter 7 looks at the application of halogenated heterocycles and discusses the chemistry and structure–activity relationships, mechanisms of action and the clinical use of halogenated heterocyclic pharmaceuticals. Chapter 8 reviews the occurrence and fate of three model halogenated heterocyclic pharmaceuticals belonging to different pharmaceutical classes in the environment. Finally, Chapter 9 reviews recent research into the direct halogenation of heterocyclic compounds with the aim of providing "greener" alternatives. Each of the methods discussed, electrophilic, nucleophilic and radical, was approached from the perspective of alternative strategies, reagents, solvents and activation.

I wish to thank all of the authors and my colleagues who contributed to the book's preparation, especially to Slovenko Polanc and Viacheslav Petrov. Finally, I would like to thank my family for their support and patience.

Ljubljana, Slovenia Jernej Iskra

Contents

Top Heterocycl Chem (2012) 27: 1–32
DOI: 10.1007/7081_2011_63

Published online: 23 September 2011

Perfluoroheteroaromatic Chemistry: Multifunctional Systems from Perfluorinated Heterocycles by Nucleophilic Aromatic Substitution Processes

Graham Sandford

Abstract Perfluoroheteroaromatic systems, such as pentafluoropyridine, tetrafluoropyrazine and tetrafluoropyridazine, are highly electrophilic due to the presence of several fluorine atoms attached to the heteroaryl rings and, consequently, react very readily with a wide range of nucleophilic species. In this chapter, we discuss reports of recent nucleophilic aromatic substitution reactions of perfluoroheteroaromatic substrates that have enabled the synthesis of, for example, a wide range of highly functionalised heterocyclic derivatives, macrocycles, ring-fused systems and glycosyl donors.

Keywords Macrocyle · Nucleophilic aromatic substitution · Pentafluoropyridine · Perfluoroheteroaromatic chemistry · Perfluoroheterocycle · Ring fused heterocycle · Scaffold

Contents

Abbreviations

d	Day
DIPEA	Di*iso*propylethylamine
DMEU	*N*, *N*-Dimethylethylene urea

G. Sandford
Department of Chemistry, Durham University, South Road, Durham DH1 3LE, UK
e-mail: Graham.Sandford@durham.ac.uk

DMF	*N,N*-Dimethyl formamide
LDA	Lithium di*iso*propylamine
MeCN	Acetonitrile
MW	Microwave irradiation
Nuc	Nucleophile
rt	Room temperature
SEM	(Trimethylsilyl)ethoxymethyl
TDAE	Tetrakis(dimethylamino)ethylene
THF	Tetrahydrofuran
THP	Tetrahydropyran
TMEDA	Tetramethylethylene diamine

1 Introduction

The chemistry of heteroaromatic systems such as pyridine, quinoline and the diazines are, of course, of fundamental importance to synthetic organic chemistry [1], and the large number of systems that possess a heteroaromatic subunit as part of their structure continue to find valuable applications in all sectors of the chemical industry including pharmaceuticals, agrochemicals and materials [2]. A relatively small part of the vast heterocyclic chemistry literature is dedicated to the chemistry of perfluoroheteroaromatic derivatives such as pentafluoropyridine, heptafluoroquinoline and the tetrafluoro-diazines, that is, systems in which all unfunctionalised ring carbon atoms are bonded to fluorine rather than hydrogen [3, 4]. However, the exploration of the chemistry of perfluoroheteroaromatic systems has led, previously, to the discovery of some remarkable new processes and molecular architectures including, for example, various stable valence bond isomers [3], and has extended our knowledge of mechanistic organic chemistry in general. Of course, in developing the chemistry of this new class of organic compound, all the usual problems of, for example, reagent use, regioselectivity, reagent compatibility, effect of substituents, kinetics, thermodynamics, solvents, and reaction conditions, must be explored and established and this process continues to the present.

The first realistic and readily scaled synthesis of pentafluoropyridine and various other perfluoroheteroaromatic systems was reported by Chambers [3, 5] in the early 1960s. Reaction of the corresponding perchlorinated heteroaromatic derivative, prepared by exhaustive chlorination of the parent hydrocarbon with either chlorine or phosphorous pentachloride, with potassium fluoride at high temperature in a sealed autoclave provides useful quantities of the perfluoroheteroaromatic starting materials and some of these systems, which are usually volatile colourless liquids, are now commercially available (Scheme 1).

The chemistry of perfluoroheteroaromatic systems has developed significantly and was summarised in an exhaustive review article by Brooke [4] in 1997 and the present chapter aims to provide an overview of the main developments in the field

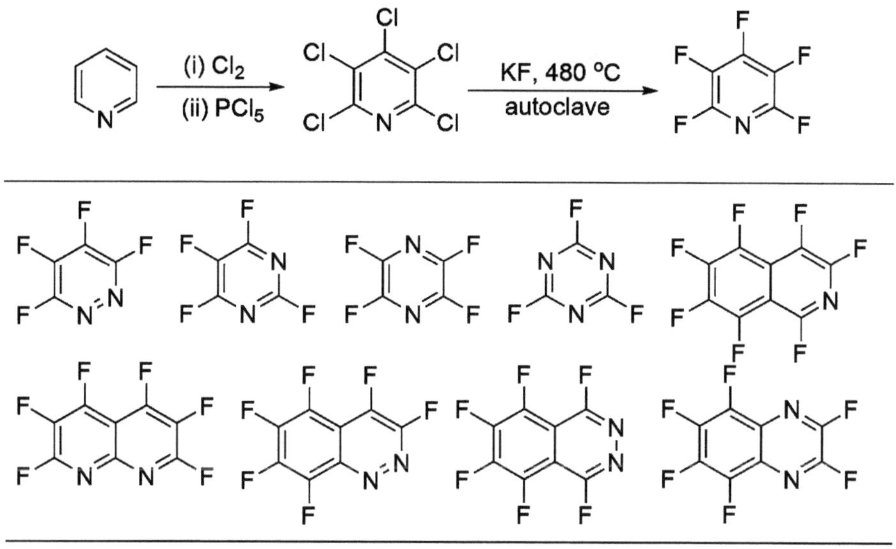

Scheme 1 Synthesis of perfluoroheteroaromatic systems [3]

since this report was published. More general discussions of the chemistry of perfluorinated heterocycles may be found in several reviews [3, 6], book chapters [7, 8] and monographs [9] concerning various aspects of organofluorine chemistry. A discussion of transition metal mediated functionalisation processes of perfluoroheteroaromatic derivatives involving C–F bond activation is not included in this chapter, and the reader is directed to publications by Braun and co-workers [10, 11] for an introduction to this emerging field.

2 Reactions of Perfluoroheteroaromatic Systems

2.1 Nucleophilic Aromatic Substitution (S_NAr) Processes

The chemistry of perfluoroheteroaromatic substrates is dominated by reactions with nucleophilic species because the presence of several highly electronegative fluorine atoms attached to the ring makes the heterocycle very susceptible towards nucleophilic attack [3, 4]. Nucleophilic aromatic substitution reactions follow the well-established two-step addition–elimination mechanism that proceed via appropriate Meisenheimer intermediates (Scheme 2), and this reactivity has been termed 'mirror image' chemistry [7], which contrasts the very well known chemistry of aromatic hydrocarbons that undergo electrophilic substitution processes [12].

Extensive discussions regarding the factors that affect the regioselectivity of nucleophilic aromatic substitution processes involving perfluoroheteroaromatic

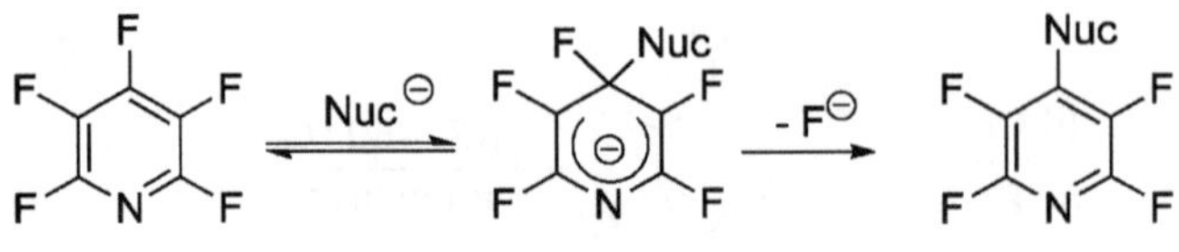

Scheme 2 Nucleophilic aromatic substitution (S_NAr) mechanism [7]

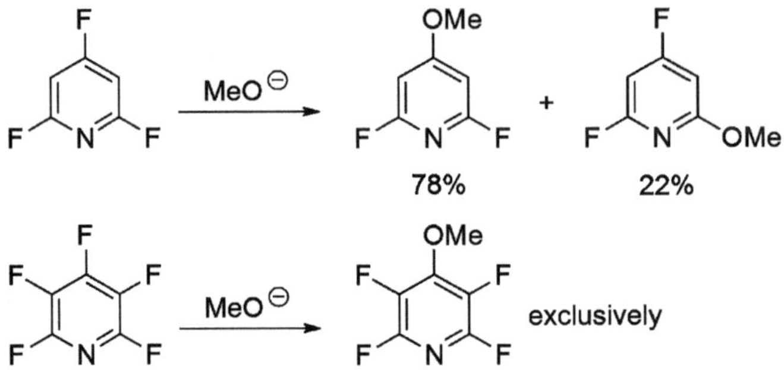

Scheme 3 Regioselectivity of S_NAr processes [7]

systems have been published in the course of developing the chemistry of pentafluoropyridine and related heteroaromatic systems [3, 4, 13] and is still a topic of some discussion [14, 15].

Comparison of the rates of reactivity of a series of perfluorinated heteroaromatic systems with nucleophiles show that the ring nitrogen exerts the dominating influence and activates *ortho* and *para* positions [7, 14]. Reactions of 2,4,6-trifluoropyridine show that the rates of nucleophilic attack at the 4- and 2-positions is in the ratio 3:1, indicating that substitution at the *para* positions is slightly favoured (Scheme 3). The fact that pentafluoropyridine reacts exclusively with nucleophiles at the 4-position, therefore, indicates that the fluorine atoms attached to the heteroaryl ring play an important role in effecting the regioselectivity of the nucleophilic substitution processes in these systems.

Kinetic studies [7, 14] have been used to ascertain the separate activating influences of fluorine attached to aromatic rings and it was observed that fluorine located at sites *ortho* and *meta* to the site of nucleophilic attack are strongly activating, whereas fluorine that is *para* to the site of nucleophilic attack is slightly deactivating. A consideration of carbanion stabilities explains the deactivating effect of *para* fluorine. In planar carbanions, such as in the Meisenheimer intermediates for heteroaryl systems, repulsion between the fluorine atom lone pairs and the negative charge are maximised and, therefore, overall destabilising. For fluorine atoms *meta* to the site of attack, the electronegative fluorine atom inductively stabilises the negatively charged intermediate thereby overall activating the system. We would expect fluorine atoms *ortho* to the site of attack to have a similar effect to *para* fluorine atoms if the stability of the Meisenheimer intermediates were the only factors to consider, but experimentally determined kinetic measurements show this is not the case. Here, *ortho* fluorine atoms influence the site of attack in the initial state, making the carbon–fluorine bond under attack more electron deficient in nature by inductive withdrawal. Consequently, nucleophilic aromatic substitution in perfluoroheteroaromatic systems occurs at sites that are preferentially *para* to ring nitrogen and maximise the number of activating *ortho* and *meta* fluorine atoms while minimising the number of fluorine atoms *para* to the site of attack. This situation occurs in reactions of pentafluoropyridine (Scheme 3) where the 4-position is clearly favoured over the 2- and 3-sites, and this is consistent with experimental observations.

These ideas have been expanded to predict and explain the orientation of nucleophilic aromatic substitution processes for a variety of perfluorinated heteroaromatic systems [7], and recent examples that are consistent with this mechanistic analysis are apparent in the sections below.

2.2 *Reactions of Perfluoroheteroaromatic Systems with Monofunctional Nucleophiles*

Reactions of a wide range of nucleophiles with pentafluoropyridine occur almost exclusively to give products arising from displacement of the most activated fluorine atom *para* to ring nitrogen and, recently, reactions with *N*,*N*-dimethylaminopyridine [16], adamantylthiazylamide salts [17], cyclam [18], phosphonate derivatives [19], benzodichalcogenophenes [20] and trifluoromethane thiolate anion [21] further confirm this well-established regioselectivity (Table 1).

In principle, 4-substituted-tetrafluoropyridine derivatives are sufficiently electrophilic to be able to react with further nucleophilic species and used as starting materials for the synthesis of a range of highly substituted polyfunctional pyridine derivatives. Potentially, substitution of fluorine at either the 2- or 3-positions or displacement of the 4-substitutent could occur and the product profile is dependent upon both the 4-substituent and the nucleophile (Scheme 4).

Reactions of a range of 4-substituted tetrafluoropyridine derivatives with diethylamine illustrate the regioselectivity of S_NAr processes depending upon the nature of the substituent located at the 4-position [22] (Table 2). When the 4-substituent is NEt_2, OEt, Br and SO_2Ph, disubstituted trifluoropyridine products are formed that arise from selective substitution of the most activated fluorine atom that is *ortho* to ring nitrogen. In contrast, if the substituent is the more labile nitro group, *ipso* substitution of the nitro group competes effectively with substitution *ortho* to ring nitrogen.

In many cases, therefore, sequential reaction of pentafluoropyridine with nucleophiles occurs in the order of displacement of fluorine firstly from the 4-then 2- followed by the 6-position. This reactivity profile has been utilised recently to synthesise a variety of polynitroxides [23] and triaminopyridine derivatives [24] (Scheme 5). All the fluorine atoms of pentafluoropyridine are, in principle, susceptible towards nucleophilic attack and, indeed, all five may be

Table 1 Reactions of pentafluoropyridine with monofunctional nucleophiles [16–21]

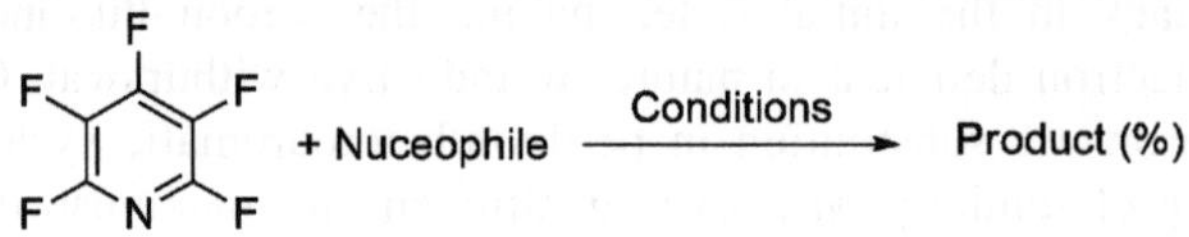

Nucleophile	Conditions	Product (%)
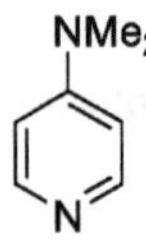	1) MeCN, rt 2) $NaBF_4$, MeCN, rt	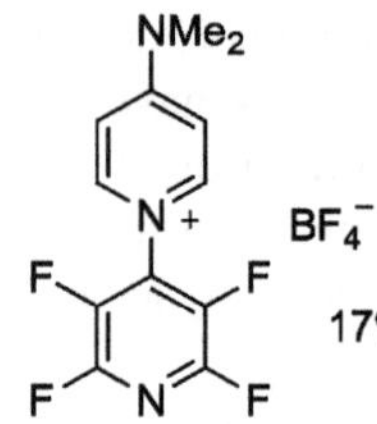 17%
	CsF, MeCN 20°C	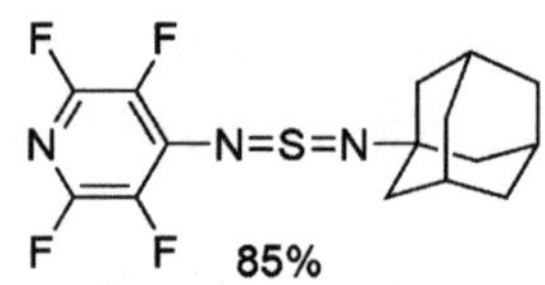 85%
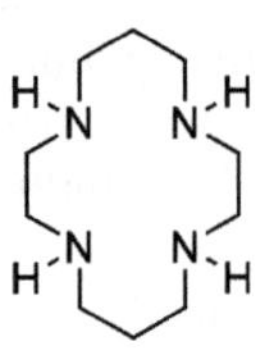	DMF, 100°C 13 h	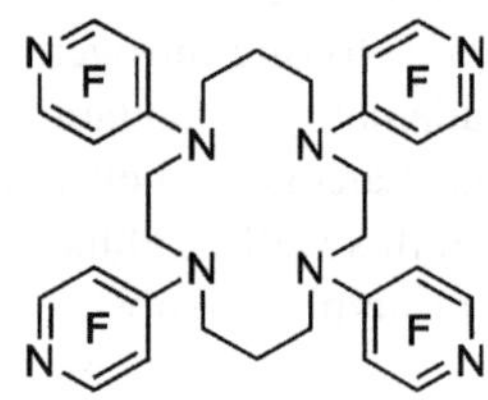 94%
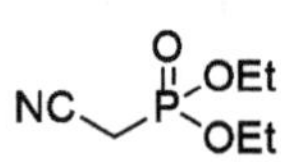	CsF, 3 h	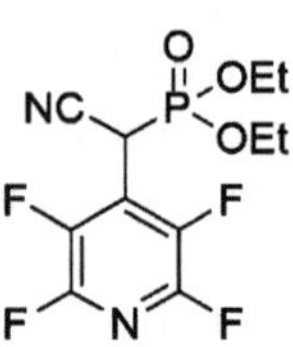 66%
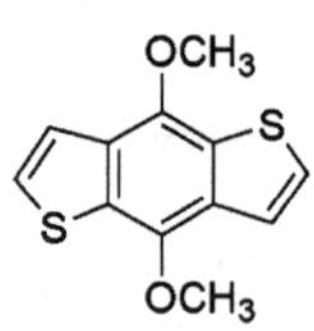	*n*-BuLi TMEDA THF	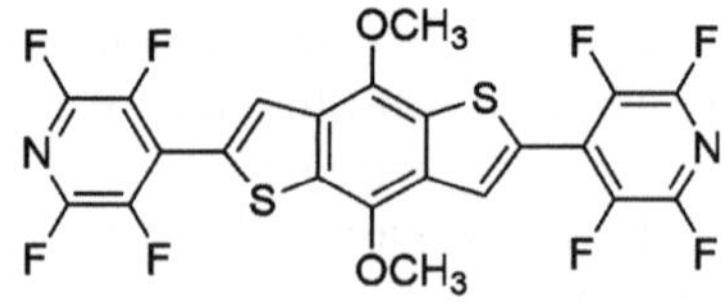 74%
Et_4NSCF_3	MeCN rt	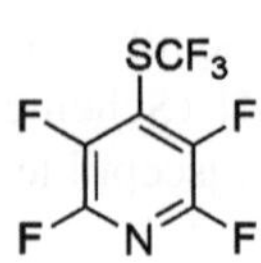

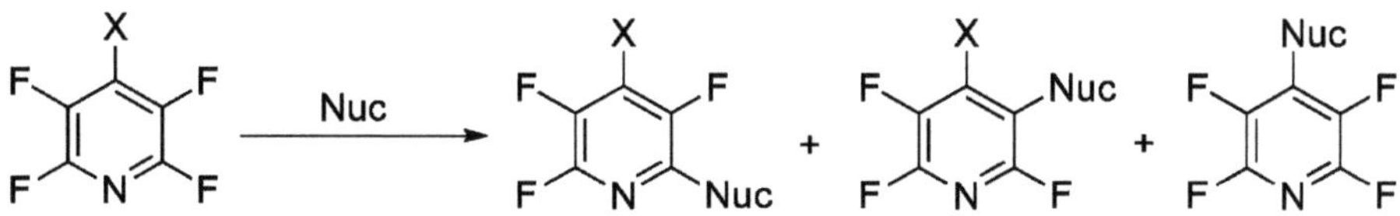

X = H, Cl, Br, OR, NR_2, Ph, NO_2, SO_2Ph, etc

Scheme 4 Possible products derived from S_NAr processes involving tetrafluoropyridine substrates

Table 2 Reactions of 4-substituted tetrafluoropyridine derivatives with diethylamine [22]

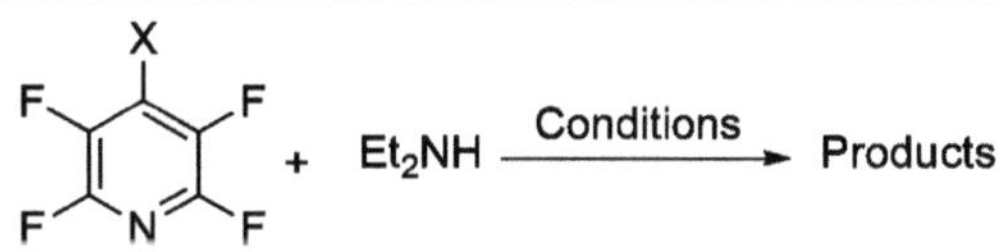

X	Product(s)	Yield
NEt_2	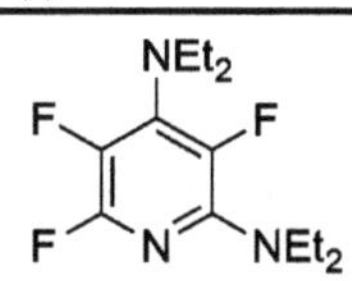	63%
OEt	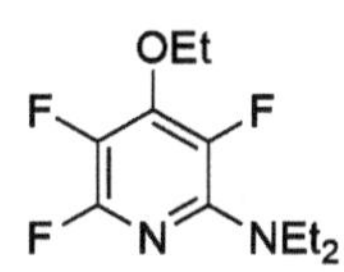	63%
Br	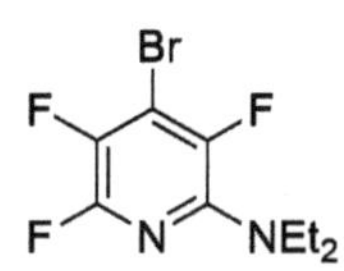	61%
SO_2Ph	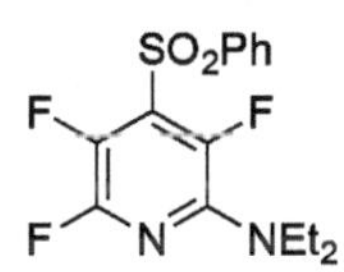	38%
NO_2	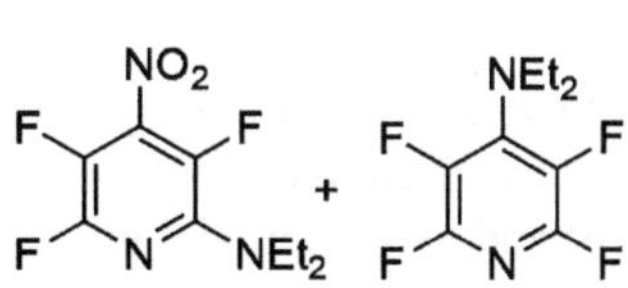	35% 2:1

displaced upon reaction with an excess of a pyrazole salt [25], demonstrating the synthetic potential of using these highly fluorinated systems for the preparation of densely functionalised pyridine derivatives.

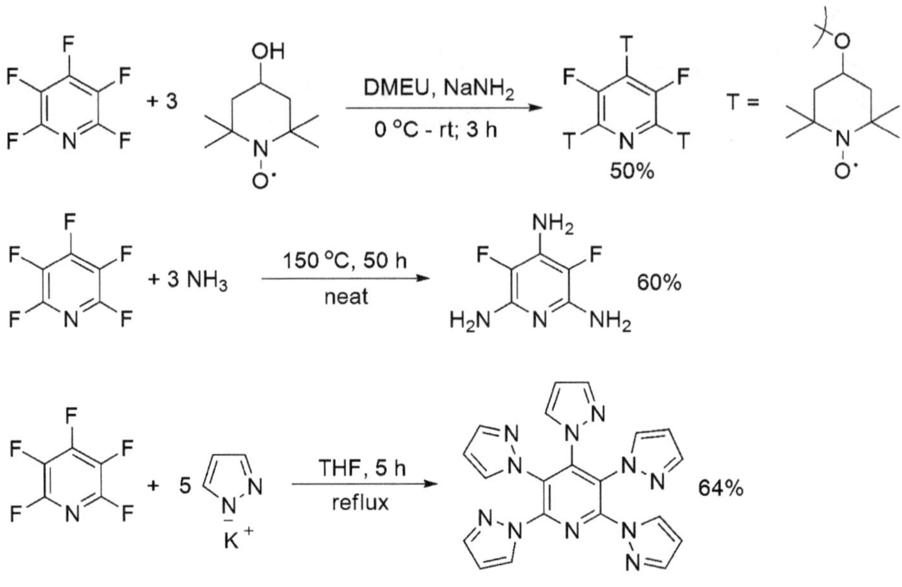

Scheme 5 Polysubstitution reactions of pentafluoropyridine [23–25]

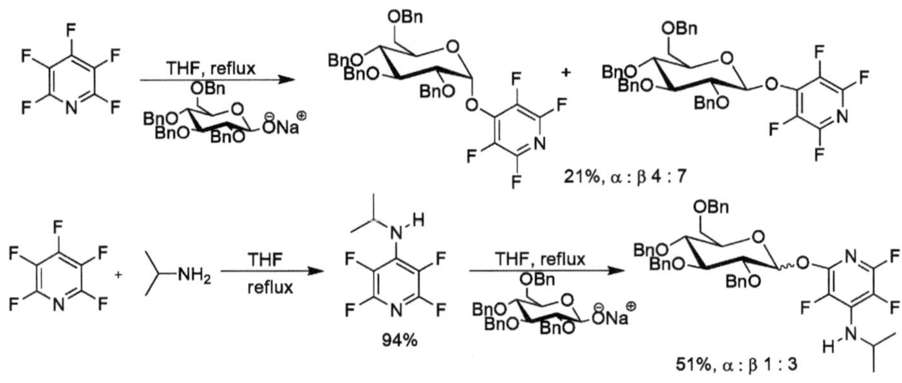

Scheme 6 Synthesis of glycosyl donors [26, 27]

By similar sequential nucleophilic substitution processes, various glycosyl donor systems have been synthesised by reaction of appropriate glucose salts with perfluoropyridine systems (Scheme 6) and used in model glycosylation reactions [26, 27].

Related perfluorobrominated scaffolds, synthesised by displacement of fluorine by bromine at the 2,4 and/or 6 positions, have more flexible functionality due to the synthetic possibilities offered by carbon–bromine bonds for use in reactions including palladium-catalysed coupling and carbanion generation processes. In this context, 4-bromo-tetrafluoropyridine [28] and 2,4,6-tribromo-3,5-difluoro-pyridine [29, 30] synthesised by reaction of pentafluoropyridine with a mixture of hydrogen

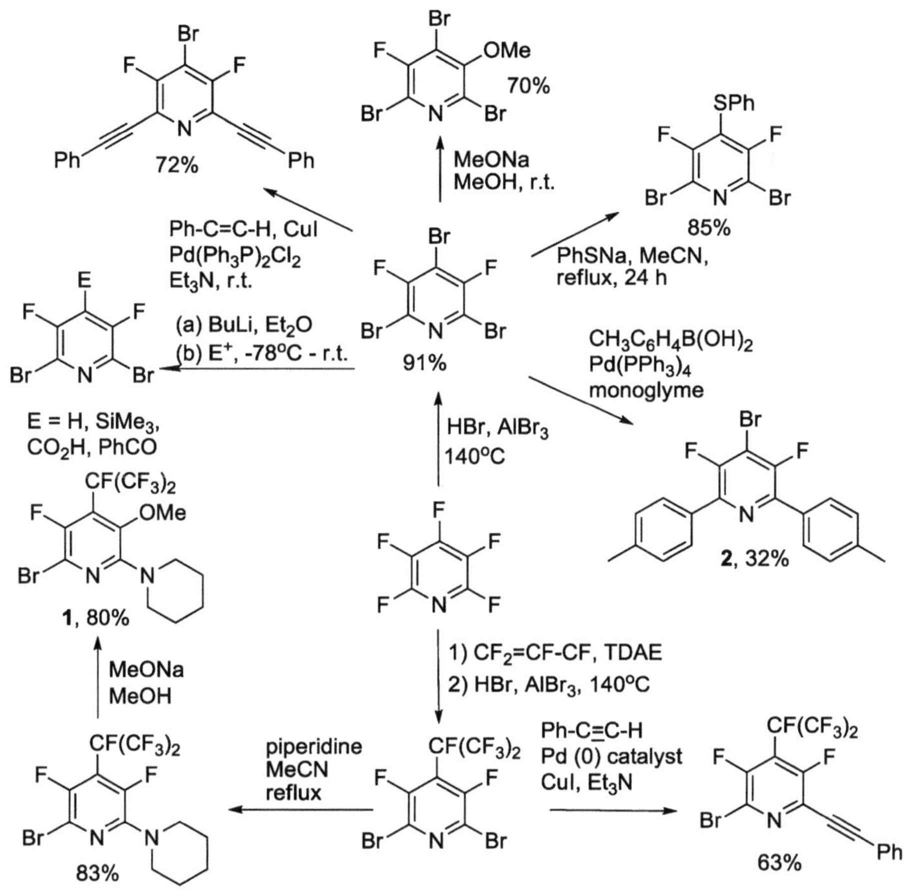

Scheme 7 Perfluorobromopyridine derivatives as core scaffolds [29, 30]

bromide and aluminium tribromide in an autoclave at 140 °C, were assessed as polyfunctional scaffolds for the synthesis of a variety of pentasubstituted pyridine derivatives (Scheme 7), some of which **1**,**2** were characterised by X-ray crystallography (Fig. 1).

Perfluorinated heterocyclic systems may be used as starting materials for rapid analogue synthesis approaches to highly functionalised biologically active systems because a range of polysubstituted systems can be derived from such scaffolds by nucleophilic aromatic substitution processes as shown above. The wide range of nucleophiles available (O, N, C and S centred) makes the theoretical number of highly functionalised heterocyclic systems that could be accessed by S_NAr methodology very large indeed and, depending on the nucleophiles chosen, could give rapid access to many drug-like molecules that fall within the 'Lipinski parameters' [31]. It is now widely accepted that molecules that are most likely to possess drug-like physiochemical properties fall within the empirical 'rules of 5' described by

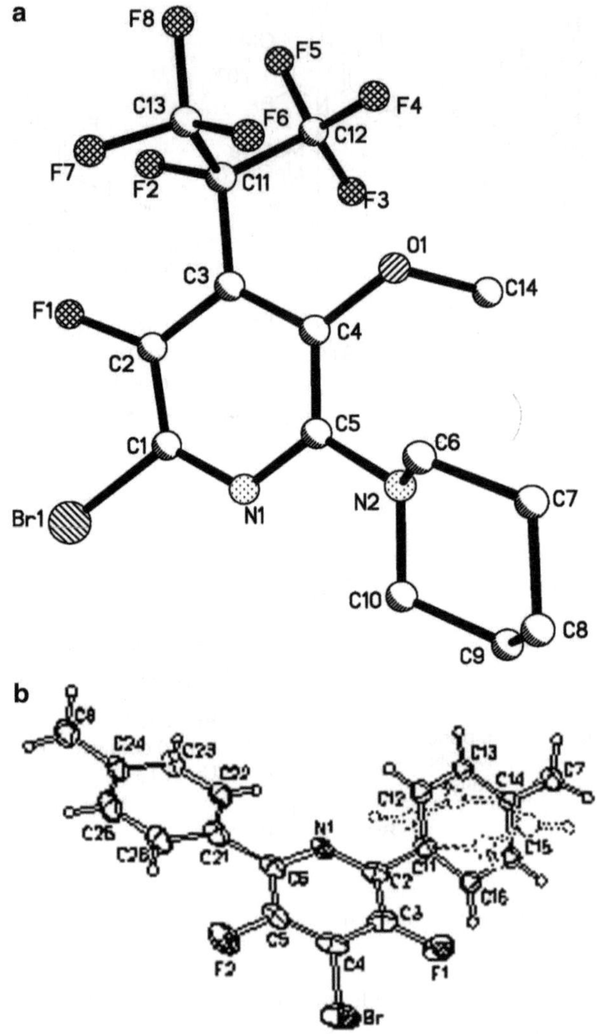

Fig. 1 Structures of polyfunctional pyridine systems **1** (**a**, *above*) and **2** (**b**, *below*) [29, 30]

Lipinski [31] and, furthermore, small molecules with these structural features are also more likely to possess good oral bioavailability. Consequently, there is a continuing requirement for the development of methodology that allows the ready synthesis of libraries of small molecules [32, 33] that may be screened against new and existing biological targets for the purpose of lead generation and the use of highly fluorinated heteroaromatic scaffolds as starting materials for the synthesis of model systems that are related to drug-like molecules has been explored recently.

Indeed, medicinal chemists were able to rapidly synthesise several small libraries of trisubstituted pyridine derivatives from pentafluoropyridine (Table 3) by sequential S_NAr processes similar to those described above. The factors VIIa/TF

Table 3 Biologically active polysubstituted systems from pentafluoropyridine [34–38]

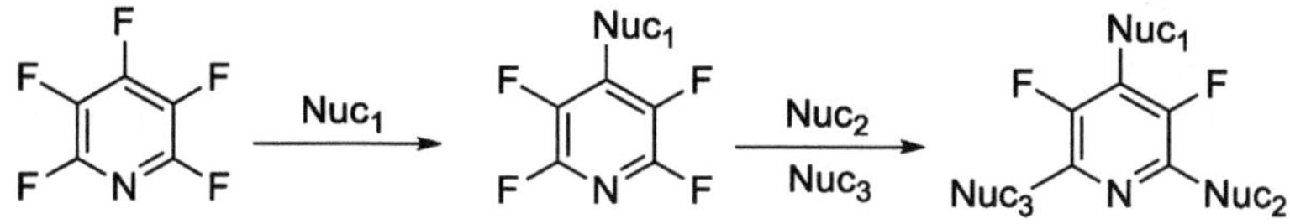

Nuc$_1$	Nuc$_2$	Nuc$_3$	Product
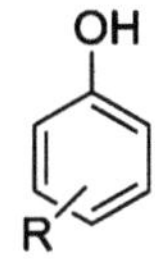	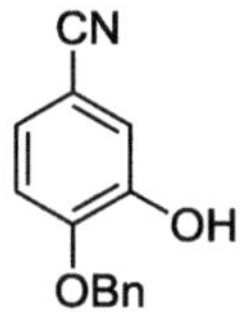	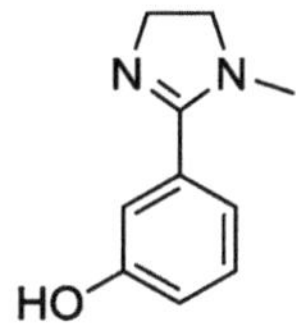	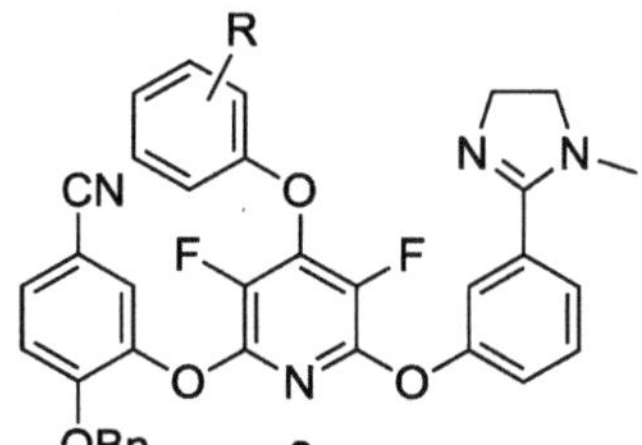 **3**
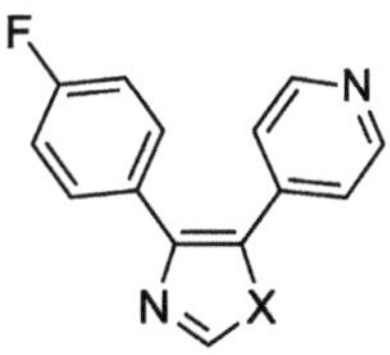	NH_3	NH_3	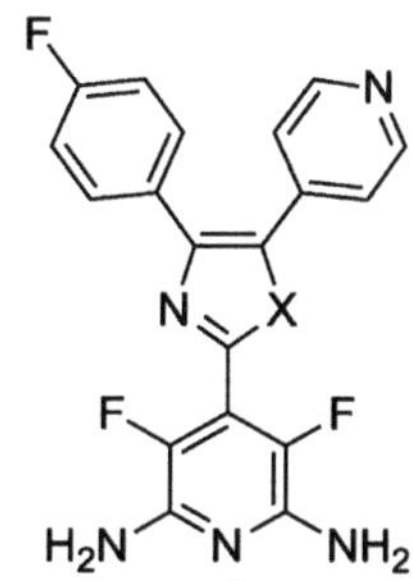 **4**

complex and Xa are proteins known to be involved in the blood coagulation cascade [34] and, as such, are validated targets in the search for novel antithrombotic drugs [35, 36]. Having established that a series of 2,6-diphenoxypyridines to be modest inhibitors of factor Xa, medicinal chemists synthesised a small library of 3,5-difluoro-triaryloxypyridine derivatives (**3**, Table 3) from pentafluoropyridine [37]. The 2,4,6-substitution pattern was obtained through sequential nucleophilic aromatic substitution by substituted phenol systems.

It is thought that inhibition of the p38 kinase protein could treat the underlying cause of chronic inflammatory diseases and a diverse set of aryl substituted pyridinylimidazoles to give potentially high affinity binding to the active site were synthesised. Deprotonation of the SEM-protected imidazole gave a carbanion that reacted as a nucleophile with pentafluoropyridine to give the expected 4-substituted pyridine. Bromination at the 4- and 5-positions of the imidazole was followed by regioselective Stille reaction to yield a pyridinylimidazole derivative while subsequent Suzuki or Stille coupling at the remaining carbon-bromine bond was followed by deprotection of the imidazole. Finally, diamination at the 2- and 6- position of the tetrafluoropyridine gave the desired biologically active 3,5-difluoropyridine system (**4**, Table 3) [38].

Tetrafluorodiazine systems are approximately 1,000 times more reactive towards nucleophiles than pentafluoropyridine and nucleophilic aromatic substitution processes occur very readily [7]. Tetrafluoropyrimidine may be used as a scaffold for the synthesis of a range of 2,4,6-trisubstituted pyrimidine derivatives

Table 4 Tetrafluoropyrimidine as a core scaffold [39]

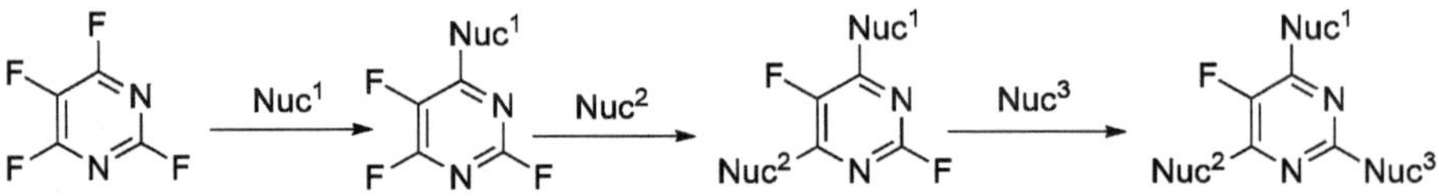

Nuc^1	Nuc^2	Nuc^3	
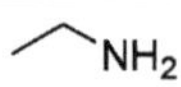 THF, DIPEA 0 °C	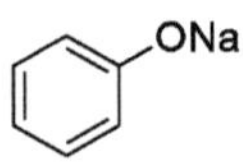 THF, r.t.	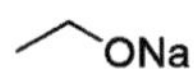 THF, MW 140 °C, 15 min	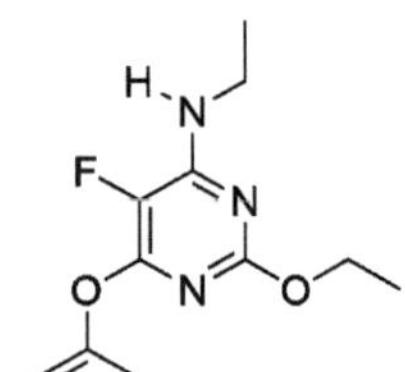 **5**
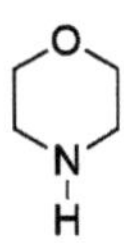 THF, DIPEA 0 °C	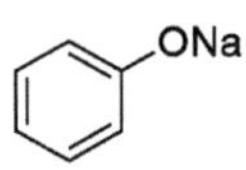 THF, r.t.	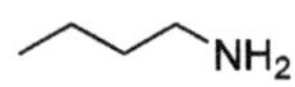 THF, MW 140 °C, 1 h	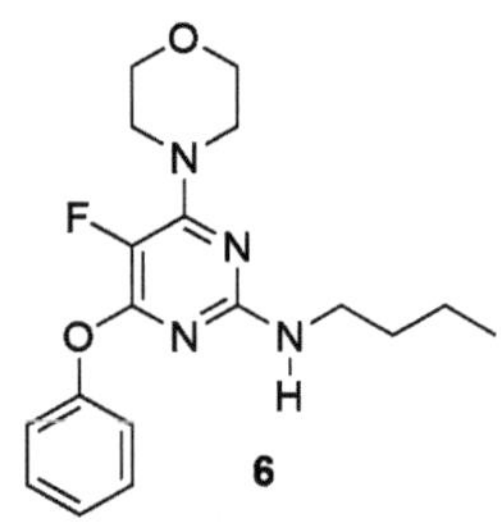 **6**
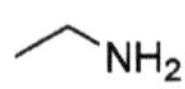 THF, DIPEA 0 °C	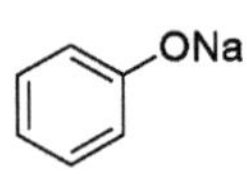 THF, r.t.	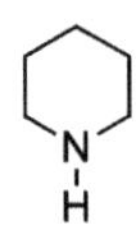 THF, MW DIPEA 140 °C, 1 h	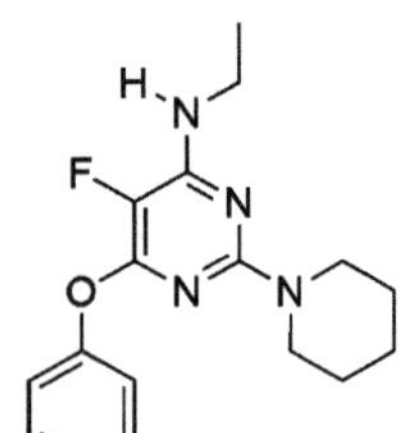
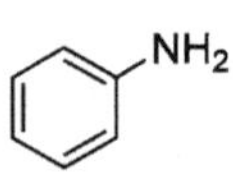 THF, DIPEA 0 °C	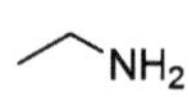 THF, DIPEA r.t.	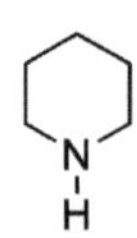 THF, MW DIPEA 140 °C, 1 h	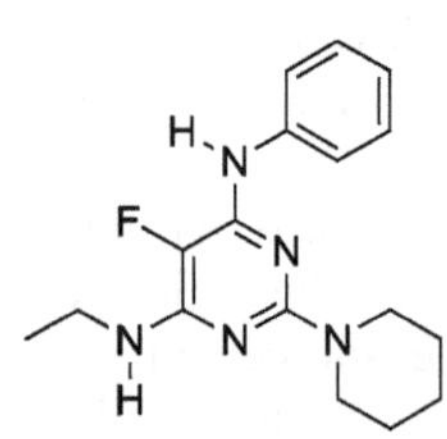

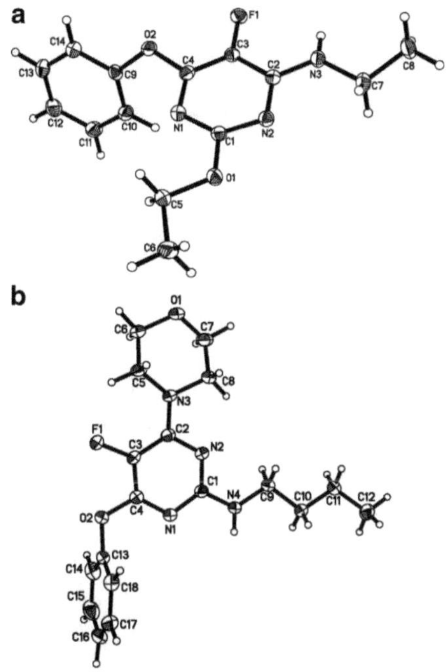

Fig. 2 Polysubstituted pyrimidine systems **5** (**a**, *above*) and **6** (**b**, *below*) [39]

[39], upon sequential displacement of the fluorine atoms attached to the strongly activated 4-, 6- and 2-positions (Table 4) and were characterised by X-ray crystallography [39] (**5**,**6**, Fig. 2). Many 5-fluoropyrimidine derivatives have valuable biological activity [39] and this rapid multiple polysubstitution strategy allows ready access to libraries of polyfunctional pyrimidine systems for bioassay screening studies.

Similarly, trifluoropyridazinone, readily synthesised by reaction of tetrafluoropyridazine in sulfuric acid, may be used as the starting material for the synthesis of a variety of 4,5-diamino-fluoropyridazinone systems [40]. Reaction of trifluoropyridazinone gives a mixture of products arising from displacement of fluorine from either the 4- or the 5-positions (Table 5), both positions activated by *para* ring nitrogen, but these isomers can be separated by column chromatography and used in subsequent S_NAr processes for the synthesis of a range of diaminated pyridazinone derivatives [40] (Table 6).

Reactions of trifluoropyridazinone with highly basic sodium methoxide or phenoxide gave complex mixtures of products and tar derived, presumably, from deprotonation of the pyridazinone ring NH and subsequent polymerisation [40]. Protection of the ring NH group as a tetrahydropyran derivative, however, allows the functionalisation of the pyridazinone core scaffold by oxygen-centred nucleophiles (Scheme 8) extending the range of functional pyridazinone systems that may be accessed by this general strategy [41].

Table 5 Reactions of trifluoropyridazinone [40]

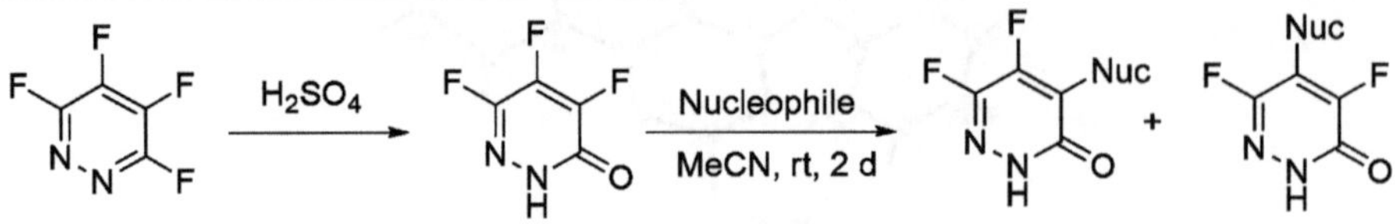

78%

Nucleophile	Yield 4-isomer[a]	Yield 5-isomer[a]	Isomer ratio[b]
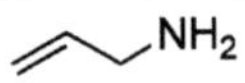	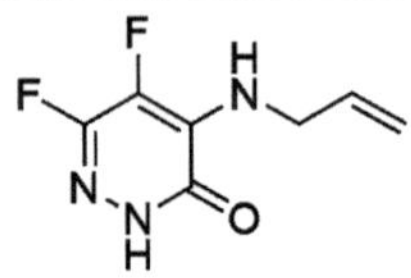 40%	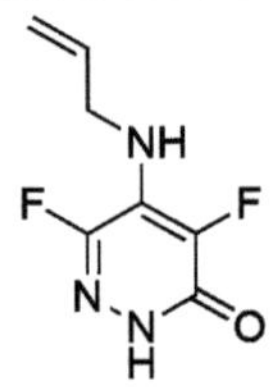 13%	60:40
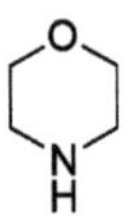	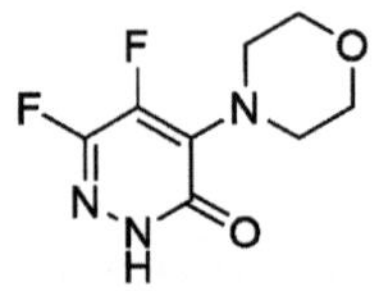 59%	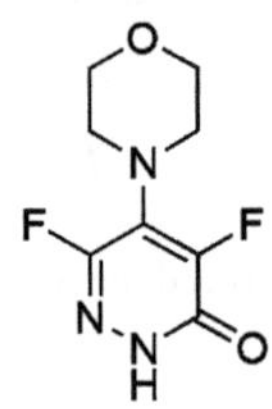 12%	77:23
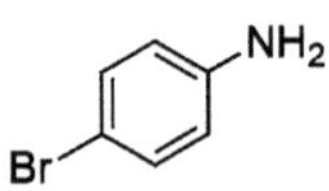	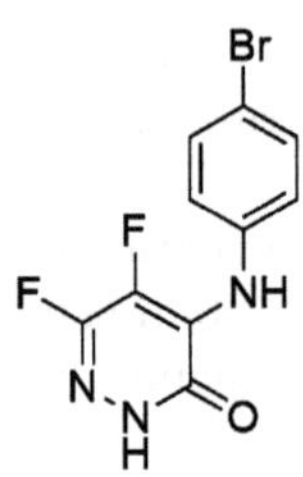 42%	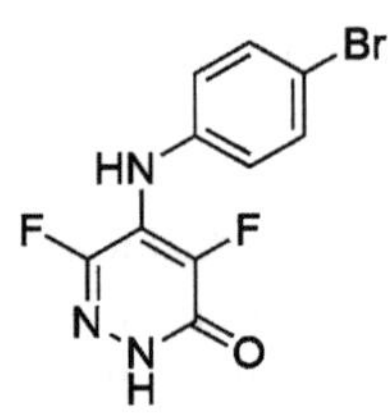 not isolated	96:4

[a]Isolated by column chromatography [b]Isomer ratio determined by ^{19}F NMR analysis of the crude reaction mixture

Octafluoro-4,4′-bipyridine also reacts efficiently with diethylamine [42] to give diaminated product (Scheme 9).

Table 6 Synthesis of disubstituted pyridazinone derivatives [40]

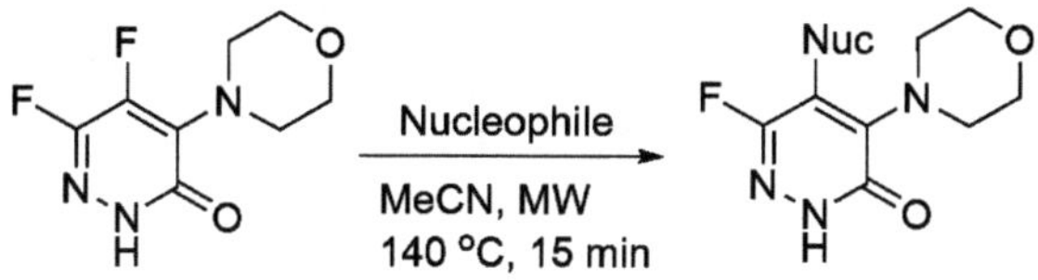

Nucleophile	Product	Yield (%)
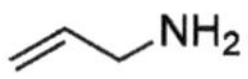	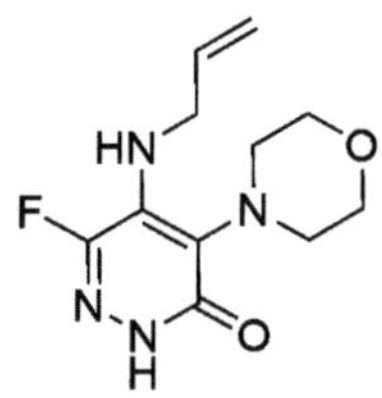	56
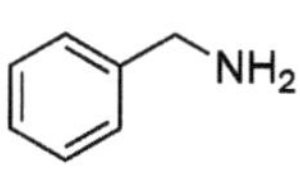	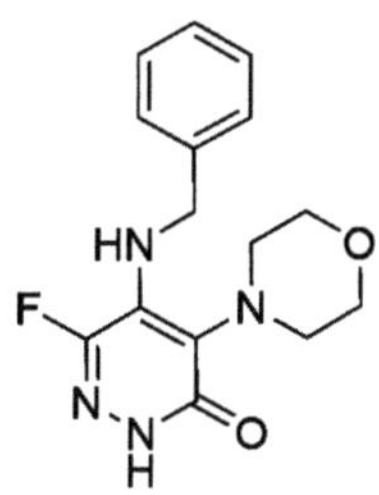	62
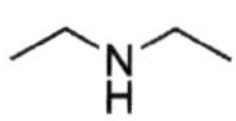	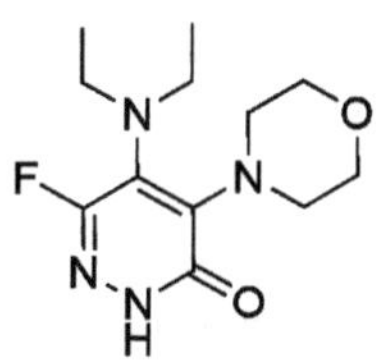	25
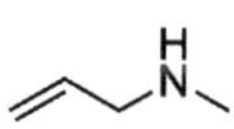	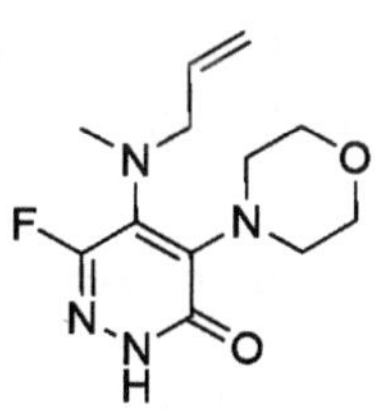	43

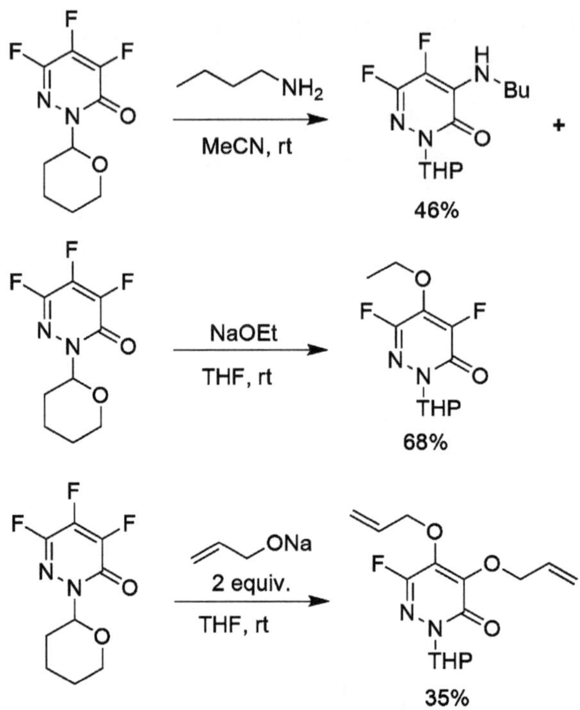

Scheme 8 Reactions of THP-protected trifluoropyridazinone [41]

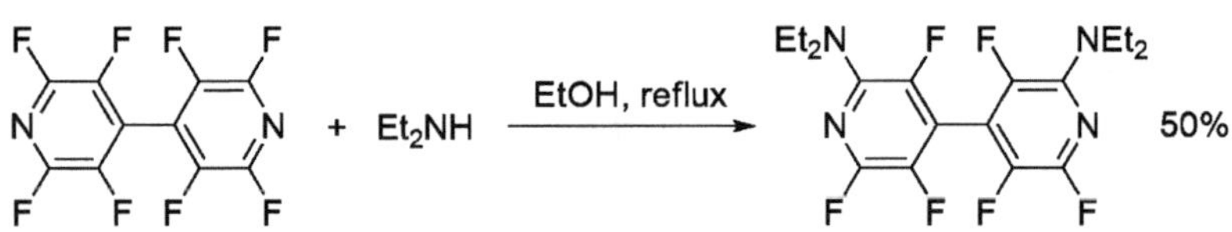

Scheme 9 Reaction of octafluoro-4,4′-bipyridine [42]

2.3 *Reactions of Perfluoroheteroaromatic Systems with Difunctional Nucleophiles*

2.3.1 Cyclisation by Intermolecular Processes

Polyfunctional building blocks are, of course, used in many synthetic applications such as for the preparation of polymers, dendrimers and biopolymers as well as for various applications in macrocyclic and supramolecular chemistry [43, 44]. As outlined above, pentafluoropyridine is a multifunctional building block that reacts sequentially with a range of nucleophiles and utilisation of this reactivity profile for the preparation of a variety of structurally diverse macrocyclic ring systems has been developed [45, 46], in which the macrocyclic rings may be constructed by

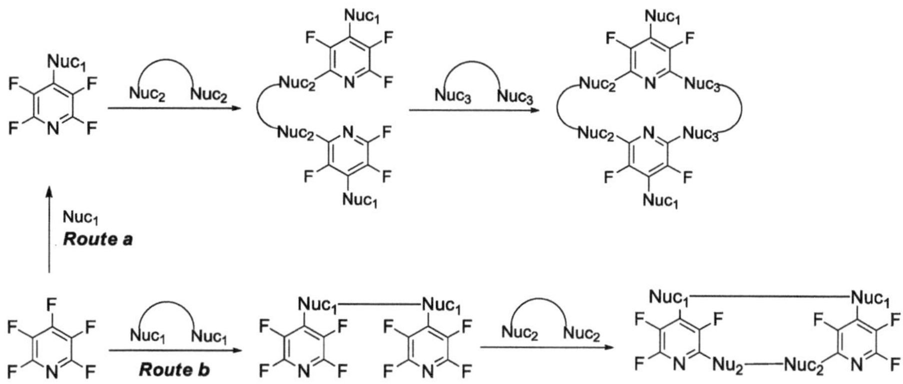

Scheme 10 Strategies for the synthesis of macrocycles from pentafluoropyridine [46]

processes involving reactions with a variety of a difunctional nucleophiles. Two strategies involving different reaction sequences with mono- and difunctional nucleophiles have been used to construct novel macrocycles from pentafluoropyridine despite the fact that syntheses of macrocycles that involve a nucleophilic aromatic substitution process as the ring forming step have not been widely adopted in supramolecular chemistry (Scheme 10).

By utilising the strategy outlined above (Route a, Scheme 10), functionalisation of pentafluoropyridine at the 4-position by reaction with various nucleophiles ($Nuc^1 = OMe, CF(CF_3)_2$, Scheme 10) and subsequent reaction with appropriate di-oxyanions, generated in situ from bis-trimethylsilyl derivatives of a variety diols and catalytic quantities of fluoride ion, or a diamine, firstly gives a bridged system, in which two pyridine sub-units are connected by a polyether or polyamine chain respectively (Scheme 11) [45, 46]. Ring closure to the desired macrocycles may be achieved by addition of a further equivalent of dinucleophile. An example of this synthetic strategy is shown in detail in Scheme 11 and X-ray crystallography confirmed the macrocyclic structure **7** (Fig. 3). Macrocycles prepared by similar synthetic methodology from 4-alkoxytetrafluoropyridine ($Nuc^1 = OCH_3$; $OCH_2CH_2C_6F_{13}$) derivatives [47], are also indicated in Scheme 11, demonstrating that this stepwise methodology allows the synthesis of many structurally diverse macrocycles from pentafluoropyridine, depending upon the choice of nucleophile and dinucleophile [46].

Analysis by X-ray crystallography reveals that the perfluoroalkylated macrocycle **8** has some unusual structural features [46, 47] (Fig. 4). Inspection of the crystal lattice indicates that, in the solid state, the individual molecules adopt positions in which the long perfluoroalkyl substituent groups overlap to form fluorine-rich environments or 'domains' within the crystal lattice, reflecting the solubility characteristics of fluorocarbon systems that has been used so effectively in fluorous phase chemistry [48, 49].

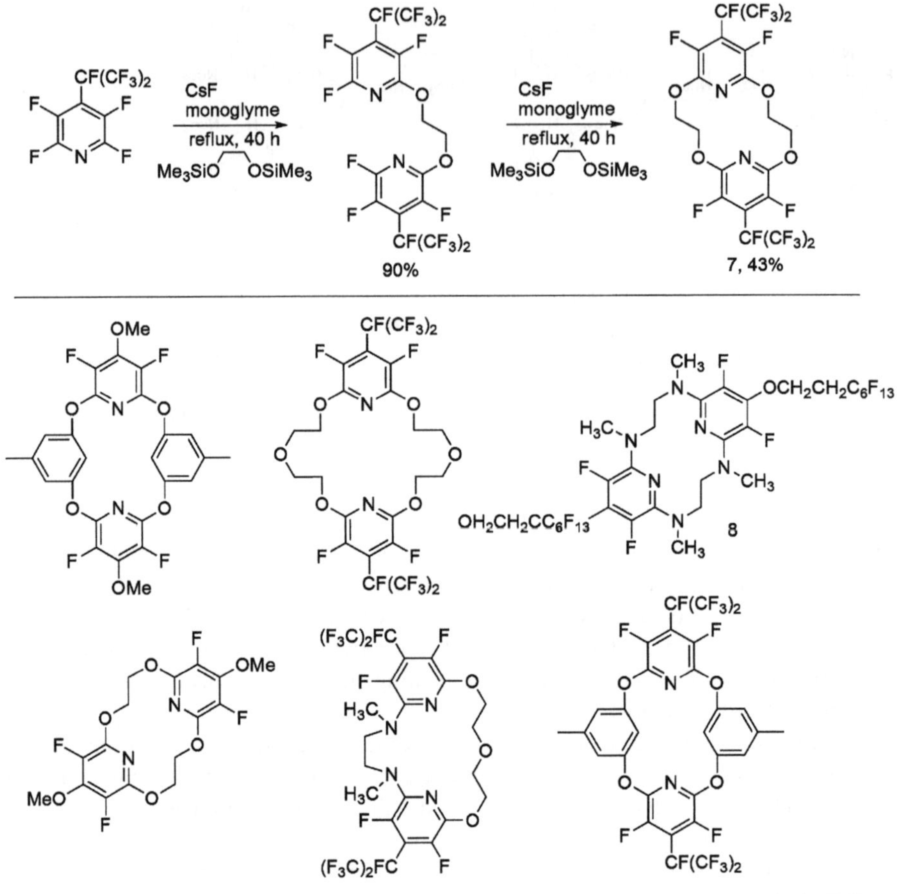

Scheme 11 Synthesis of macrocycles [45, 46]

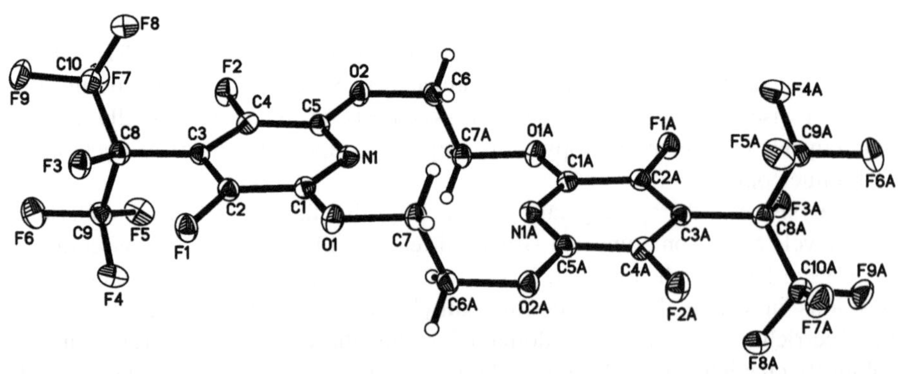

Fig. 3 Molecular structure of macrocycle **7** [46]

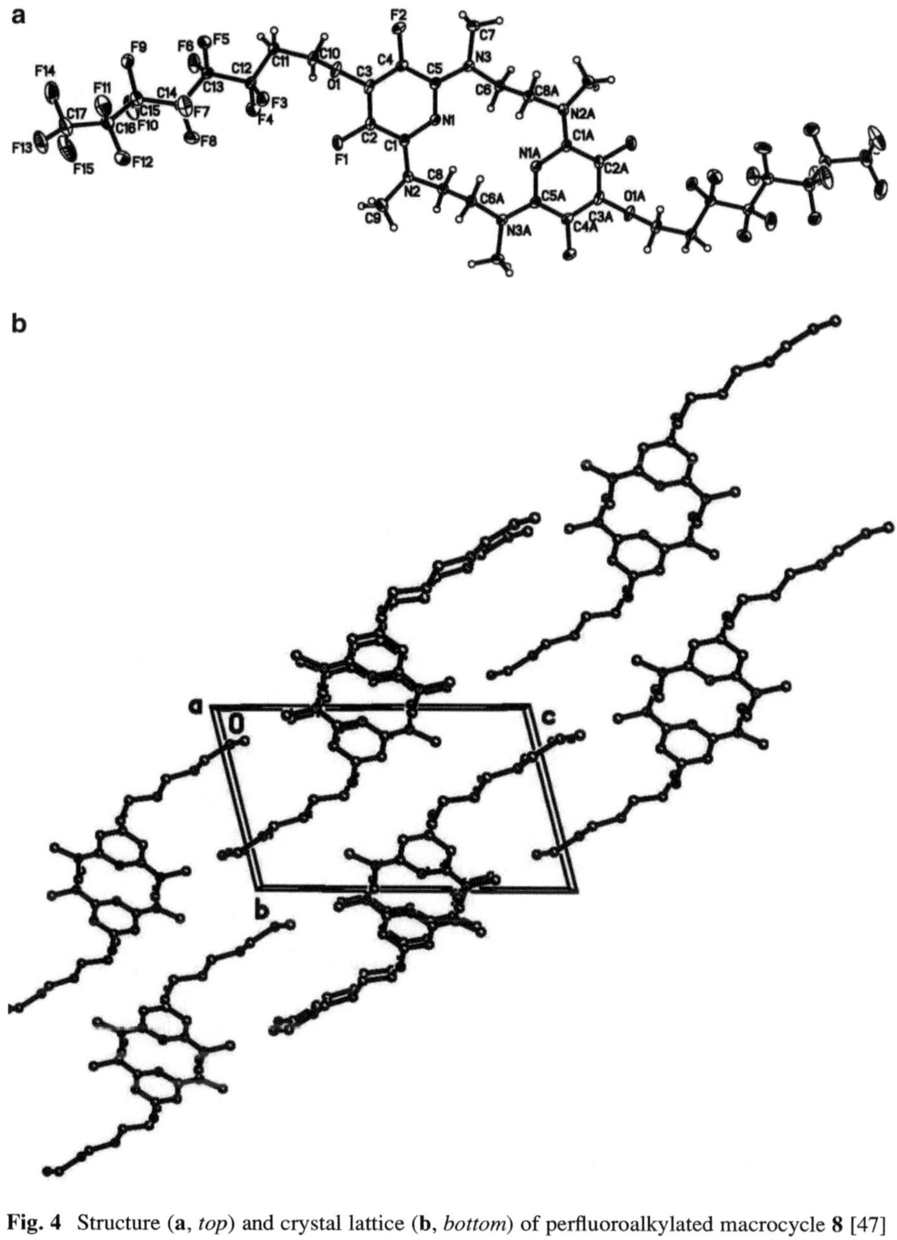

Fig. 4 Structure (**a**, *top*) and crystal lattice (**b**, *bottom*) of perfluoroalkylated macrocycle **8** [47]

By a related process (Route b, Scheme 10), reaction of pentafluoropyridine with an appropriate diamine gave a bridged derivative in which two pyridine units are connected via the 4-pyridine positions. Cyclisation by reaction with a further diamine unit, which displaces the fluorine atoms attached to the most reactive 2-positions, completes the macrocycle synthesis. Similarly, macrocycle **9**, derived

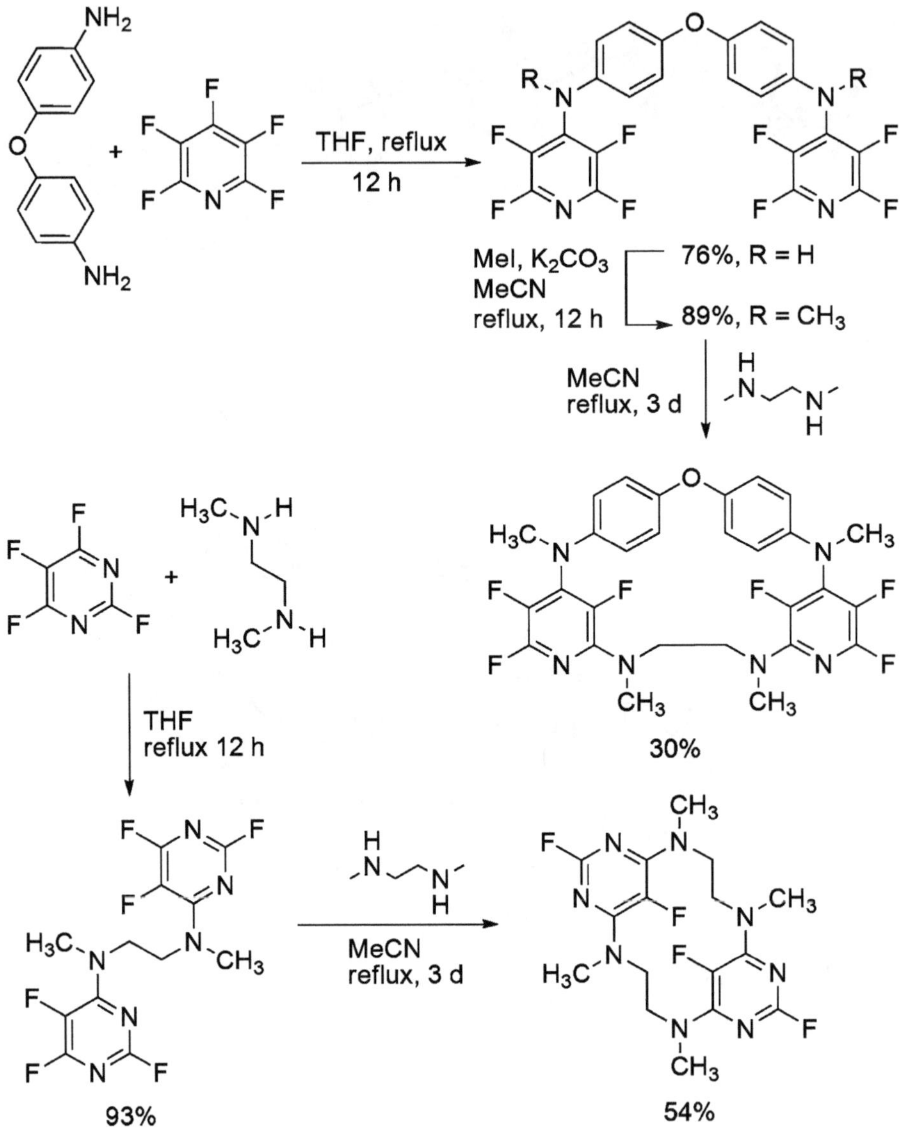

Scheme 12 Macrocycles from pentafluoropyridine and tetrafluoropyrimidine [50]

from tetrafluoropyrimidine, has been synthesised by a corresponding route (Scheme 12) and confirmed by X-ray crystallography (Fig. 5) [50].

Of course, in general, macrocyclic systems are used as effective complexing agents for a variety of ions and mixing a solution of macrocycle **7** with a mixture of aqueous sodium chloride, bromide and iodide led to the observation of macrocycle/anion complexes by negative ion electrospray mass spectrometry. Systems that

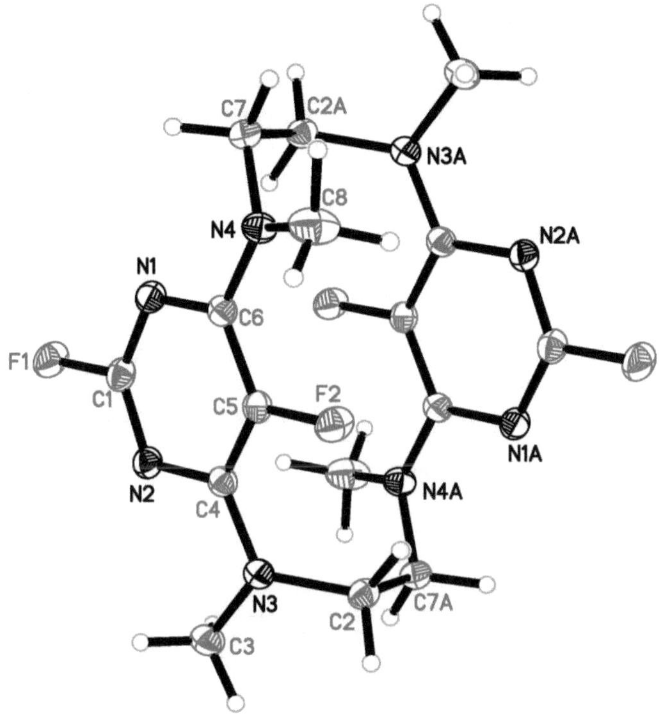

Fig. 5 Macrocycle **9** derived from tetrafluoropyrimidine [50]

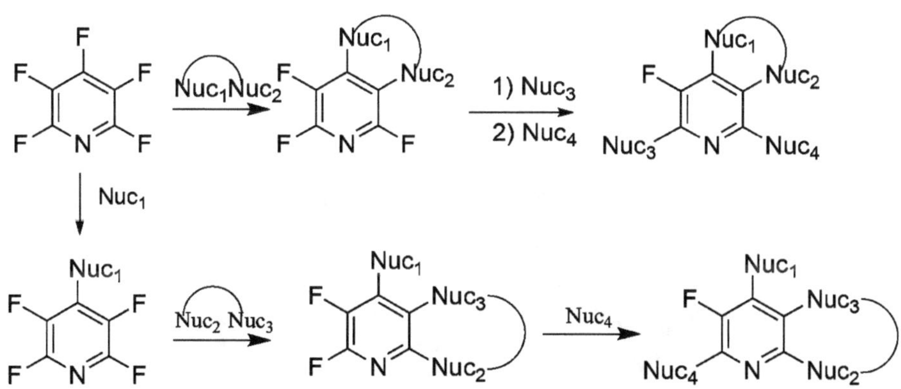

Scheme 13 Synthetic strategy for the synthesis of ring fused systems from pentafluoropyridine [22, 52]

form complexes with anionic guests, such as polyammonium, guanidinium and pyrrole-based systems [51] usually contain sites that are capable of directed hydrogen bonding, a situation that is not possible in this case.

2.3.2 Cyclisation by Intramolecular Processes

The order of nucleophilic substitution for the displacement of fluorine atoms in pentafluoropyridine [4], as outlined above, is generally in the order 4 > 2 > 3, but this reactivity can be altered by reaction with appropriate bifunctional nucleophiles. Substitution of the most activated 4-position may be followed by attack at the adjacent 3-position due to the geometric constraints of the heteroaromatic system as outlined in Scheme 13. Similarly, tetrafluoropyridine derivatives bearing substituents at the 4-position can, in principle, react with

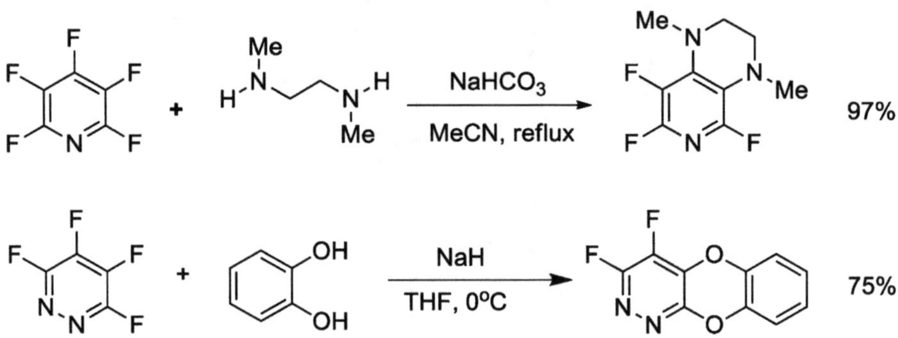

Scheme 14 Synthesis of [6,6]-ring fused systems [52, 53]

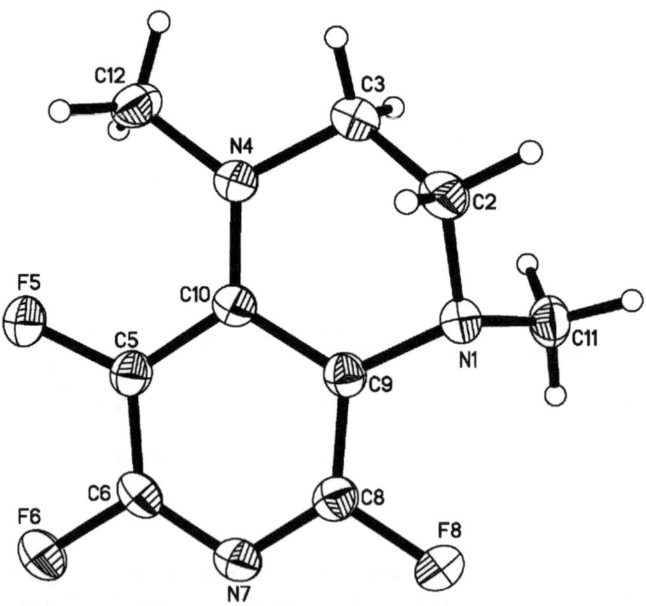

Fig. 6 Pyrido-pyrazine system **10** [52]

Table 7 Synthesis of [5,6]-ring fused systems [54–57]

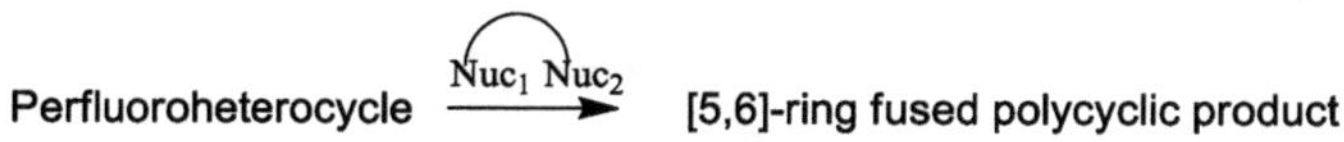

Perfluoroheterocycle	Nuc_1 Nuc_2	Conditions	Product
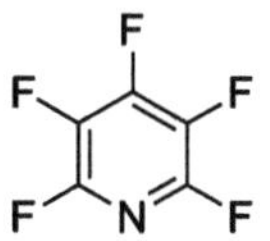	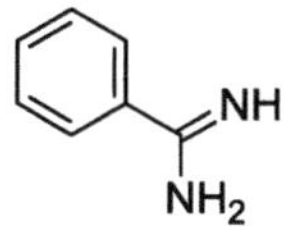	1) $NaHCO_3$ MeCN, reflux 2) LDA, THF, −78°C−rt	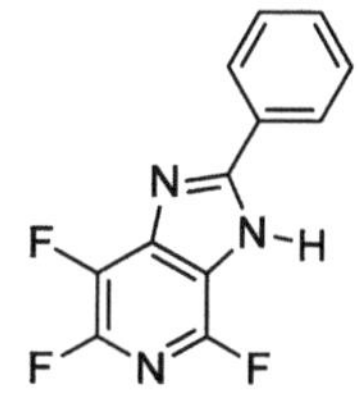 91%
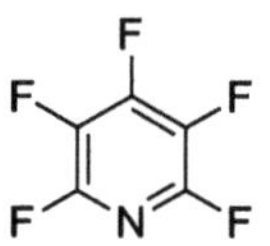	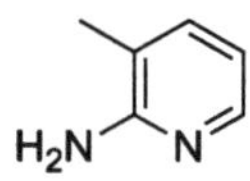	$NaHCO_3$ MeCN reflux, 70 h	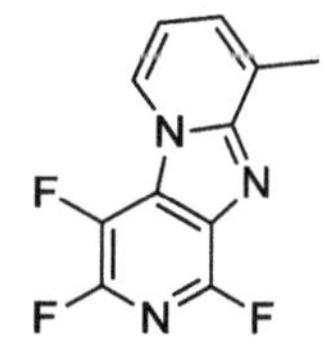 82%
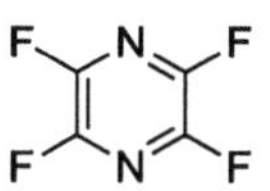	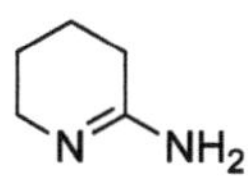	MeCN, reflux $NaHCO_3$	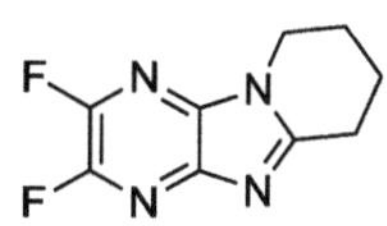 79%
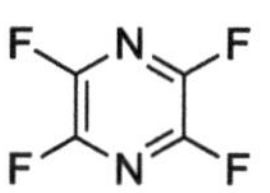	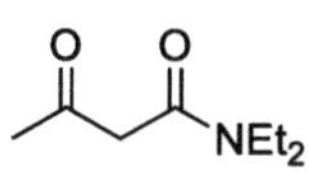	NaH, THF reflux, 1 d	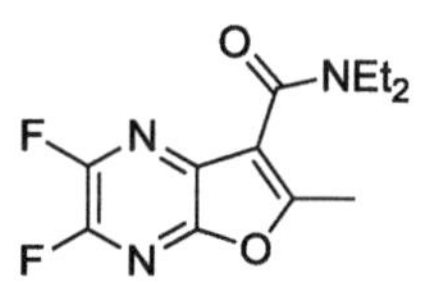 60%
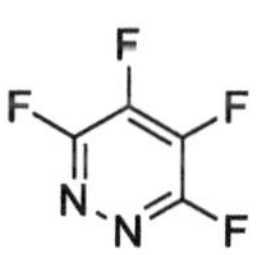	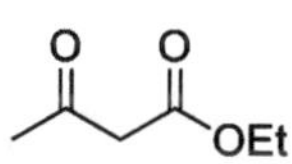	DIPEA, THF rt, 4 d	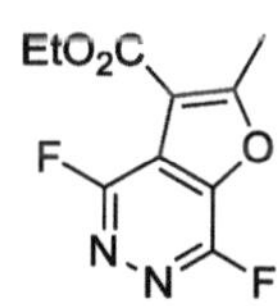 40%

appropriate difunctional nucleophiles to give polyfunctional annelated systems by an intramolecular process [22, 52] (Scheme 13).

Using the synthetic strategy outlined in Scheme 13, various [6,6]-ring fused systems have been prepared by reaction of appropriate difunctional

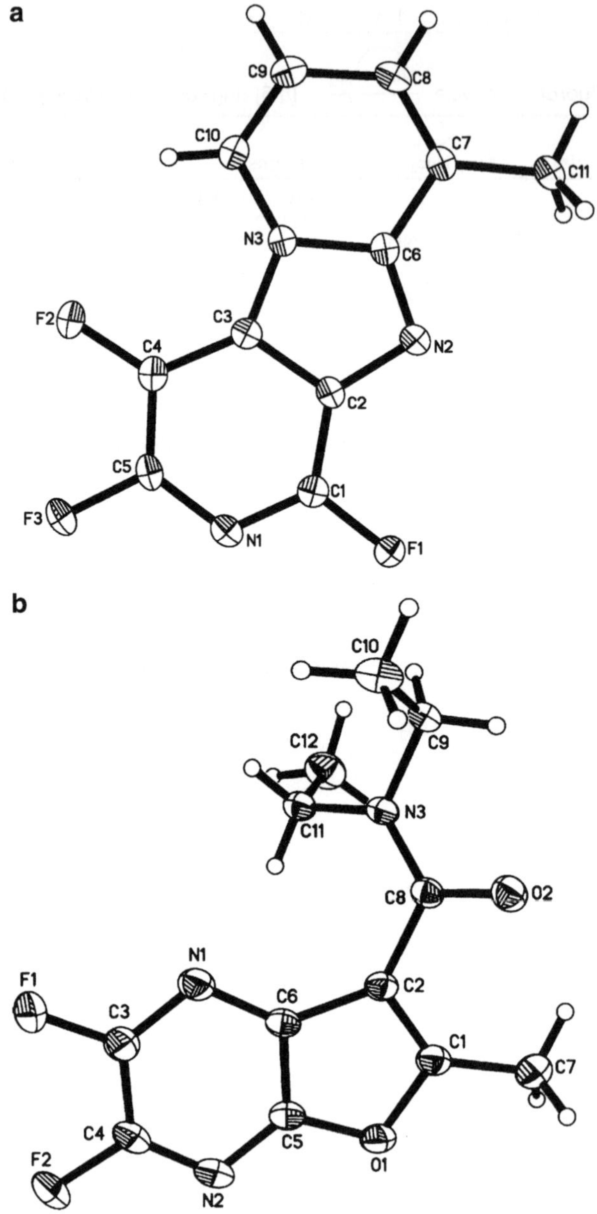

Fig. 7 Structures of [5,6]-ring fused systems **11** (**a**, *above*) and **12** (**b**, *below*) [55, 57]

nucleophiles and either pentafluoropyridine[52] or tetrafluoropyridazine [53] (Scheme 14). For example, reaction of pentafluoropyridine with *N,N*-dimethylaminoethylene diamine with pentafluoropyridine in the presence of base in acetonitrile at reflux temperature or under microwave irradiation gave

Table 8 Synthesis of ring fused systems from tetrafluoropyridine derivatives [22, 54, 57, 58]

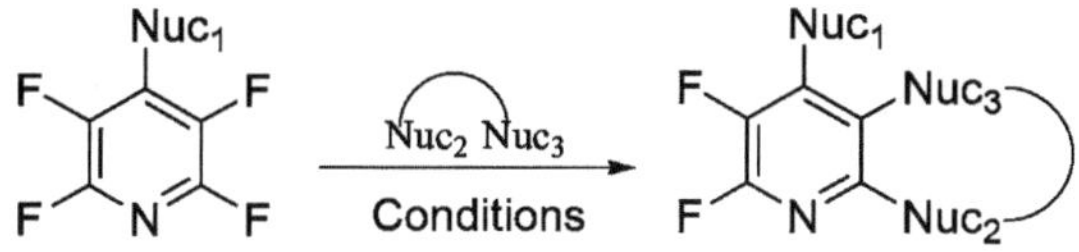

Nuc_1	Nuc_2 Nuc_3		Product	
$PhSO_2$	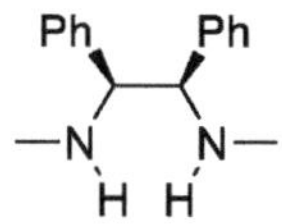	$NaHCO_3$, MeCN reflux	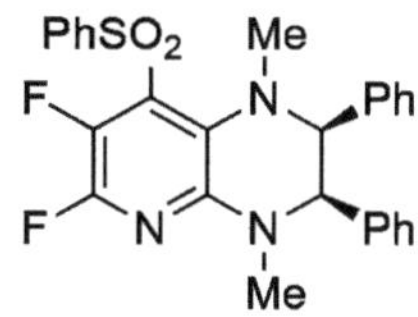	34%
$PhSO_2$	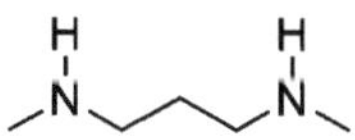	$NaHCO_3$, MeCN reflux	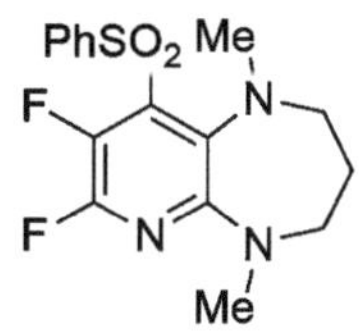	62%
$PhSO_2$	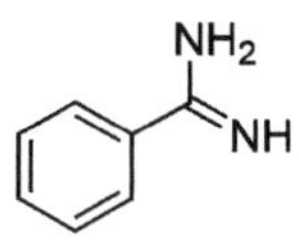	$NaHCO_3$, MeCN reflux	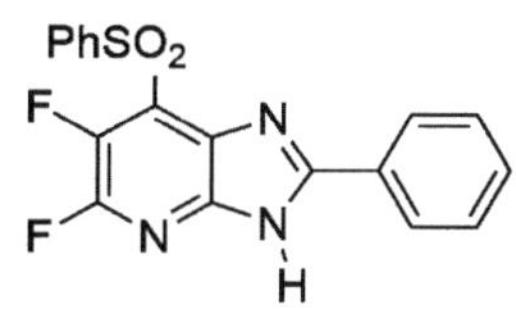	36%
CN	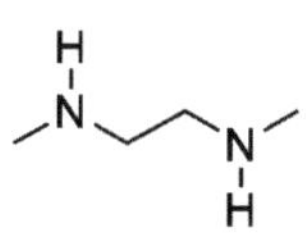	$NaHCO_3$, MeCN reflux	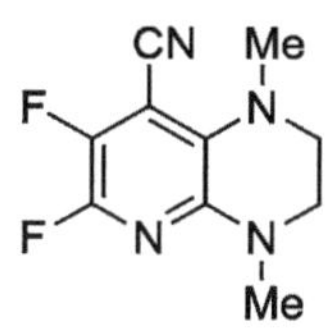	90%
CN	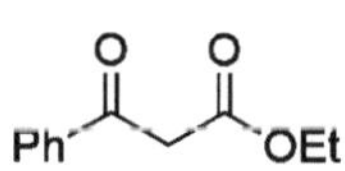	NaH, THF rt, 1 h	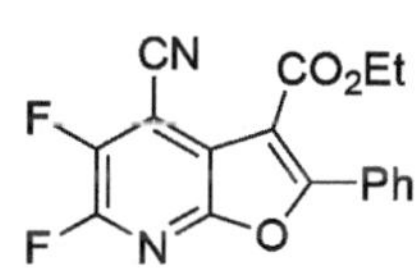	59%

the pyridopyridazine system **10** as shown by X-ray crystallography (Fig. 6) in high yield arising from initial substitution of fluorine located at the 4-position followed by cyclisation and displacement of fluorine at the less activated 3-position. Similarly, reaction of catechol with tetrafluoropyridazine gave a tricyclic ring-fused system [53].

In related processes, [5,6]-ring fused systems may be synthesised by reaction of appropriate difunctional nucleophiles such as benzamidine or 2-aminopyridine

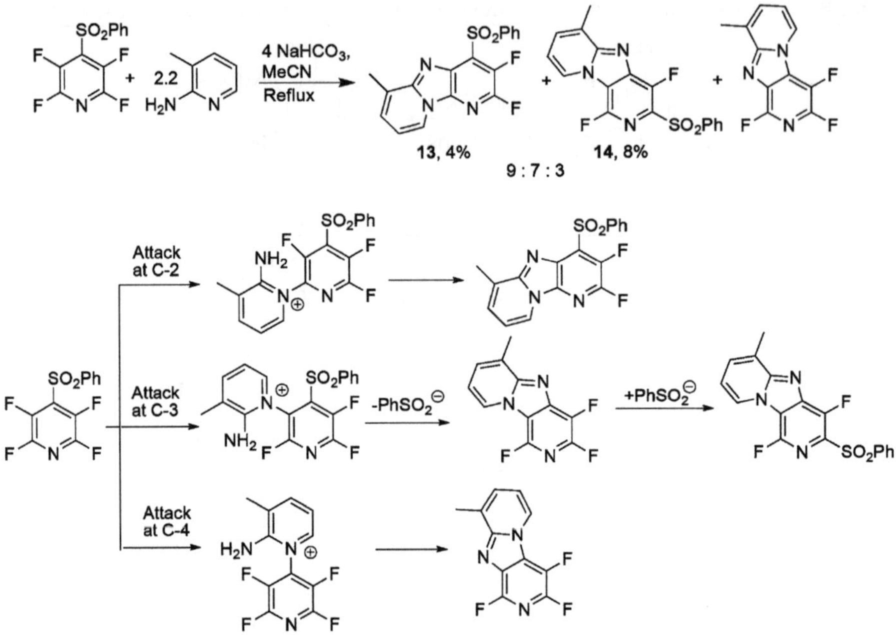

Scheme 15 Reaction of 4-phenylsulfonyl-tetrafluoropyridine with 2-amino-3-picoline [54]

derivatives, with pentafluoropyridine [54, 55] (Table 7). For example, heating a mixture of 2-amino-3-picoline with pentafluoropyridine in acetonitrile readily gives the [6,5,6]-tricyclic system **11** by sequential intramolecular S_NAr processes involving initial nucleophilic attack by the ring nitrogen atom of the picoline dinucleophile and subsequent cyclisation via the pendant amino group. In related processes, tetrafluoropyrazine [56] and tetrafluoropyridazine [57] react very efficiently with difunctional nucleophiles such as picoline, ethyl acetoacetate and benzamidine derivatives, to allow the synthesis of corresponding [5,6]-ring fused systems such as **12** (Table 7) and the structures of some of these very structurally unusual heterocyclic systems were confirmed by X-ray crystallography (Fig. 7).

As discussed above, 4-subsituted tetrafluoropyridine derivatives are still sufficiently electrophilic to react with nucleophiles and, in these cases, react with difunctional nucleophiles to give ring fused systems [22, 58] upon displacement of fluorine from the 2-positon followed, in general, by cyclisation by displacement of fluorine from the most geometrically accessible 3-postion (Table 8).

Reaction of 4-cyano- and 4-phenylsulfonyl-tetrafluoropyridine can, however, lead to some unexpected products upon reaction with appropriate difunctional nucleophiles (Scheme 15) [54]. Reaction of the 4-phenylsulfonyl tetrafluoropyridine with 2-amino-3-picoline gave a mixture of products, some of which **13**,**14** were characterised by X-ray crystallography [54] (Fig. 8), arising from displacement of fluorine from the 2- and 3-positions as well as the phenylsulfonyl

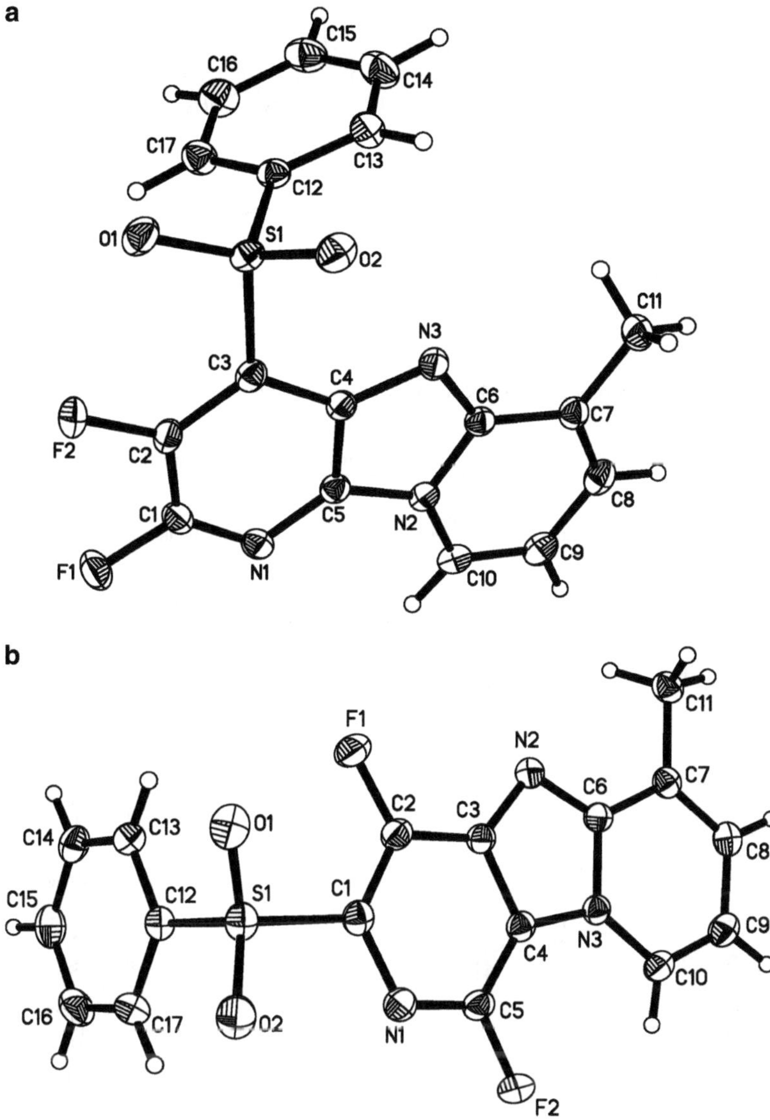

Fig. 8 Ring fused systems **13** (**a**, *above*) and **14** (**b**, *below*) from 4-phenylsulfonyl-tetrafluoropyridine [54]

group, followed by subsequent cyclisation and routes to all three products obtained have been formulated (Scheme 15).

Reaction of 4-cyanotetrafluoropyridine with benzamidine leads to a pyridopyrimidine system as the major product arising from substitution of fluorine at the 3-position followed by annelation involving the pendant cyano group [54] (Scheme 16).

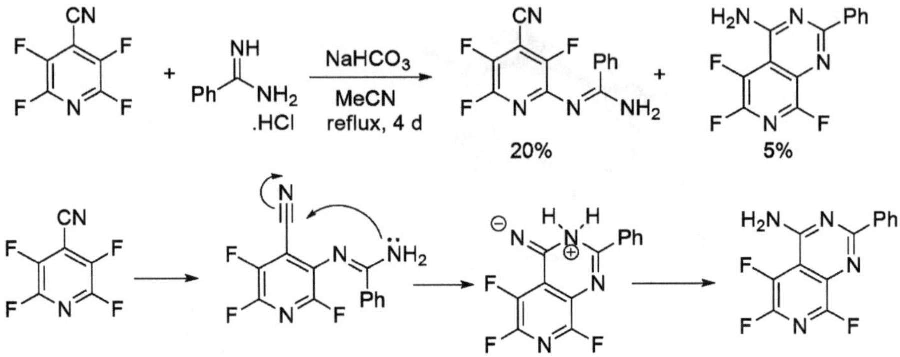

Scheme 16 Cyclisation processes of 4-cyano-tetrafluoropyridine [54]

Table 9 Annelation reactions of trifluoro- and difluoro-pyridazinone substrates [40, 41]

Perfluoroheterocycle —(Nuc1 Nuc2)→ [5,6]-ring fused polycyclic product

Perfluoroheterocycle	Nuc1 Nuc2	Conditions	Product
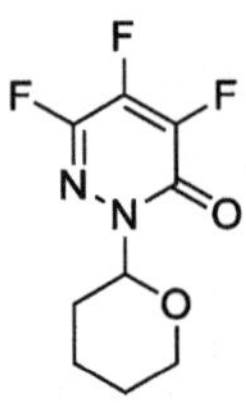	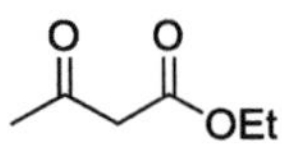	NaH, THF reflux, 4 h	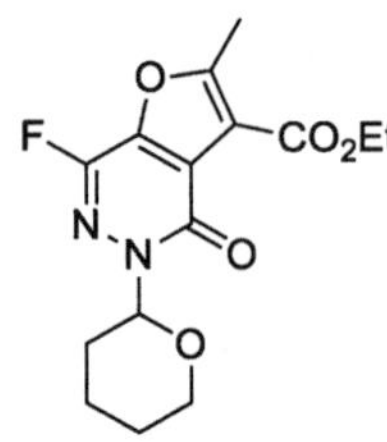 87%
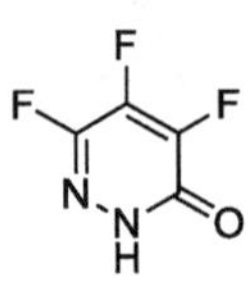	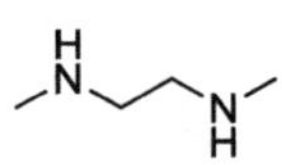	MeCN rt	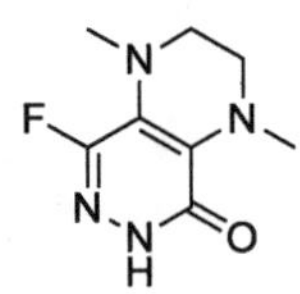 82%
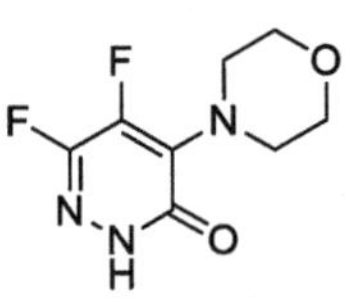	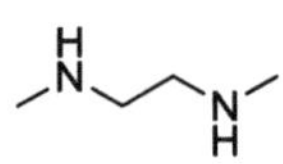	MeCN MW, 20 min, 150°C	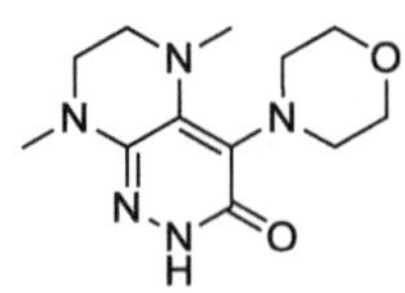 79%
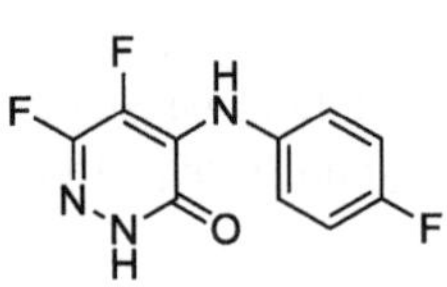	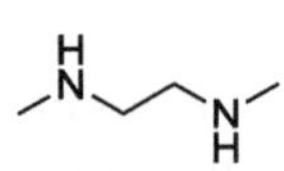	MeCN MW, 30 min, 150°C	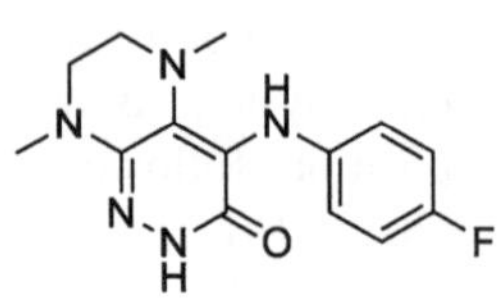 67%

Similarly, fused ring systems arising from annelation reactions of trifluoro- and difluoro-pyridazinone substrates have been reported [40, 41] (Table 9).

3 Conclusions

Recent research into the nucleophilic aromatic substitution chemistry of perfluoroheteroaromatic systems such as pentafluoropyridine, tetrafluoropyridazine and trifluoropyridazinone has focused on the use of these substrates as polyfunctional scaffolds for the synthesis of a range of ring fused derivatives, macrocycles and polyfunctional heterocyclic products. The wide range of structures that can be very difficult to access using conventional 'hydrocarbon' heterocyclic chemistry provides exciting new opportunities to exploit the unique reactivity profiles of perfluoroheteroaromatic substrates for organic synthesis.

References

1. Katritzky AR, Rees CW (1984) Comprehensive heterocyclic chemistry. Pergamon, Oxford
2. Pozharskii AF, Soldantenkov AT, Katritzky AR (1997) Heterocycles in life and society. Wiley, New York
3. Chambers RD, Sargent CR (1981) Adv Heterocycl Chem 28:1–71
4. Brooke GM (1997) J Fluor Chem 86:1–76
5. Chambers RD, Hutchinson J, Musgrave WKR (1964) J Chem Soc:3573–3576
6. Silvester MJ (1994) Adv Heterocycl Chem 59:1–37
7. Chambers RD (2004) Fluorine in organic chemistry. Blackwell, Oxford
8. Baasner B, Hagemann H, Tatlow JC (2000) Houben-Weyl organo fluorine compounds, vol E10. Thieme, Stuttgart
9. Uneyama K (2006) Organofluorine chemistry. Blackwell, Oxford
10. Shaub T, Fischer P, Steffen A, Braun T, Radius U, Mix A (2008) J Am Chem Soc 130: 9304–9317
11. Braun T, Wehmeier F (2011) Eur J Inorg Chem: 613–625.
12. Olah GA, Molnar A (1995) Hydrocarbon Chemistry. Wiley, New York
13. Chambers RD, Waterhouse JS, Williams DLH (1977) J Chem Soc Perkin 2:585–588
14. Chambers RD, Martin PA, Sandford G, Williams DLH (2008) J Fluor Chem 129:998–1002
15. Baker J, Muir M (2010) Can J Chem 88:588–597
16. Murray CB, Sandford G, Korn SR, Yufit DS, Howard JAK (2005) J Fluor Chem 126:571–576
17. Lork E, Mews R, Shakirov MM, Watson PG, Zibarev AV (2002) J Fluor Chem 115:165–168
18. Beletskaya IP, Artamkina GA, Ivushkin VA, Guilard R (2000) Tetrahedron Lett 41:313–316
19. Artamkina GA, Tarasenko EA, Lukashev NV, Beletskaya IP (1998) Tetrahedron Lett 39:901–904
20. Wang Y, Parkin SR, Watson MD (2008) Org Lett 10:4421–4424
21. Tyrra W, Naumann D, Hoge B, Yagupolskii YL (2003) J Fluor Chem 119:101–107
22. Hargreaves CA, Sandford G, Slater R, Yufit D, Howard JAK, Vong A (2007) Tetrahedron 63:5204–5211

23. Zeika O, Li Y, Jockusch S, Parkin G, Sattler A, Sattler W, Turro NJ (2010) Org Lett 12: 3696–3699
24. Kusov SZ, Rodionov VI, Vaganova TA, Shundrina IK, Malykhin EV (2009) J Fluor Chem 130:461–465
25. Cornago P, Escolastico C, Santa Maria MD, Claramunt RM, Fernandez-Castano C, Foces-Foces C, Fayet JP, Elguero J (1996) Tetrahedron 52:11075–11094
26. Huchel U, Schmidt C, Schmidt RR (1998) Eur J Org Chem 7:1353–1360
27. Hargreaves CA, Sandford G, Davis BG (2007) Fluorinated synthons. In: Soloshonok VA (ed) ACS Symposium Series 949. ACS, Washington DC, p 323–336
28. Christopher JA, Brophy L, Lynn SM, Miller DD, Sloan LA, Sandford G (2008) J Fluor Chem 129:447–454
29. Chambers RD, Hoskin PR, Sandford G, Yufit DS, Howard JAK (2001) J Chem Soc Perkin Trans 1:2788–2795
30. Benmansour H, Chambers RD, Sandford G, McGowan G, Dahaoui S, Yufit DS, Howard JAK (2001) J Fluor Chem 112:349–355
31. Lipinski CA, Lombardo F, Dominy BW, Feeney PJ (1997) Adv Drug Deliv Rev 23:3–25
32. Collins I (2000) J Chem Soc Perkin Trans 1:2845–2861
33. Collins I (2002) J Chem Soc Perkin Trans 1:1921–1940
34. Davie EW, Fujikawa K, Kisiel W (1991) Biochemistry 30:10363–10370
35. Kranjc A, Kikelj D, Peterlin-Masic L (2005) Curr Pharm Des 11:4207–4227
36. Walenga JM, Jeske WP, Hoppensteadt D, Fareed J (2003) Curr Opin Invest Drugs 4:272–281
37. Ng HP, Buckman BO, Eagen KA, Guilford WJ, Kochanny MJ, Mohan R, Shaw KJ, Wu SC, Lentz D, Liang A, Ho LTE, Smith D, Subramanyam B, Vergona R, Walters J, White KA, Sullivan ME, Morrissey MM, Phillips GB (2002) Bioorg Med Chem 10:657–666
38. Revesz L, Padova FED, Buhl T, Feifel R, Gram H, Hiestand P, Manning U, Wolf R, Zimmerlin AG (2002) Bioorg Med Chem Lett 12:2109–2112
39. Parks EL, Sandford G, Yufit DS, Howard JAK, Christopher JA, Miller DD (2010) Tetrahedron 66:6195–6204
40. Pattison G, Sandford G, Yufit DS, Howard JAK, Christopher JA, Miller DD (2009) J Org Chem 74:5533–5540
41. Pattison G, Sandford G, Yufit DS, Howard JAK, Christopher JA, Miller DD (2009) Tetrahedron 65:8844–8850
42. Gutov AV, Rusanov EB, Ryabitskii AB, Chernega AN (2010) J Fluor Chem 131:278–281
43. Lehn JM, Atwood JL, Davies JED, MacNicol DD, Votgle F (1996) Comprehensive supramolecular chemistry, vol 1–11. OUP, Oxford
44. Schneider HJ, Yatsimirsky AK (2000) Principles and methods in supramolecular chemistry. Wiley, New York
45. Sandford G (2003) Chem Eur J 9:1464–1469
46. Chambers RD, Hoskin PR, Kenwright AR, Khalil A, Richmond P, Sandford G, Yufit DS, Howard JAK (2003) Org Biomol Chem 1:2137–2147
47. Chambers RD, Khalil A, Richmond P, Sandford G, Yufit DS, Howard JAK (2004) J Fluor Chem 125:715–720
48. Hope EG, Stuart AM (2000) Adv Inorg Fluorides:403–436.
49. Horvath IT (1998) Acc Chem Res 31:641–650
50. Ranjbar-Karimi R, Sandford G, Yufit DS, Howard JAK (2008) J Fluor Chem 129:307–313
51. Beer PD, Gale PA (2001) Angew Chem Int Ed Engl 40:486–516
52. Sandford G, Slater R, Yufit DS, Howard JAK, Vong A (2005) J Org Chem 70:7208–7216
53. Pattison G, Sandford G, Yufit DS, Howard JAK, Christopher JA, Miller DD (2010) Beilstein J Org Chem 6:45
54. Cartwright MW, Sandford G, Yufit DS, Howard JAK, Christopher JA, Miller DD (2007) Tetrahedron 63:7027–7035
55. Cartwright MW, Convery L, Kraynck T, Sandford G, Yufit DS, Howard JAK, Christopher JA, Miller DD (2010) Tetrahedron 66:519–529

56. Parks EL, Sandford G, Yufit DS, Howard JAK, Christopher JA, Miller DD (2010) J Fluor Chem 131:1086–1090
57. Cartwright MW, Parks EL, Pattison G, Slater R, Sandford G, Wilson I, Yufit DS, Howard JAK, Christopher JA, Miller DD (2010) Tetrahedron 66:3222–3227
58. Baron A, Sandford G, Slater R, Yufit DS, Howard JAK, Vong A (2005) J Org Chem 70:9377–9381

Top Heterocycl Chem (2012) 27: 33–64
DOI: 10.1007/7081_2011_58

Published online: 3 November 2011

Monofluorinated Heterocycles

Andrei A. Gakh

Abstract The chapter presents an overview of synthetic chemistry of ring-fluorinated heterocycles containing only one fluorine atom (monofluorinated heterocycles). Particular attention is given to modern nucleophilic and electrophilic fluorination methods, including catalytic reactions, chiral and microwave-assisted synthesis, carbene and hypervalent chemistry, utilization of ionic intermediates and ionic liquids, and others. One of the major emphases of the chapter is identification of the remaining "white spots" as opportunities for the future research effort.

Keywords Fluorination · Heterocycles

Contents

A.A. Gakh (✉)
National Institutes of Health, NIDDK, 8 Center Drive, Bethesda, MD 20892, USA
e-mail: andrei.gakh@nih.gov

Abbreviations

Accufluor®	1-Fluoro-4-hydroxy-1,4-diazoniabicyclo[2.2.2] octane bis-tetrafluoroborate
CEFOX	Cesium fluoroxysulphate
DAST	Diethylaminosulfur trifluoride
Deoxo-Fluor®	Bis(2-methoxyethyl)aminosulfur trifluoride
EMEF	*N*-Ethyl(hexamethylenetetraammonium) fluoride
NBS/Et_3N•3HF	*N*-Bromsuccinimide/Et_3N•3HF
NFSI	*N*-Fluorobenzenesulfonimide
"Proton sponge"/Et_3N•3HF	1,8-Bis(dimethylamino)naphthalene/Et_3N•3HF
Selectfluor®	1-Chloromethyl-4-fluoro-1,4-diazoniabicyclo[2.2.2] octane bis-tetrafluoroborate
TBAF (Bu_4NF)	Tetrabutylammonium fluoride
TMAF (Me_4NF)	Tetramethylammonium fluoride

1 Introduction

Ring-fluorinated heterocycles containing only one fluorine atom (monofluorinated heterocycles) constitute an important family of organic compounds with a wide array of applications ranging from drugs to multi-ton industrial intermediates [1, 2]. Although the first representatives of monofluorinated heterocycles, including 2-fluoropyridine [3], were synthesized almost 100 years ago, major developments in this field were made during the last two decades. Despite general recognition of their academic and practical importance, the family of monofluorinated heterocycles has significant gaps in synthetic methodology, availability, and basic physical–chemical data. It is expected that these gaps would eventually be filled. The main purpose of this review is not only to explore the field of synthetic chemistry of the monofluorinated heterocycles but also to identify the remaining gaps as opportunities for the future research effort.

2 Three-Membered Monofluoroheterocycles

The monofluorinated derivatives are expected to be stable enough for isolation and characterization for almost all possible three-membered heterocycles, including marginally thermally stable 3-fluorodiazirine **13** [4]. In practice, however, many parent monofluorinated heterocycles **1–15** have not been reported yet, whereas their derivatives have been synthesized successfully. This is an unfortunate common feature for almost all of the three member-ring heterocycles presented below (Fig. 1).

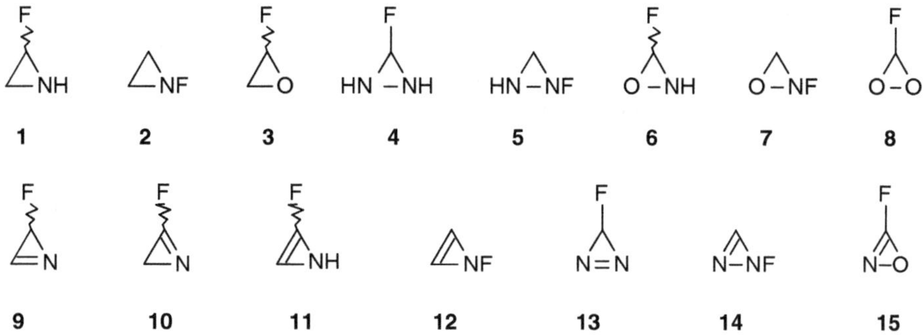

Fig. 1 Three-membered monofluoroheterocycles

A reliable method of synthesis was reported only for the parent 2-fluorooxirane (**3**). It has been prepared in four steps from chlorotrifluoroethylene according to the procedure presented below (Scheme 1). The compound **3** is of theoretical interest as one of the smallest chiral compounds [5].

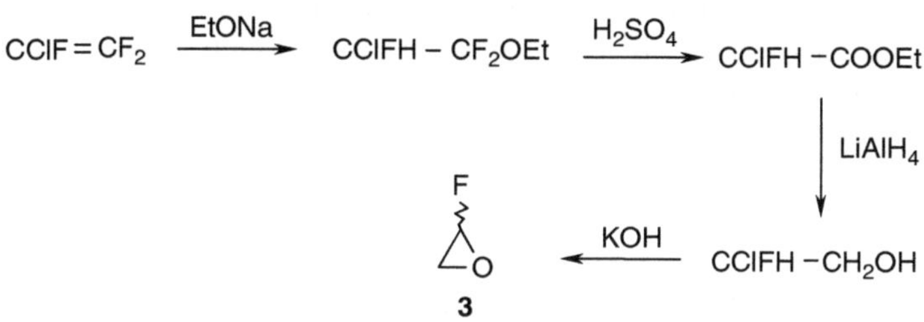

Scheme 1 Synthesis of 2-fluorooxirane (**3**)

Parent 2-fluoroaziridine (**1**) is also chiral. The compound is expected to be stable, as well as its nonchiral *N*-fluoro isomer **2**. The derivatives of both compounds **1** and **2** are known and are of considerable academic interest due to their unusual geometry and substantial strain.

One of the recently reported syntheses of substituted 2-fluoroaziridines **1a** is based on ring closure of β-fluorinated β-chloroamines (Scheme 2). These β-fluorinated β-chloroamines can be prepared via reduction of corresponding chlorofluoroacetic acid amides [6].

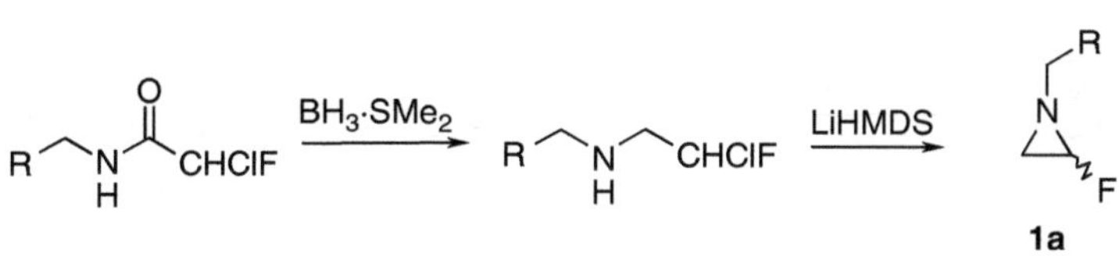

Scheme 2 Synthesis of 2-fluoroaziridines **1a** from β-fluorinated β-chloroamines

Derivatives of 2-fluoroaziridine (**1**) can also be prepared via fluorocarbene addition to a C=N double bond. An example of this approach is presented below (Scheme 3). In this case, the ester of dibromofluoroacetic acid serves as a source of fluorocarbene-type intermediate, which is described in terms of "the Reformatsky-type aza-Darzens reaction" [7]. The reaction proceeds with reasonable stereoselectivity and produces a mixture of syn/anti products in better than 5:1 ratio. It is reported that the major syn isomers of **1b** are actually less thermodynamically stable than the minor anti-isomers [7].

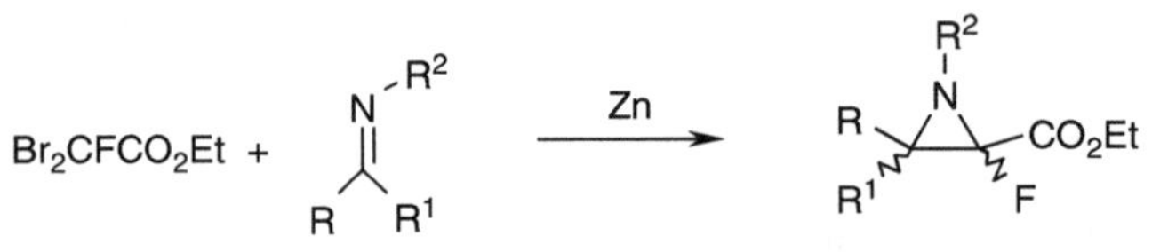

1b

Scheme 3 Synthesis of 2-fluoroaziridines **1b** from β-fluorinated β-chloroamines

The simplest fluorocarbene, CHF: (generated under ultrasound conditions from $CHFBr_2$ and "activated" Pb in the presence of tetrabutylammonium bromide, TBAB) can be trapped by imines to yield 2-fluoroaziridines **1c** and **1d** (Scheme 4). The reaction likely proceeds via formation of azometine ylides [8].

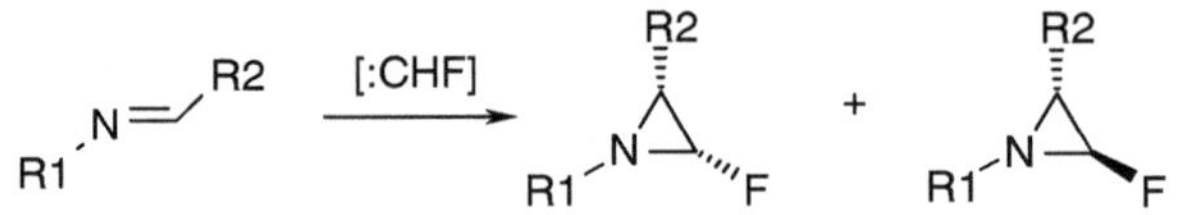

1c **1d**

Scheme 4 Synthesis of 2-fluoroaziridines **1c** and **1d**

N-fluorinated carboxyaziridines **2a** are also reported. These N-fluorinated heterocycles **2a** are of theoretical interest due to the exceptionally high configuration stability of the nitrogen atom (the barrier of interconversion sometimes exceeds 30 kcal/mol) [9]. One of the modern synthetic approaches entails stereoselective direct fluorination of the *N*-aminomethyl derivatives leading to **2a** (Scheme 5).

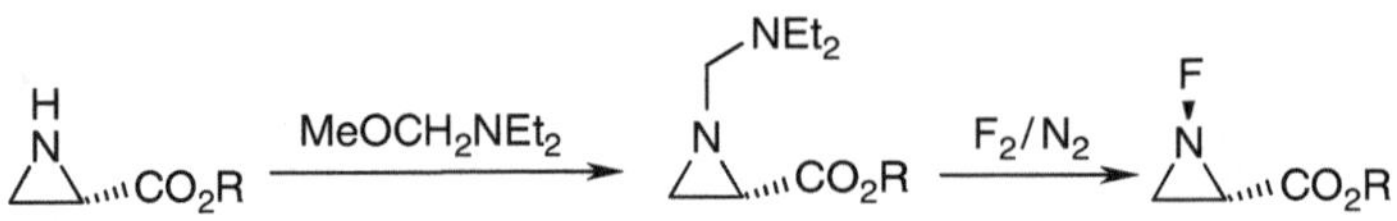

2a

Scheme 5 Synthesis of *N*-fluorocarboxyaziridines **2a**

Only polyfluorinated derivatives of fluorodiaziridines **4** and **5** were reported so far, including perfluorodiziridine (**4a**), but the employed synthetic procedures are specific for polyfluorinated organic compounds and cannot be used for the preparation of the parent monofluorinated heterocycles **4** and **5** (Scheme 6) [10]. The same is true for fluorooxaziridines **6** and **7**.

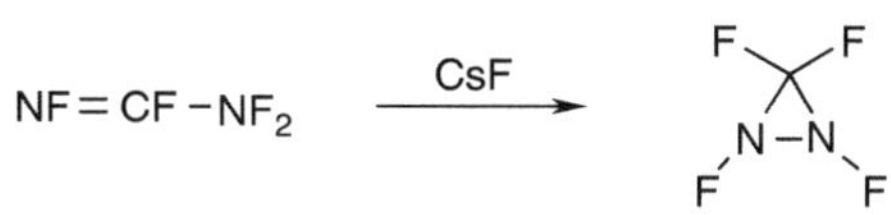

4a

Scheme 6 Preparation of perfluorodiziridine (**4a**)

Unlike compounds **4–12** and **14–15**, the derivatives of monofluorinated diazirine **13** are well represented in the available literature and were a subject of extensive computational studies related to their stability in respect to the diazirine-diazo ring-opening process [4]. Several methods of synthesis were reported. One of them is based on nucleophilic fluorination of the corresponding 3-chloro derivatives of **13** using various sources of fluoride anion, such as Bu_4NF (Scheme 7) [11]. It was suggested that the reaction leading to fluorodiazirines **13a** might involve a diazirinium cation/anion pair. Indeed, some nucleophilic substitution reactions of 3-chlorodiazirines are first order reactions [12].

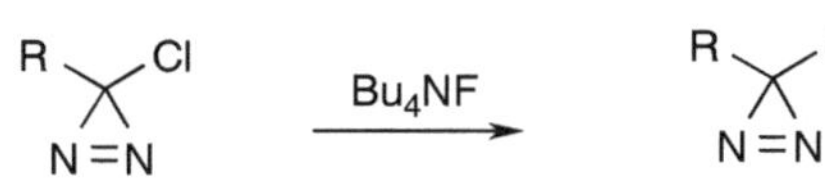

13a

Scheme 7 Nucleophilic fluorination of 3-chlorodiazirines

The 2,4-dinitrophenoxy group can also be used as the leaving group (e.g., compound **13b**) in the synthesis of fluorodiazirines **13c** and **13d**. This leaving group allows introduction not only of fluorine but also of chlorine using Bu_4NCl instead of Bu_4NF (Scheme 8) [13].

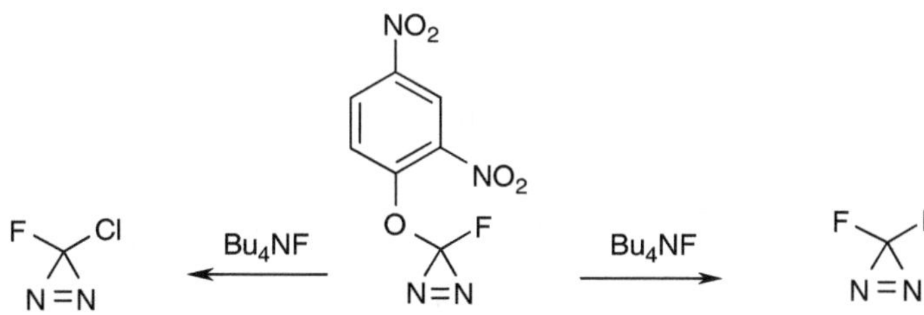

13c **13b** **13d**

Scheme 8 Preparation of chlorofluorodiazirine **13c**

Perhaps one of the most valuable properties of monofluorodiazirines is their facile thermolysis leading to corresponding monofluorocabenes via N_2 elimination. Unlike their chloro- and bromo- analogs, monofluorocarbenes are stable enough to be captured with a variety of electron rich substrates, including double bonds [13]. The latter reaction yields monofluorocyclopropanes (Scheme 9) [13, 14].

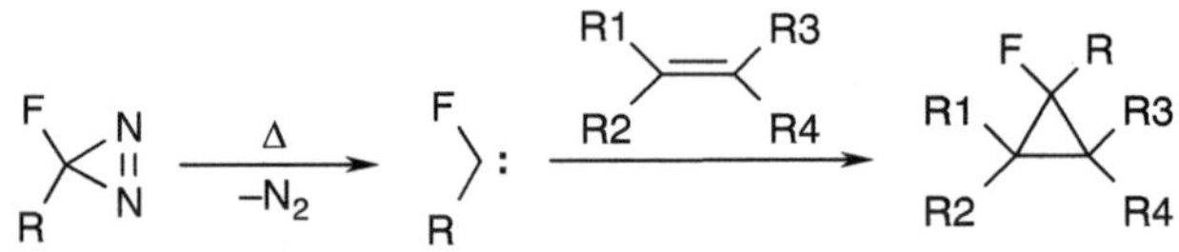

Scheme 9 Generation of monofluorocabenes from monofluorodiazirines

3 Four-Membered Monofluoroheterocycles

Four-membered monofluorinated heterocycles **16–21** are in general far more abundant than the three-membered ones, but this only applies to the easier-to-make 3-fluoroheterocycles **17**, **19**, and **21**. As expected, these small ring compounds also carry substantial strain due to the unfavorable stretching of their valence angles from the ideal tetrahedral arrangement to the almost square geometry. Some of the most common monofluorinated four-membered heterocycles are presented below (Fig. 2).

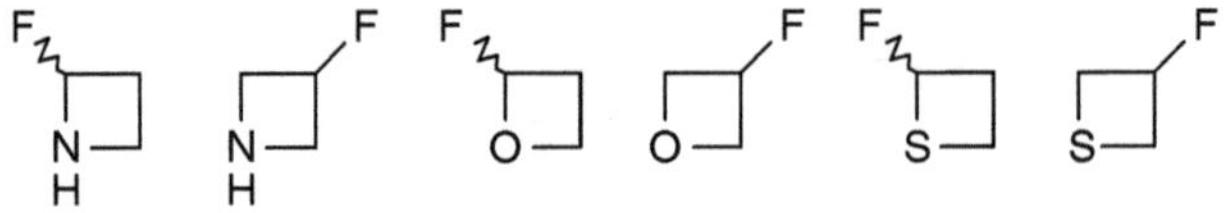

16 **17** **18** **19** **20** **21**

Fig. 2 Four-membered monofluoroheterocycles

Unlike 2-fluoroazetidine (**16**), 3-fluoroazetidine (**17**) is now commercially available. Typical synthetic procedures for the synthesis of **17** derivatives are based on the cyclization of appropriate fluorinated aliphatic precursors. This methodology was successfully employed for the preparation of Boc-protected 3-fluoroazetidine-3-carboxylic acid (**17a,** Scheme 10) [15].

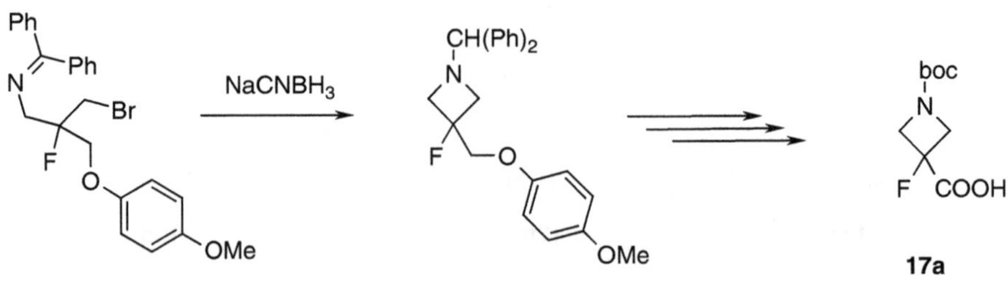

17a

Scheme 10 Preparation of Boc-protected 3-fluoroazetidine-3-carboxylic acid (**17a**)

A similar approach has been used for the synthesis of other 3-substiuted 3-fluoroazetidines (**17b**) of biological interest. In this case, the introduction of the fluorine atom has been achieved via bromofluorination of a double bond with the NBS/Et_3N•3HF system (Scheme 11) [16].

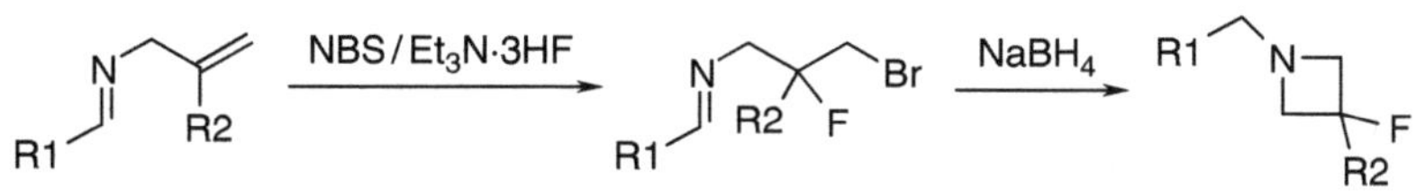

Scheme 11 Synthesis of 3-substiuted 3-fluoroazetidines **17b** via bromofluorination

Other known methods for the synthesis of 3-fluoroazetidines **17c** include direct fluorination of azetidines [17] and the reaction of 1-azabicyclo[n.1.0]alkanes with liquid HF or the pyridine•10HF complex (Scheme 12) [18].

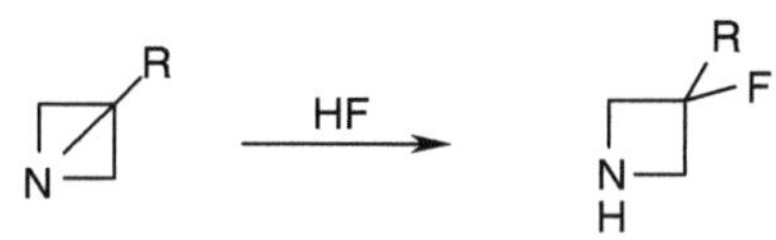

Scheme 12 Synthesis of 3-substiuted 3-fluoroazetidines **17c** via 1-azabicyclo[n.1.0]alkanes

Stereoselective electrophilic fluorination of substituted azetidinones (β-lactams) with *N*-fluorobenzenesulfonimide (NFSI) can be used for the synthesis of 3-fluoroazetidinones **17d** (Scheme 13). These compounds are of interest for the development of carbapenem antibiotics [19].

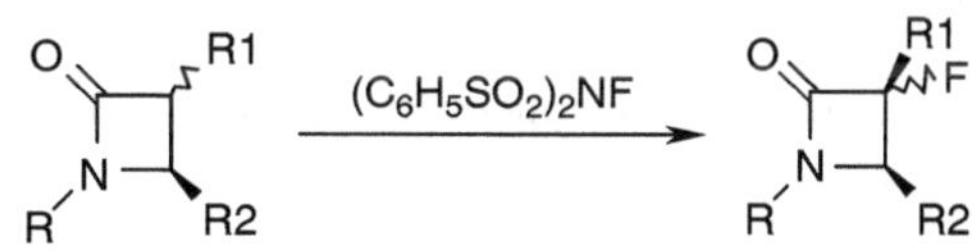

Scheme 13 Stereoselective electrophilic fluorination of β-lactams

Similar to 2-fluoroazetidine (**16**) and 3-fluoroazetidine (**17**), derivatives of 3-fluorooxetane (**19**) are better known and more studied than 2-fluorooxetane (**18**). Parent 3-fluorooxetane (**19**) can be easily prepared from 3-oxetyl tosylate and KF (Scheme 14) [20].

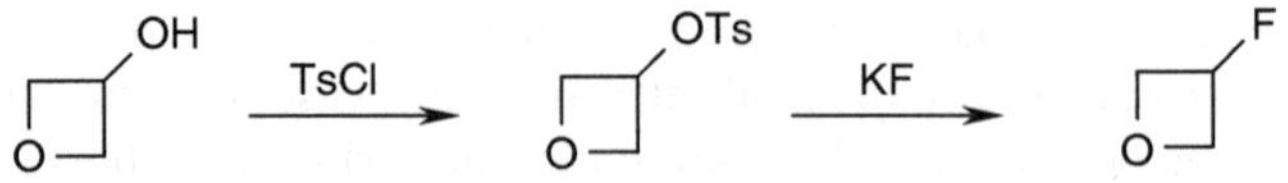

19

Scheme 14 Synthesis of 3-fluorooxetane (**19**)

Alternatively, 3-oxetanols can be directly converted to 3-fluorooxetanes **19a** using DAST as a fluorinating agent (Scheme 15). This method was used for the preparation of 3-fluorooxetane δ-amino acids [21].

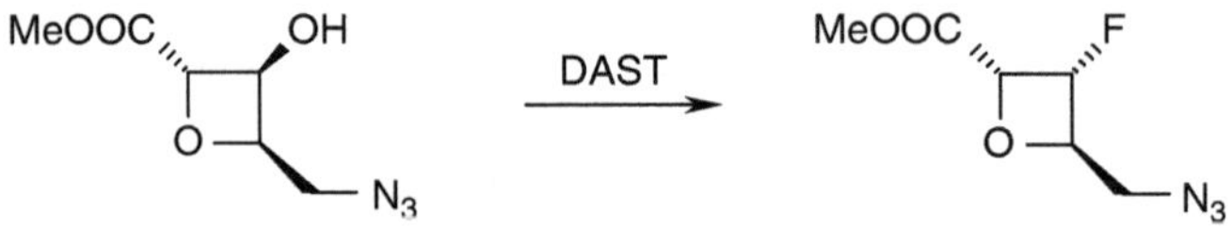

19a

Scheme 15 DAST fluorination

Keto-derivatives of 3-fluorooxetane (fluorinated β-lactones **19b**) are also known. They can be prepared via cyclization of α-fluoro-β-hydroxycarboxylic acids. In some cases, corresponding 1-fluoroalkenes were also detected in the reaction mixture. These 1- fluoroalkenes are the primary thermal decomposition products of the fluorinated β-lactones **19b** (CO_2 elimination, Scheme 16) [22].

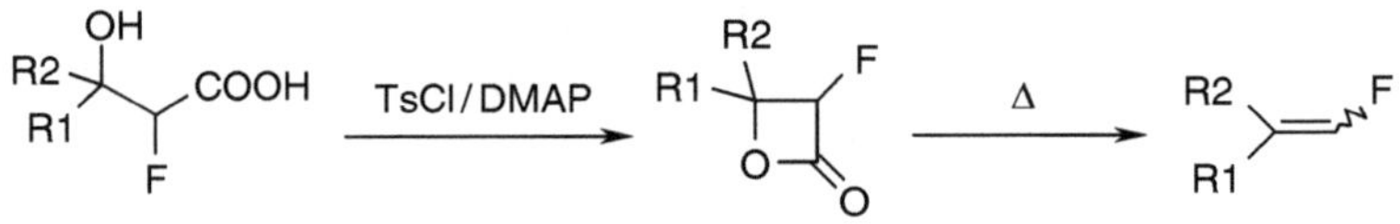

19b

Scheme 16 Synthesis and thermal decomposition of fluorinated β-lactones **19b**

A derivative of 2-fluorothietane (**20**) was recently prepared and successfully used in a glycosidation reaction leading to the S-analog of the natural antibiotic Oxetanocin A. The synthesis of this 2-fluorothietane derivative **20a** was performed using direct fluorination of substituted thietane with DAST or Deoxo-Fluor® in 2-position (Scheme 17) [23].

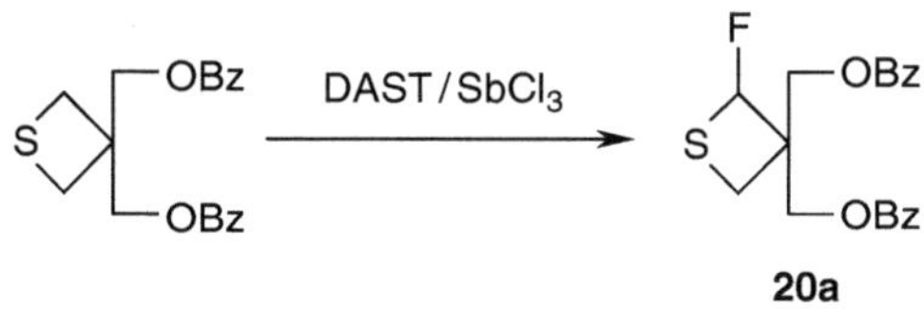

Scheme 17 Synthesis of the 2-fluorothietane derivative **20a**

4 Five-Membered Monofluoroheterocycles

Synthesis and reactions of five-membered monofluoroheterocycles are frequently reported in the literature. Some of the popular fluorinated heterocycles of this class are presented below (Fig. 3).

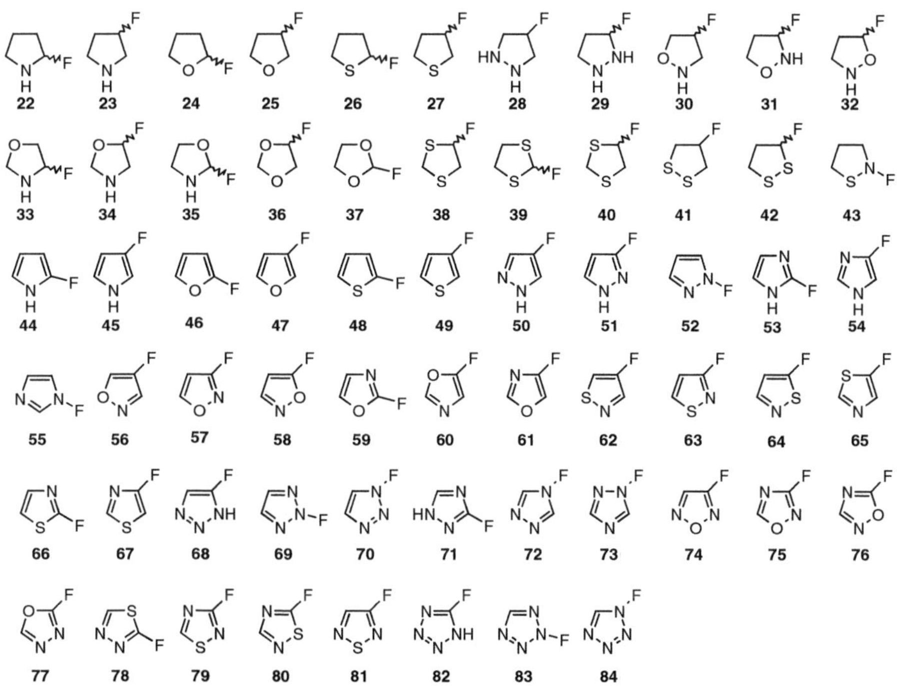

Fig. 3 Five-membered monofluoroheterocycles

Unlike 2-fluoropyrrolidine (**22**), both stereoisomers of the parent 3-fluoropyrrolidine (**23**) are now available commercially. They can be prepared from corresponding enantiomerically pure 3-tosylates using KF as a nucleophilic fluorinating agent (Scheme 18) [24]. Alternatively, stereoselective fluorination of 3-hydroxypyrrolidines can be accomplished with DAST [25].

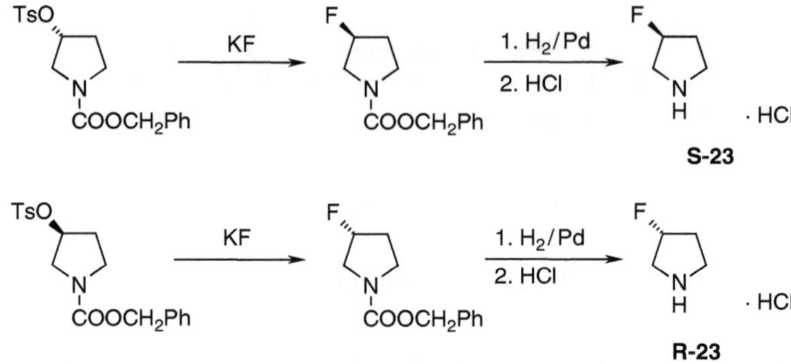

Scheme 18 Synthesis of *R*-and *S*-3-fluoropyrrolidines

Substituted 3-fluoropyrolidines are extensively studied, primarily in the form of 2-carboxy derivatives (fluorinated prolines), which are valuable nonnatural amino acids capable of improving the conformational stability of collagen [26, 27]. A practical synthesis of the 4-fluoroproline diastereomer (**23a**) from 4-hydroxyproline was recently reported (Scheme 19) [27].

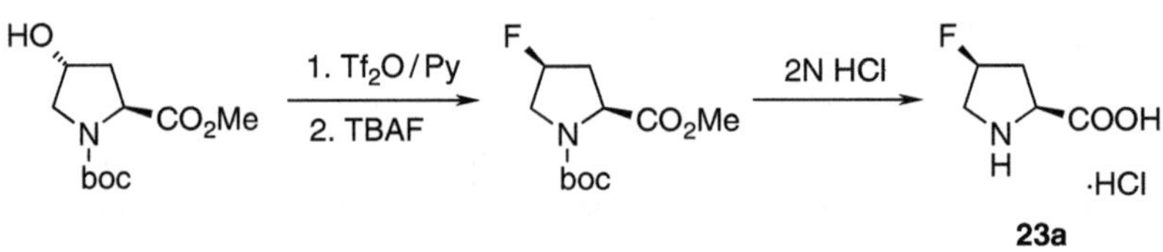

Scheme 19 Synthesis of 4-fluoroproline (**23a**)

Parent 2-fluorotetrahydrofuran (2-fluorooxolan, **24**) was prepared by the anode oxidation of THF using Et_3N•5HF as a fluorine source (Scheme 20). The synthesis can be performed in solvent-free conditions [28].

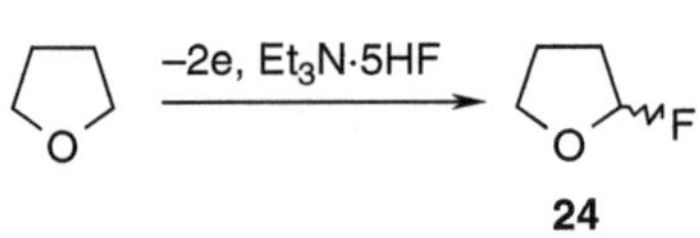

Scheme 20 Synthesis of 2-fluorotetrahydrofuran (**24**)

Direct fluorination of γ-butyrolactone leads to a mixture of predominantly monofluorinated γ-butyrolactones **24b**, **24c**, and **24d** (Scheme 21). Regioselectivity of the fluorination process can be improved by adding NaF as a HF scavenger [29].

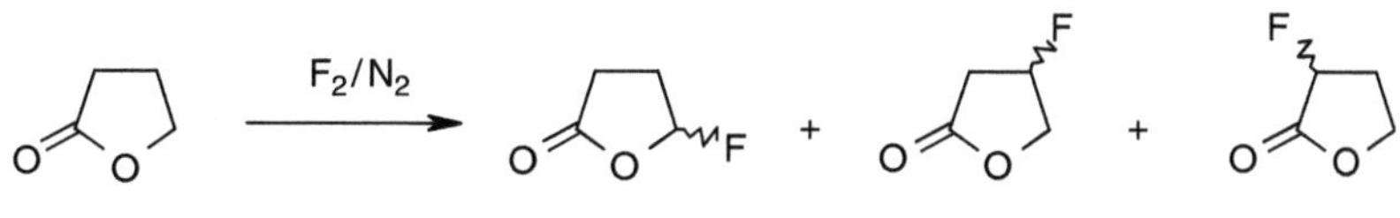

Scheme 21 Direct fluorination of γ-butyrolactone

Regioselective electrophilic fluorination of γ-butyrolactone can be accomplished using *N*-fluoro-*o*-benzenedisulfonimide via silyl enol ether (Scheme 22) [30].

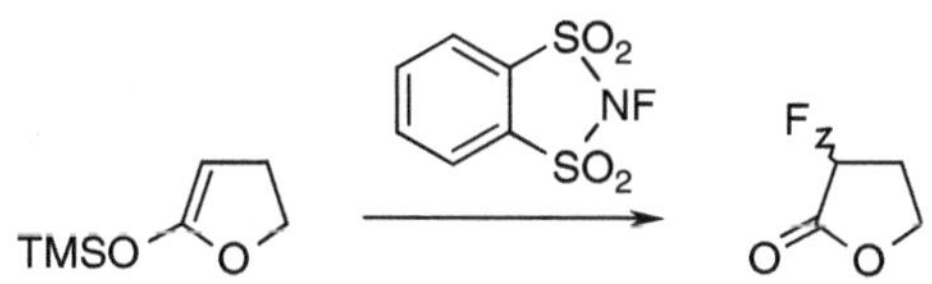

Scheme 22 Electrophilic fluorination of γ-butyrolactone silyl enol ether

Polyhydroxy derivatives of 3-fluorooxolane (**25**) are a very important class of heterocycles, which are primarily found among fluorinated carbohydrates and nucleosides. Several important drugs contain a fluorinated oxolan fragment (Fig. 4). Among them are the Hepatitis B drug Clevudine (L-FMAU, 1-[(2*S*,3*R*,4*S*,5*S*)-3-fluoro-4-hydroxy-5-(hydroxymethyl)oxolan-2-yl]-5-methylpyrimidine-2,4-dione) [31] and the leukemia drug Clofarabine (5-(6-amino-2-chloro-purin-9-yl) -4-fluoro-2- (hydroxymethyl)oxolan-3-ol) [32].

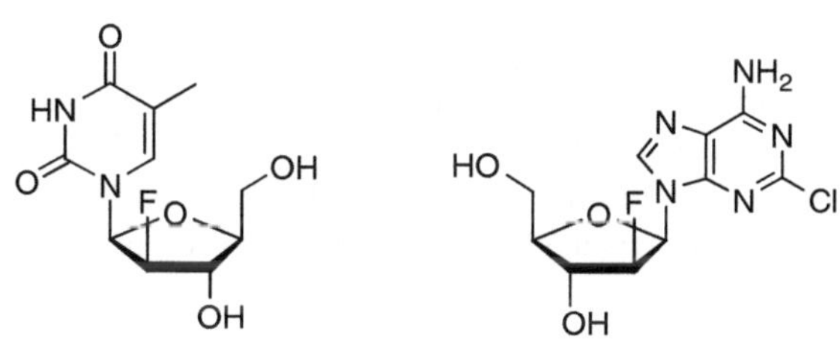

Fig. 4 Clevudine and Clofarabine

Synthesis of these monofluorinated nucleosides can be accomplished via coupling of fluorinated ribofuranose derivatives with appropriate nucleoside bases. One practical method to synthesize Clevudine involves fluorination of 1,3,5-tri-*O*-benzoyl-2-*O*-imidazolylsulfonyl-α-L-ribofuranose with $Et_3N{\bullet}3HF$ [33]. The nucleophilic substitution reaction is facilitated by the use of the potent imidazolylsulfonyl leaving group (Scheme 23).

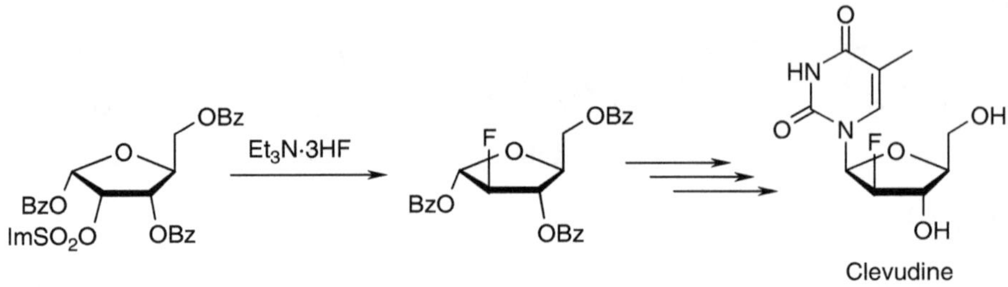

Scheme 23 Synthesis of Clevudine

The introduction of fluorine into a ribose ring can also be accomplished in appropriately activated nucleosides, but the synthetic utility of this approach is somewhat limited by the availability of parent nucleosides with a proper stereoconfiguration of the leaving groups. In some cases, a base itself can act as a leaving group (anhydro-nucleoside approach). An example of this route is presented below (Scheme 24) [34].

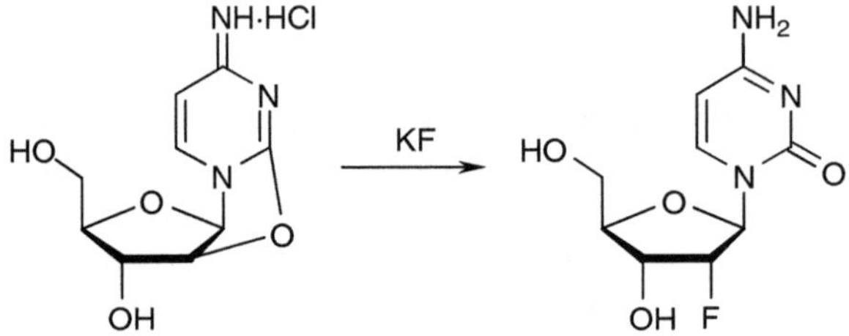

Scheme 24 Fluorination via anhydro-nucleoside approach

Both 2- and 3-fluoropyrroles (**44** and **45**) as well as their derivatives are valuable precursors in the synthesis of fluorinated porphyrins [35]. Parent 2-fluoropyrrole (**44**) has been synthesized and fully characterized only recently. The compound was produced using a "soft" Lewis acid-catalyzed fluorination reaction of pyrrole with NFSI (Scheme 25) [36].

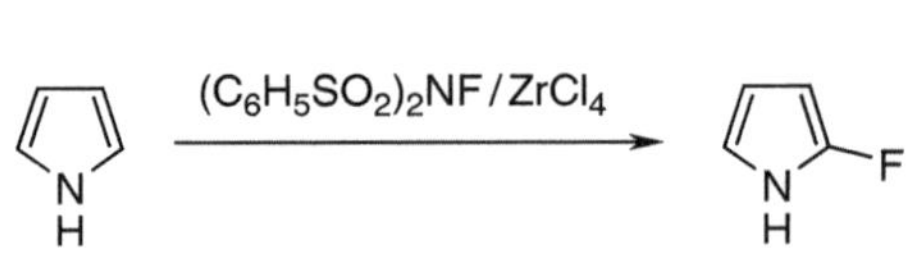

44

Scheme 25 Synthesis of 2-fluoropyrrole (**44**)

Attempts to develop an efficient method of synthesis of 3-fluoropyrrole (**45**) have been met with limited success [35]. Derivatives of 3-fluoropyrrole can be produced using several well-established methods, including the Shiemann-type photochemical fluorination via diazonium salts [37]. An efficient method for the synthesis of 3-fluoropyrrole derivatives **45a** was reported recently (Scheme 26) [38]. The method is based on KF-induced dehydrofluorination and aromatization of substituted 3,3-difluoropyrrolidines.

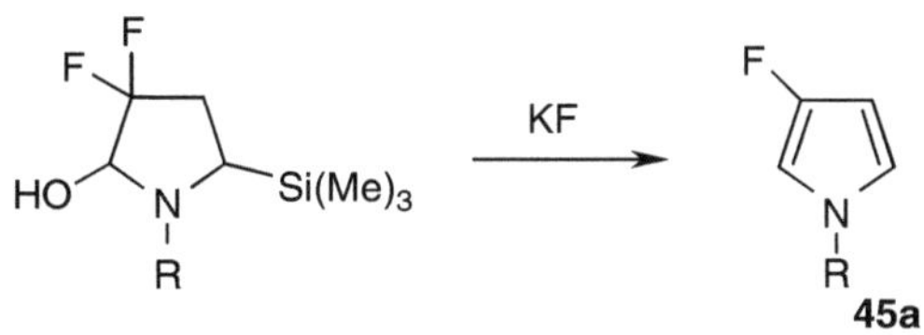

45a

Scheme 26 Synthesis of 3-fluoropyrrole derivatives **45a**

Another method for the synthesis of substituted 3-fluoropyrroles **45b** requires the use of $AuCl_3$ as a catalyst (Scheme 27) [39].

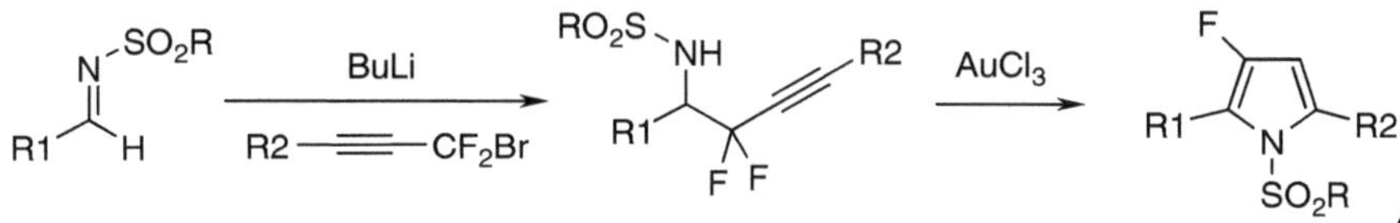

45b

Scheme 27 Synthesis of 3-fluoropyrroles **45b** via $AuCl_3$ – catalyzed heterocyclization

Preparative synthesis of *N*-fluoropyrrole was not found in the available literature. The inversion process of *N*-fluoropyrrole was recently investigated in detail using ab initio and density functional techniques [40].

Convenient preparative methods for the synthesis of parent 2- and 3-fluorofurans (**46** and **47**) have not been documented. Some thermochemical parameters of 2-fluorofuran (**46**) were calculated using ab intio G3 method [41]. Derivatives of both 2- and 3-fluorofurans (**46** and **47**) are known. 2-Fluorofurans **46a** can be prepared by the direct fluorination of furan derivatives with NFSI (Scheme 28) [42].

46a

Scheme 28 Synthesis of 2-fluorofurans **46a**

Fluorodecarboxylation of 2-furancarboxylic acids with Selectfluor® (1-chloromethyl-4-fluoro-1,4-diazoniabicyclo[2.2.2]octane bis-tetrafluoroborate) can also be used for the preparation of 2-fluorofurans (e.g., **46b**, Scheme 29) [43].

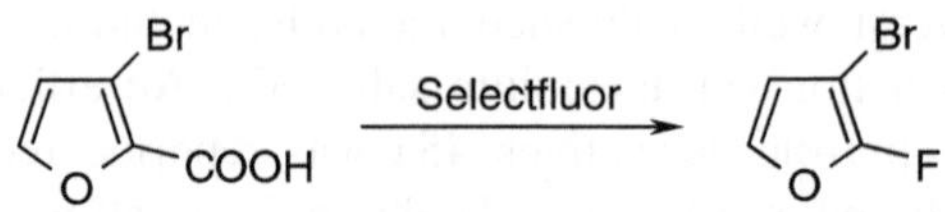

46b

Scheme 29 Synthesis of 3-bromo-2-fluorofuran **46b**

Heterocyclization of appropriate fluorinated acyclic compounds was successfully used for the synthesis of 3-fluorofurans (**47a** and **47b**). One of the recently developed methods employs iodocyclization of *gem*-difluorohomopropargyl alcohols (Scheme 30) [44].

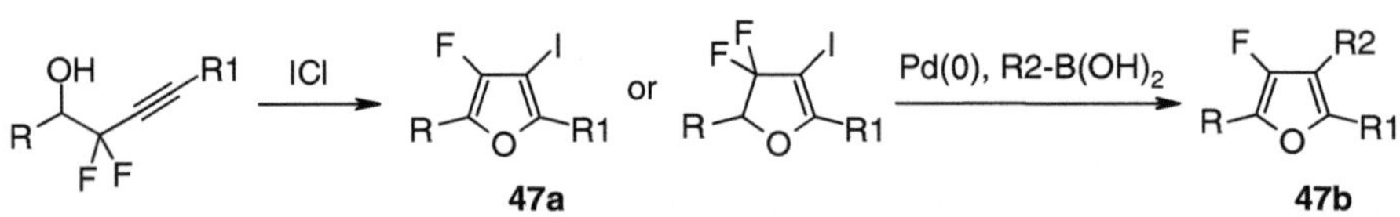

47a 47b

Scheme 30 Iodocyclization of *gem*-difluorohomopropargyl alcohols

Nonfluorinated acyclic compounds can be converted to 3-fluorofurans **47c** via a sequence of fluorination and heterocyclization in the presence of a gold(I)/silver catalyst (Scheme 31) [45].

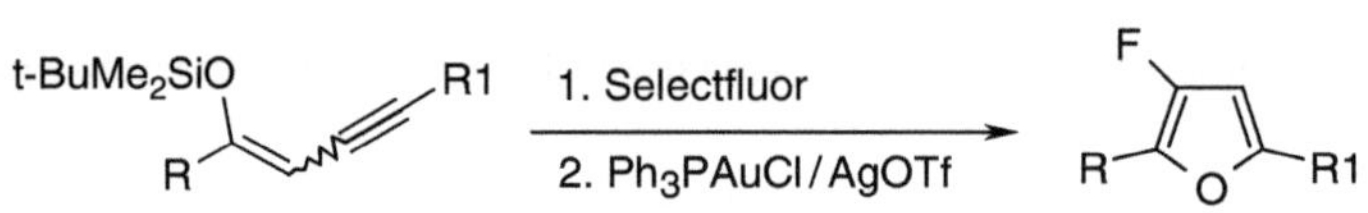

47c

Scheme 31 Fluorination and heterocyclization with gold(I)/silver catalyst

Parent 2- and 3-fluorothiophenes (**48** and **49**) and their derivatives are known. 2-Fluorothiophene (**48**) was first prepared from 2-iodothiophene more than 60 years ago (Scheme 32) [46].

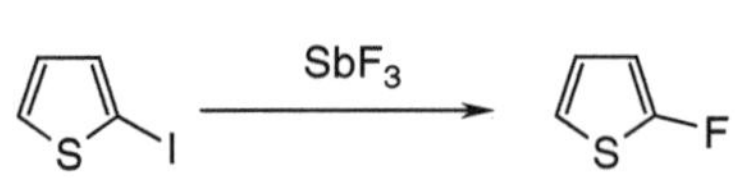

48

Scheme 32 Synthesis of 2-fluorothiophene (**48**)

Modern methods for the synthesis of 2-fluorothiophene (**48**) and its derivatives (**48a**) entail the use of electrophilic fluorinating agents, such as $FClO_3$ [47] or NFSI (Scheme 33) [42] Regioselective fluorination of 2-thienyllithium can also be achieved with *N*-fluoroquinuclidinium fluoride [48].

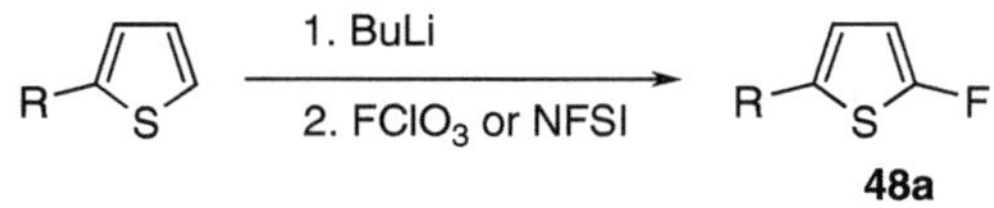

Scheme 33 Regioselective fluorination of 2-thienyllithium

Nucleophilic fluorination of iodonium salts can be used for the synthesis of 2-fluorothiophene (**48**, Scheme 34). This reaction is sensitive to the nature of the counter-anion. The best results were achieved with di-2-thienyliodonium hexafluorophosphate [49].

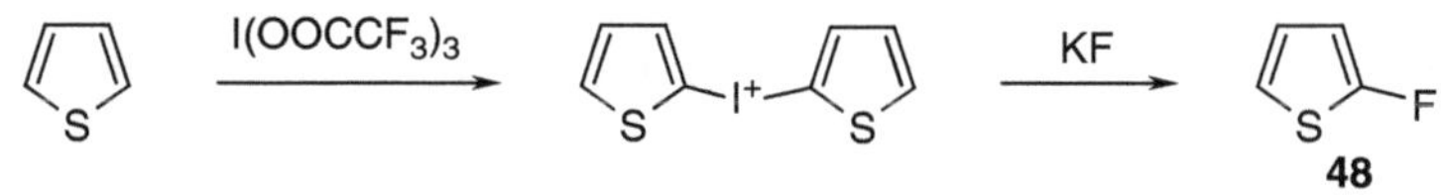

Scheme 34 Nucleophilic fluorination of iodonium salts

Radiolabeled [18F] 3-fluorothiophene (**49**) together with 2-fluorothiophene (**48**) were detected in the reaction mixture produced by direct fluorination of thiophene with [18F] elemental fluorine [50]. Pure samples of 3-fluorothiophene (**49**) can be prepared using a four-step synthetic procedure starting from methyl 3-aminothiophene-2-carboxylate (Scheme 35) [51].

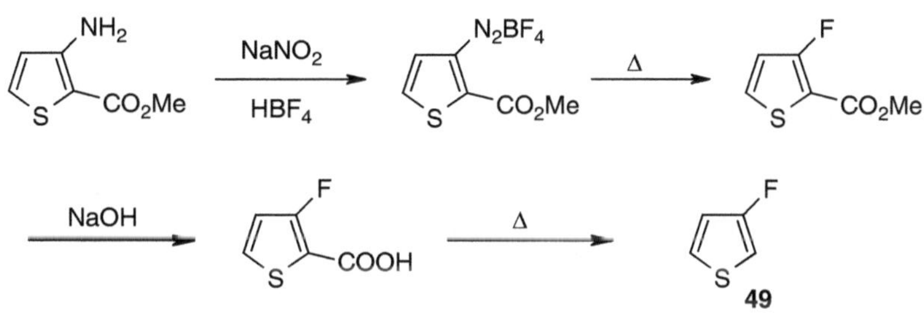

Scheme 35 Synthesis of 3-fluorothiophene (**49**)

Both C- and N-monofluorinated imidazoles are known. 2-Fluoroimidazole (**53**) was prepared via the Schiemann-type photochemical decomposition of the corresponding diazonium tetrafluoroborate in water [52]. The synthesis can be performed in solvent-free conditions using ionic liquids [53]. The compound **53** is not stable and trimerizes spontaneously to produce so-called "tri-imidazole" (aka the imidazole cyclic trimer, Scheme 36) [54]. This compound can be prepared more conveniently by the thermal decomposition of copper imidazolate [55].

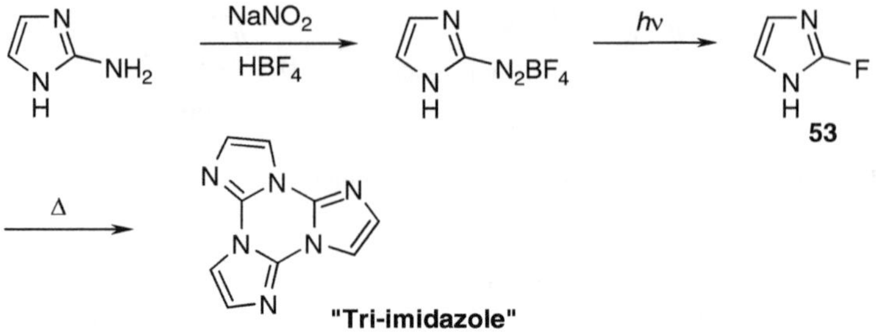

Scheme 36 Synthesis of 2-fluoroimidazole (**53**) and "tri-imidazole"

Nucleophilic fluorination of certain 2-bromoimidazole derivatives leads to 2-fluoroimidazoles (**53a**, Scheme 37). The success of this reaction depends on the proper protection of the NH group and the presence of additional electron withdrawing groups, which facilitate the reactions. Spray-dried KF with 18-crown-6 was successfully used in the synthesis of 1-methyl-2-fluoro-4,5-dicyanoimidazole [56]. Tetraalkylammonium fluorides, such as anhydrous tetramethylammonium fluoride (TMAF) and terabutylammonium fluoride (TBAF), are also effective sources of fluoride ion. The best yields were achieved in dry DMF [57].

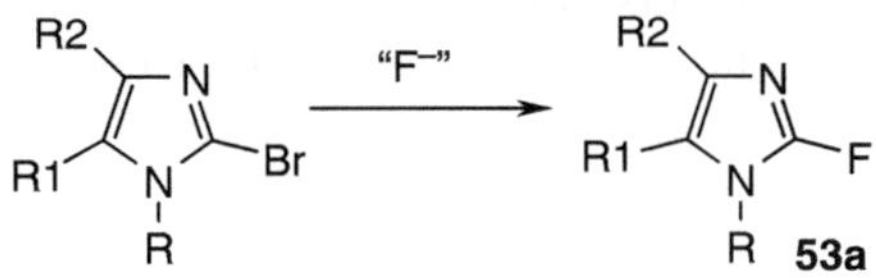

Scheme 37 Nucleophilic fluorination of 2-bromoimidazoles

4(5)-Fluoroimidazole (**54**) was prepared similarly to 2-fluoroimidazole (**53**) via the photochemical decomposition of diazonium tetrafluoroborate [52]. Modern methods for the synthesis of 4(5)-fluoroimidazole derivatives (such as **54a**) include nucleophilic fluorination with anhydrous TBAF [58] (Scheme 38).

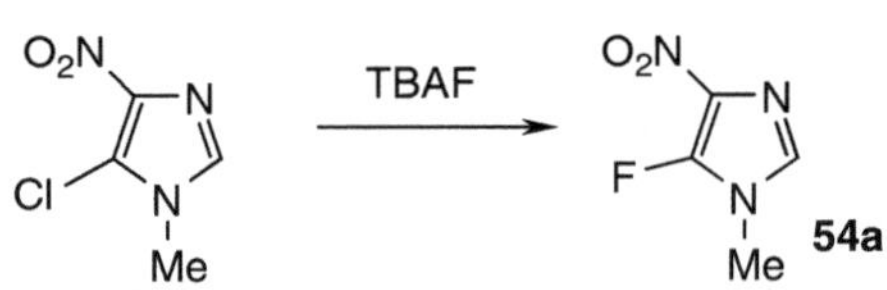

Scheme 38 Synthesis of 5-fluoro-1-methyl-4-nitroimidazole (**54a**)

2,4- and 4,5-Dinitro derivatives of *N*-fluoroimidazole (**55**) were prepared by direct fluorination of corresponding dinitroimidazoles with elemental fluorine (Scheme 39). As expected, 1-fluoro-2,4-dinitroimidazole (**55a**) is a potent electrophilic fluorinating agent [59].

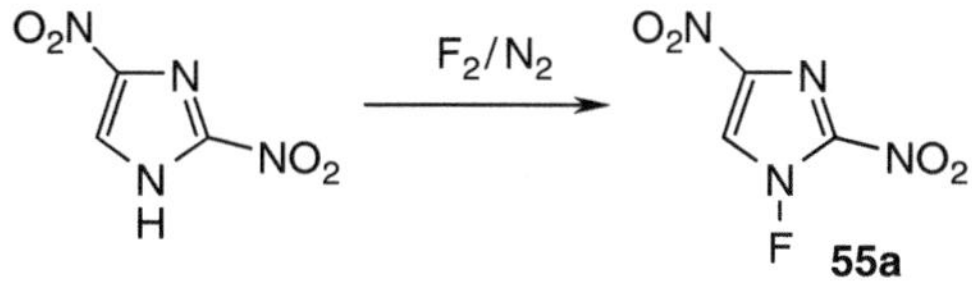
55a

Scheme 39 Synthesis of 1-fluoro-2,4-dinitroimidazole (**55a**)

Parent 4-fluoropyrazole (**50**) was initially prepared in several steps from fluoroacetic acid [60, 61]. An improved two-step synthesis of 4-fluoropyrazole (**50**) was recently reported (Scheme 40) [62]. The synthetic utility of this approach is somewhat limited due to the toxicity of fluoroacetic acid and its derivatives.

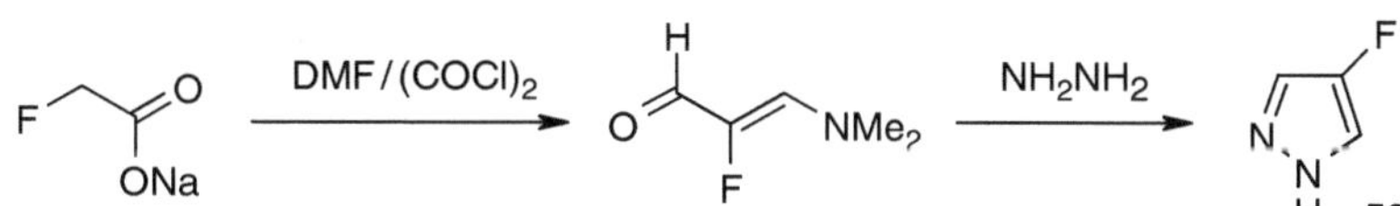
50

Scheme 40 Synthesis of 4-fluoropyrazole (**50**)

Derivatives of 4-fluoropyrazole (**50**) can be prepared by a variety of methods. Recent reports include the synthesis of versatile 5-tributylstannyl-4-fluoropyrazole (**50a**) and parent 4-fluoropyrazole (**50**) from 1,1-difluoroethylene (Scheme 41) [63].

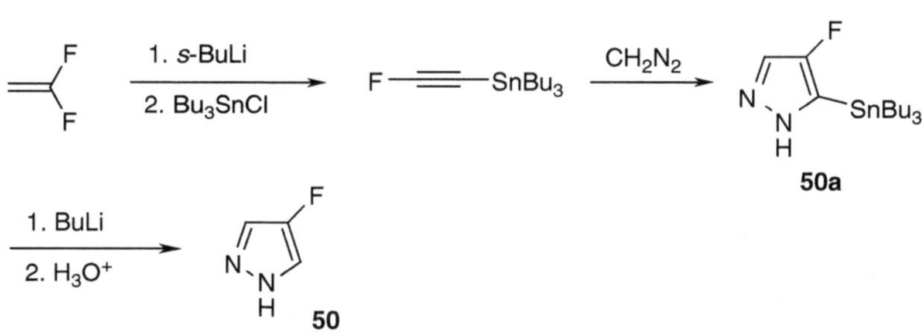
50a

50

Scheme 41 Synthesis of 5-tributylstannyl-4-fluoropyrazole (**50a**) and 4-fluoropyrazole (**50**)

A new synthesis of 4-fluoropyrazoles (**50b**) from the reaction of fluoro- and difluoroacetonitrile derivatives with hydrazine was recently described (Scheme 42). In the latter case, the formation of the heterocyclic ring was accompanied by the elimination of one fluorine atom [64].

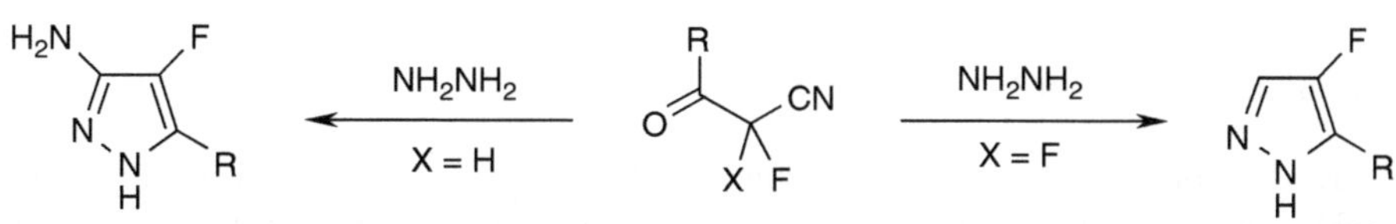
50b

Scheme 42 Heterocyclization of fluoro- and difluoroacetonitrile derivatives

Direct fluorination of 3,5-disubstituted pyrazoles having open 4 positions with diluted elemental fluorine is usually regioselective and yields mainly 4-fluoropyrazoles [65]. Microwave-induced fluorination was successfully employed for the preparation of substituted 4-fluoropyrazoles (**50c**) with Selectfluor® (Scheme 43). Defluorinated Selectfluor® catalyzes the heterocyclization of 1,3-dicarbonyl compounds with hydrazines, thus allowing one-pot synthesis of the desired 4-fluoropyrazoles directly from 1,3-diketones and substituted hydrazines [66].

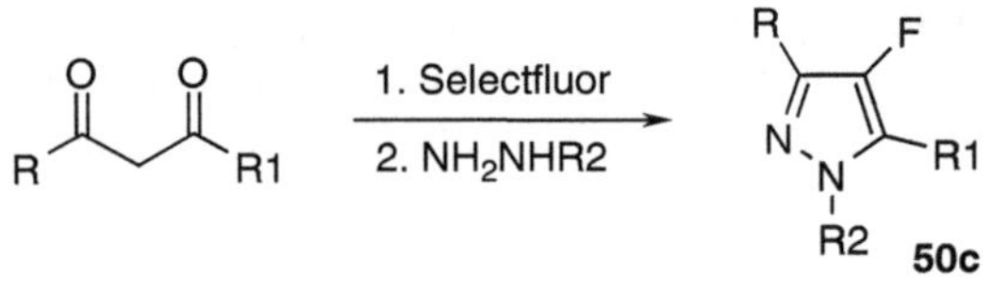

Scheme 43 One-pot synthesis of 4-fluoropyrazoles **50c**

3(5)-Fluoropyrazoles can be prepared by a variety of methods described above. In addition, substituted ethyl pyrazole-4-carboxylates **50d** can be regioselectively fluorinated in the 5-position using electrochemical fluorination methods (Scheme 44) [67].

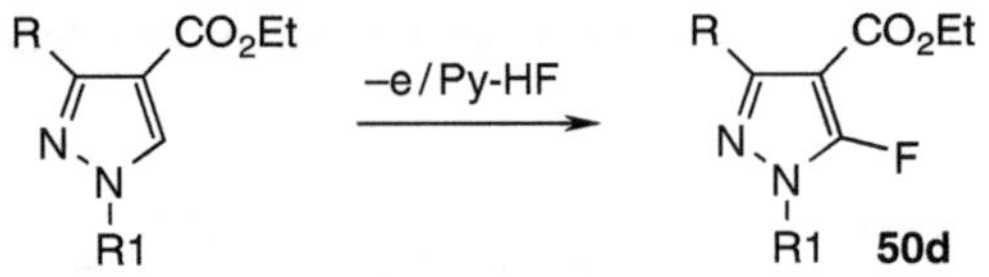

Scheme 44 Electrochemical fluorination of ethyl pyrazole-4-carboxylates

Synthesis of *N*-fluoropyrazole (**52**) derivatives was recently reported. The compounds were prepared by direct N-fluorination of nitropyrazoles with elemental fluorine (Scheme 45). N-fluorination of nitropyrazoles is a regioselective process and produces only one isomer with the fluorine atom in the most remote position from the nitro group. *N*-fluoronitropyrazoles **52a** are of interest as energetic materials [68].

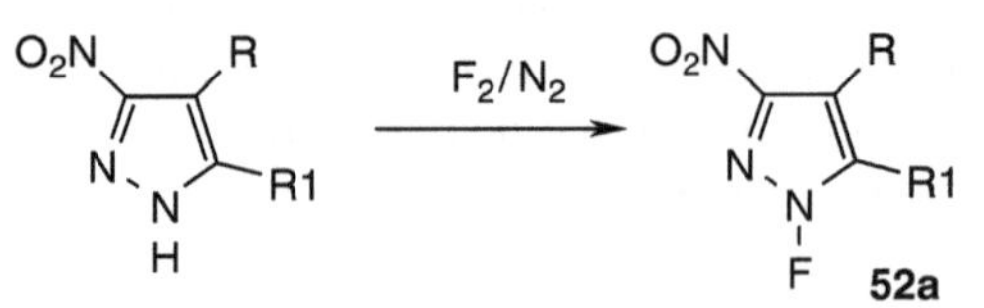

Scheme 45 Synthesis of *N*-fluoronitropyrazoles **52a**

Ring-fluorinated oxazoles are known. They easily react with nucleophiles and therefore could serve as valuable intermediates for the synthesis of polymers [69]. The more chemically stable 5-fluorothiazoles (**65a**) can be prepared by direct fluorination of substituted thiazoles with Accufluor® (1-fluoro-4-hydroxy-1, 4-diazoniabicyclo[2.2.2]octane bis(tetrafluoroborate), Scheme 46) [70].

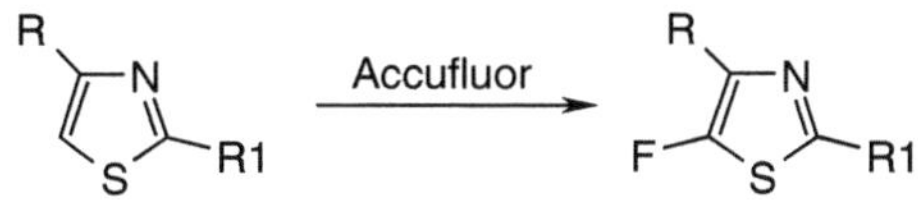

65a

Scheme 46 Fluorination of thiazoles with Accufluor®

The synthesis of parent 4-fluoroisoxazole (**56**) from fluoroacetic acid is similar to the synthesis of 4-fluoropyrazole (**50**) [60, 61]. Substituted 4-fluoroisoxazoles were prepared from 2-fluoro-1,3-dicarbonyl compounds and hydroxylamine [71]. Direct fluorination of 3,5-disubstituted isoxazoles with Selectfluor® leads to corresponding 3,5-disubstituted 4-fluoroisoxazoles **56a** (Scheme 47) [72].

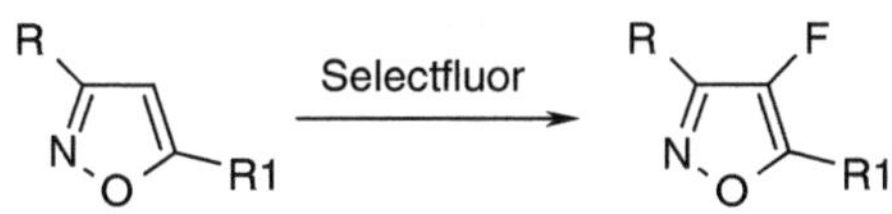

56a

Scheme 47 Fluorination of 3,5-disubstiuted isoxazoles with Selectfluor®

Fluorination of 4-nitroisoxazoline salts with $FClO_3$, followed by the elimination of HNO_2, represents another approach to the synthesis of substituted 4-fluoroisoxazoles **56b** (Scheme 48) [73].

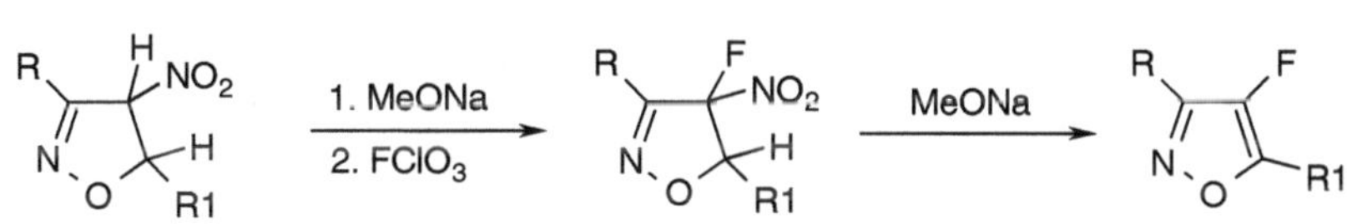

56b

Scheme 48 Synthesis of substituted 4-fluoroisoxazoles **56b**

N-fluoroisothiazolidines (**43**, predominantly known in the form of 1,1-oxide derivatives, or *N*-fluorosultams, including chiral *N*-fluorosultams **43a**, Scheme 49) are important electrophilic fluorinating agents [74]. Their applications in the enantioselective synthesis of chiral fluoroorganic compounds were recently reviewed [75].

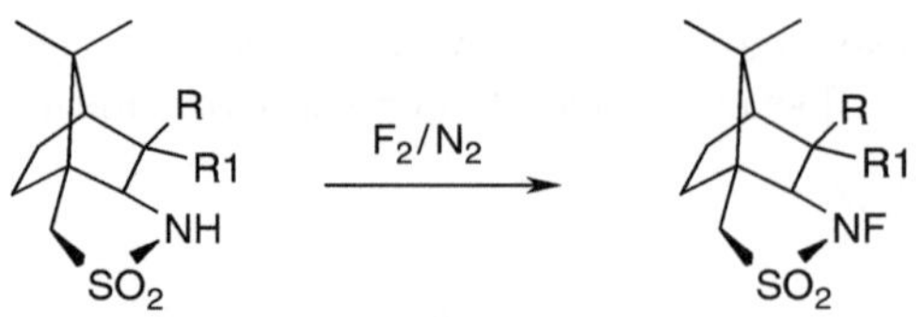

43a

Scheme 49 Synthesis of chiral *N*-fluorosultams **43a**

Parent 3(5)-fluoro-1,2,4-triazole (**71**) was prepared almost 40 years ago by fluorodenitration of 3(5)-nitro-1,2,4-triazole with HF at elevated temperatures [76]. This procedure provides a route to other 3(5)-fluoro-1,2,4-triazole derivatives **71a** (Scheme 50).

71a

Scheme 50 Synthesis of 3(5)-fluoro-1,2,4-triazoles **71a**

Halogen exchange reactions were successfully employed in the synthesis of 5-fluoro-1,2,4-triazoles (e.g., **71b**, Scheme 51) [77]. The substitution reaction is assisted by the enhanced reactivity of the 5-position in the 1,2,4-triazole ring.

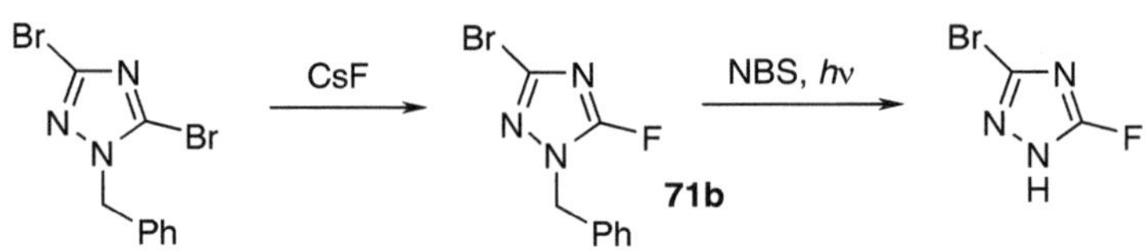

Scheme 51 Nucleophilic fluorination of 5-bromo-1,2,4-triazoles

Unlike 3(5)-fluoro-1,2,4-triazole (**71**), the preparative synthesis of parent 4(5)-fluoro-1,2,3-triazole (**68**) was not reported in the available literature. Substituted 4(5)-fluoro-1,2,3-triazoles are known. One of the recently reported procedures utilizes fluorovinyl sulfones as fluoroacetylene equivalents for cycloaddition with azides leading to substituted 4- and 5-fluoro-1,2,3-triazole derivatives **68a** and **68b**. As expected, the cycloaddition reaction is not regioselective (Scheme 52) [78].

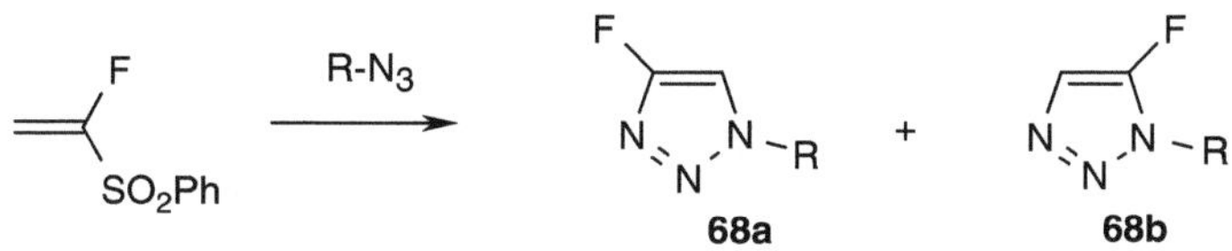

Scheme 52 Synthesis of 4- and 5-fluoro-1,2,3-triazoles **68a** and **68b**

N-fluorinated 1,2,3-triazoles (**69** and **70**) appear to be stable enough to be isolated, although the reported examples so far were limited to 1-fluoro-benzo-1,2,3-triazole (**69a**). The compound was first prepared using an anionic electrophilic fluorinating agent, cesium fluoroxysulphate (CEFOX, Scheme 53), but the use of other conventional fluorinating agents, such as diluted elemental fluorine, is also feasible [79].

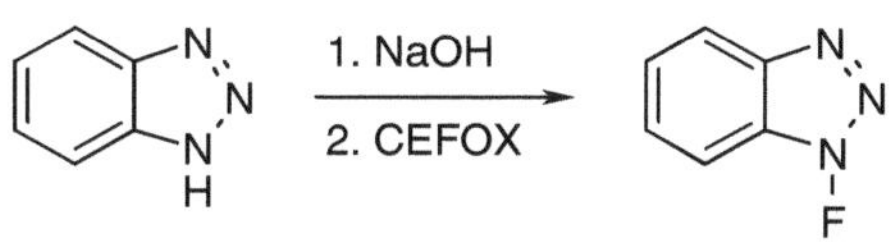

Scheme 53 Synthesis of 1-fluoro-benzo-1,2,3-triazole (**69a**)

Reports concerning monofluorinated oxadiazoles and thiadiazoles appear to be very limited. Recently, substituted 3(4)-fluoro-1,2,5-oxadiazoles (fluorofurazanes, **74a**) were successfully prepared by fluorodenitration of the corresponding nitrofurazanes (Scheme 54) [80].

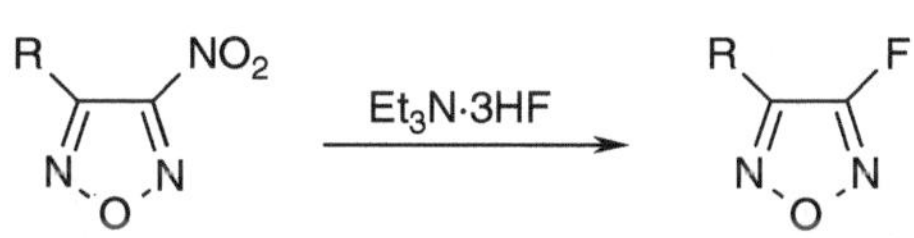

Scheme 54 Synthesis of fluorofurazanes **74a** via fluorodenitration

Derivatives of 2-fluoro-1,3,4-thiadiazole (**78**) as well as isomeric 5-fluoro-1,2,4-thiadiazole (**80**) were prepared almost 50 years ago by the action of AgF on the corresponding halothiadiazoles (Scheme 55) [81]. The synthesis of the parent fluorothiadiazoles **78, 79** and **80** has so far received little attention.

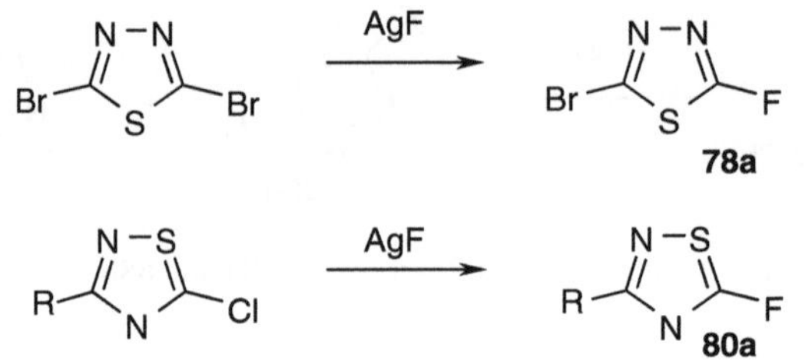

Scheme 55 Nucleophilic fluorination of halo-thiadiazoles

5 Six-Membered Monofluoroheterocycles

The family of six-membered monofluoroheterocycles includes several prominent compounds, such as parent 2-fluoropyridine (**105**) and 5-fluorouracil, a derivative of 5-fluoropyrimidine (**113**, Fig. 5).

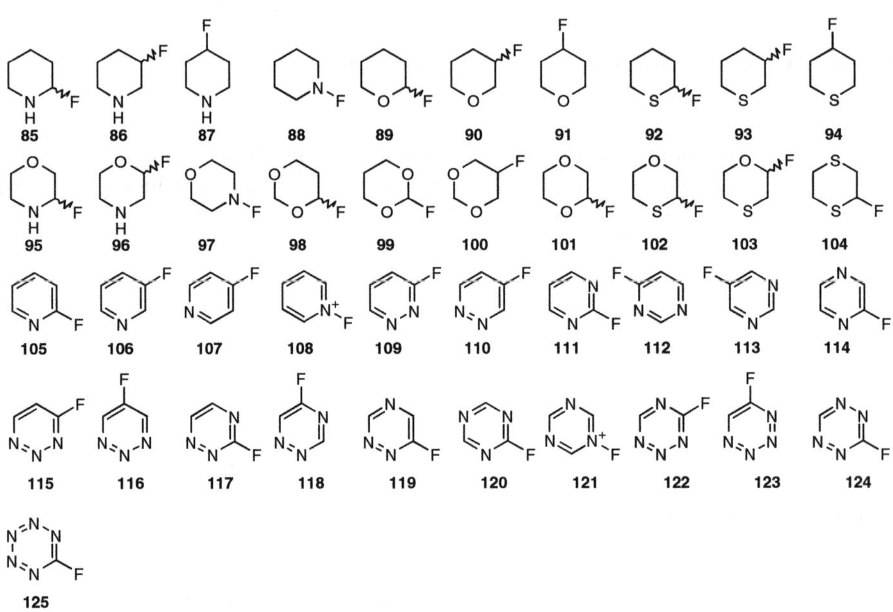

Fig. 5 Six-membered monofluoroheterocycles

Saturated six-membered monofluorinated heterocycles are well represented in the recent literature. For example, 2-fluorodioxane (**89**) was prepared in satisfactory yield by electrochemical fluorination of dioxane in Et_3N•5HF (Scheme 56) [28].

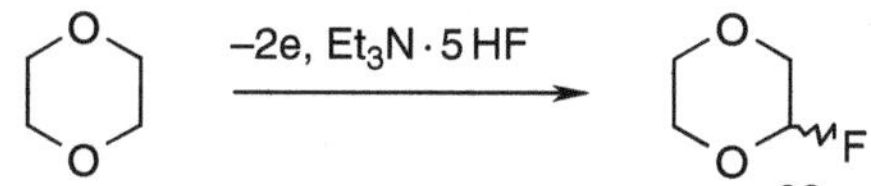

89

Scheme 56 Electrochemical fluorination of dioxane

Availability and literature precedence for monofluorinated azines with one, two, three, four, and five nitrogen atoms decline progressively. NMR spin–spin coupling constants of higher monofluorinated azines, including elusive 6-fluoropentazine (**125**), were calculated only recently [82, 83].

Fluoropyridines with one fluorine atom are probably the most explored subgroup of monofluorinated heterocycles. All monofluoropyridines (2-,3-,4-, and *N*-fluoropyridines, **105**, **106**, **107**, and **108**, respectively) are now commercially available. Halogen exchange using KF as the fluorine source and decomposition of diazonium salts in HF are equally effective in the preparation of C-fluoropyridines. Although initial attempts to prepare 4-fluoropyridine (**107**) via the diazotation-fluorination route were not successful, the compound was eventually produced under optimized reaction conditions [84]. Unlike 3-fluoropyridine (**106**), 4-fluoropyridine (**107**) is not chemically stable and easily produces the self-condensation product, *N*-(4′-pyridyl)-4-fluoropyridinium fluoride (Scheme 57), similar to 2-fluoroimidazole (**53**) [54]. Salts of 4-fluoropyridine (**107**) are more chemically stable [84].

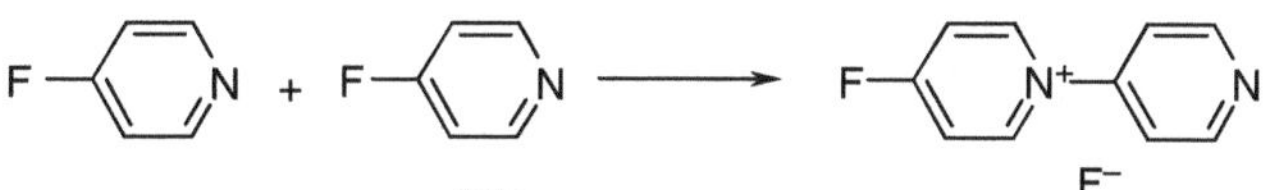

107 107

N-(4′-pyridyl)-4-fluoropyridinium fluoride

Scheme 57 Self-condensation of 4-fluoropyridine (**107**)

2-Fluoropyridines were successfully prepared by fluorodenitration of 2-nitropyridines with TBAF [85] and from 2-chloropyridines using the "proton sponge"/$Et_3N{\bullet}3HF$ system [86]. Fluorodenitration is also effective in the synthesis of 3-fluoropyridines [85]. Recent advances in the synthesis of poly-substituted 2-fluoropyridines (**105a**) include the use of a positively charged sulfonium fragment as a leaving group (Scheme 58) [87] The required cationic intermediates, 2-pyridylsulfonium salts, can be prepared from available pyridine-2(1H)-thiones.

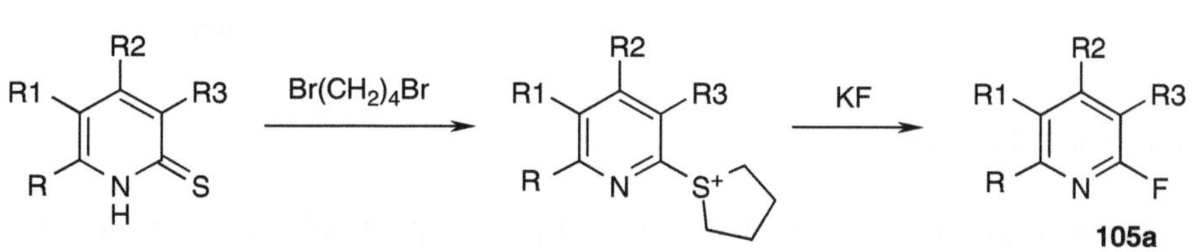

105a

Scheme 58 Nucleophilic fluorination of 2-pyridylsulfonium salts

Nucleophilic fluorination of 2-halopyridines usually requires anhydrous conditions due to the inferior reactivity of the hydrated fluoride anion. Microwave heating presents an opportunity to overcome this limitation. Affordable KF hydrate ($KF{\cdot}2H_2O$) proved to be as effective as spray-dried KF under microwave condition (Scheme 59) [88]. It is speculated that preferential heating and subsequent dehydration of $KF{\cdot}2H_2O$ crystals under microwave conditions is responsible for this activity enhancement in the synthesis of 2-fluoropyridines **105b**.

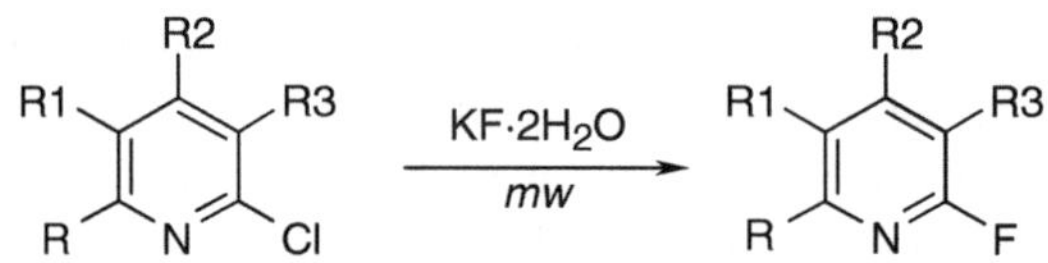

105b

Scheme 59 Microwave-assisted synthesis of 2-fluoropyridines **105b**

Microwavc hcating also provides notable rate enhancements in the synthesis of 2-fluoropyridyl derivatives of [3,2-c]pyrazolo-corticosteroids **105c** via nucleophilic fluorination (Scheme 60). A relative rate ratio of 3:1 for microwave versus conventional heating was obtained by kinetic experiments [89].

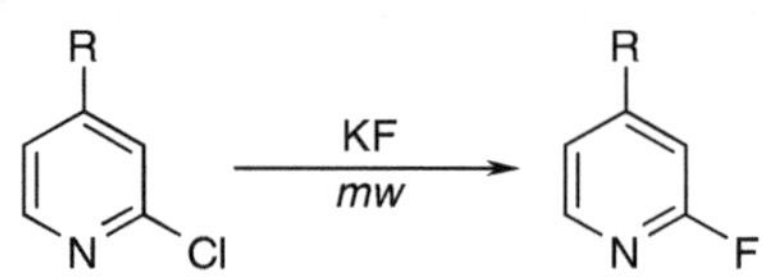

105c

Scheme 60 Microwave-assisted nucleophilic fluorination of 2-chloropyridines

The reaction of pyridine with $C_{60}F_{48}$ and other highly fluorinated fullerenes leads to 2-fluoropyridine (**105**). In this case, $C_{60}F_{48}$ acts as both an activator and a source of fluoride ions. The reaction is likely to proceed via the initial formation of a complex between pyridine and $C_{60}F_{48}$ (Scheme 61) [90].

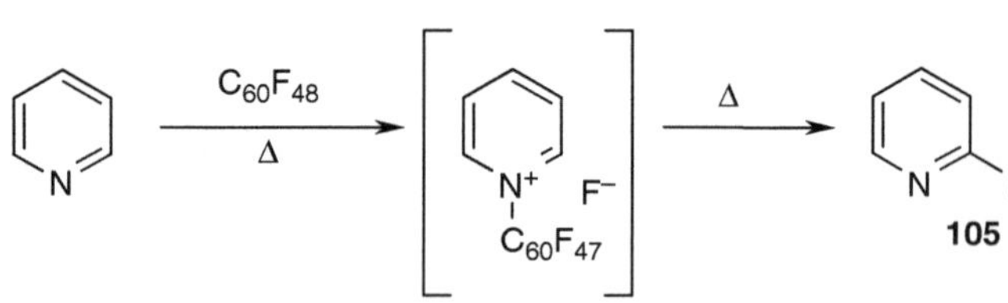

105

Scheme 61 Fluorination of pyridine with $C_{60}F_{48}$

N-fluoropyridinium salts (**108**) were originally prepared from pyridine and elemental fluorine, still a method of choice for the commercial production of these salts. Other electrophilic fluorinating agents, including CEFOX and fluorine nitrate ($FONO_2$), are also effective [79, 91]. *N*-fluoropyridinium salts (**108**) have a rich chemistry, and can act as both electrophilic fluorinating agents and precursors of reactive intermediates, such as 2-pyridyl cations/carbenes [92]. Addition of certain nucleophiles at the 2- and 4-positions via cine/tele nucleophilic substitution of hydrogen was also reported (Scheme 62) [93].

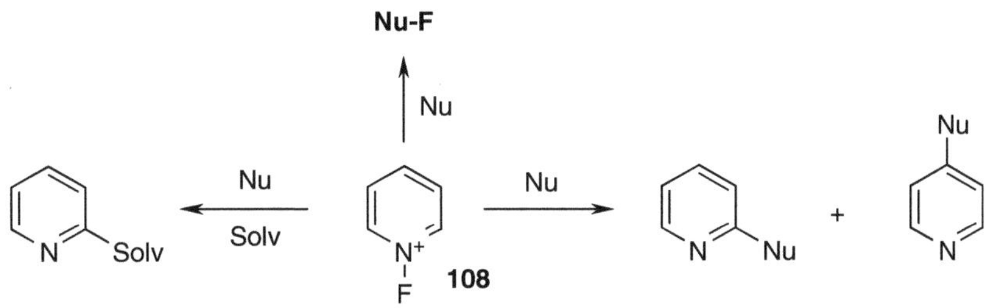

Scheme 62 Chemical transformations of *N*-fluoropyridinium salts (**108**)

N-fluoropyridinium triflate (**108a**) was successfully used as a catalyst in the reaction between N-substituted imines and ethyl diazoacetate (Scheme 63). The authors reported excellent yields of N-substituted aziridines [94].

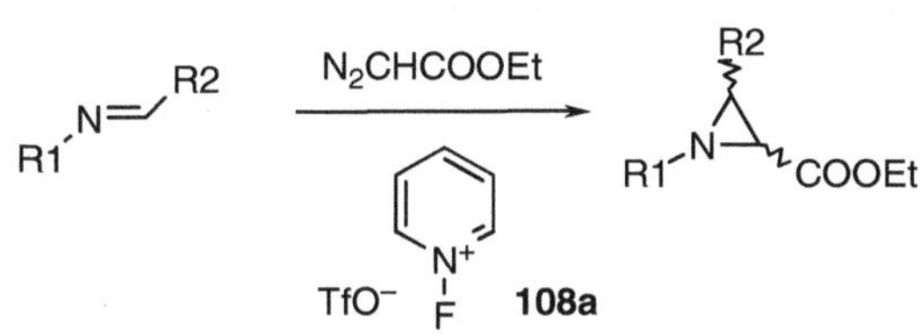

Scheme 63 Catalytic properties of *N*-fluoropyridinium triflate (**108a**)

Monofluorinated pyrimidines are of special importance among other fluorinated azoles due to the anticancer properties of 5-fluorouracyl, commonly known as **5-FU** [95]. Several methods for the synthesis of **5-FU** are known, including the industrially important direct fluorination of uracyl (Scheme 64), and the more laboratory-convenient fluorination with Selectfuor® [96]. Other monofluorinated pyrimidines can be prepared by nucleophilic fluorination of corresponding chloropyrimidines. A pyrimidine ring is substantially more activated compared to a pyridine ring so these substitution reactions usually proceed in relatively mild conditions [97].

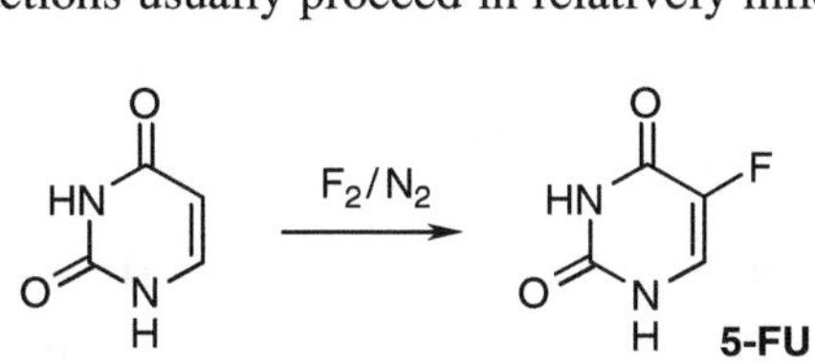

Scheme 64 Preparation of 5-fluorouracyl (**5-FU**)

Ionic liquids were successfully used in the preparation of substituted 4(5)-fluoropyridazines (**110a**, Scheme 65). The yields were moderately better under the microwave-heating conditions. The results were indicative of an assumption that the 4(5)-positions in these monochlorinated pyridazines are more active toward nucleophilic fluorination compared to the 3(6)-positions [98]. A different activity pattern was observed for polychlorinated pyridazines: the 4(5)-positions were found to be less active than the 3(6)-positions [99].

110a

Scheme 65 Nucleophilic fluorination of 4(5)-chloropyridazines

Substituted 3(6)-fluoropyridazines (**109a**) were prepared from corresponding 3 (6)-chloropyridazines and KF in the presence of a phase transfer agent (Scheme 66). The use of a phase transfer agent for chlorine–fluorine exchanges improves the yields of fluoropyridazines [100]. The "proton sponge"/Et_3N•3HF system is also an efficient fluorinating reagent for the synthesis of substituted 3(6)-fluoropyridazines (**109a**) via halogen exchange [86, 101].

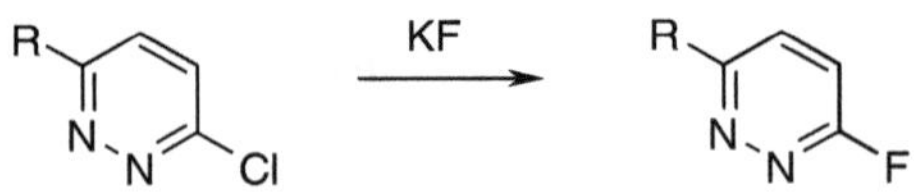

109a

Scheme 66 Nucleophilic fluorination of 3(6)-chloropyridazines

Various 4-fluoropyridazinones **109b** were prepared by nucleophilic fluorination with a novel fluorinating reagent, *N*-ethyl(hexamethylenetetraammonium) fluoride (EMEF, Scheme 67). Good regioselectivity was observed in the presence of the two chlorine atoms [102].

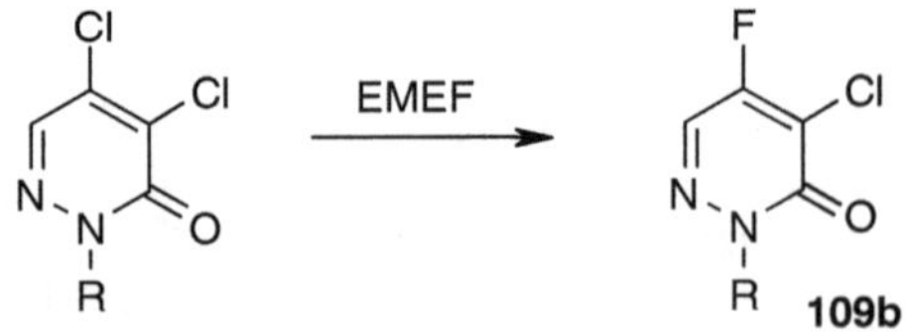

109b

Scheme 67 Synthesis of 4-fluoropyridazinones **109b**

The simplest method for the synthesis of the parent 2-fluoropyrazine (**114**) entails nucleophilic fluorination of 2-chloropyrazine with KF in *N*-methylpyrrolidone (Scheme 68) [103]. This method, with some variations, is applicable for the preparation of a variety of substituted 2-fluoropyrazines [104].

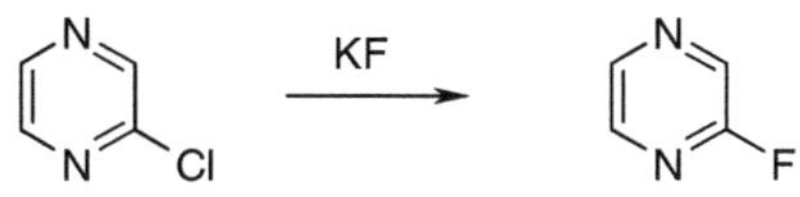

114

Scheme 68 Synthesis of 2-fluoropyrazine (**114**)

Derivatives of C-fluorinated triazines (**115**, **116**, **117**, **118**, **119**, and **120**) can be prepared by nucleophilic fluorination of corresponding chlorinated triazines [105, 106]. The fluorine atom in C-fluorinated triazines is activated by the presence of three nitrogen atoms and allows facile synthesis of other functionalized triazines [105, 106]. Photolysis of C-fluorinated 1,2,3-triazines **115a** leads to transient formation of unstable fluoroazetes via the loss of N_2 (Scheme 69) [107].

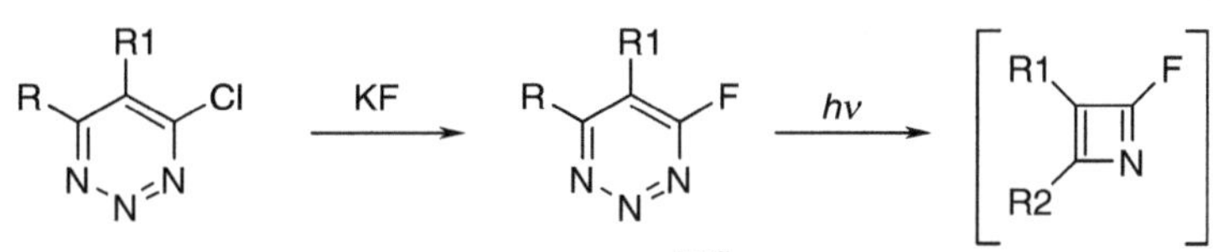

115a

Scheme 69 Synthesis and photolysis of 4(6)-fluoro-1,2,3-triazines **115a**

Derivatives of the N-fluorinated 1,3,5-triazinium cation (**121**) are also known. They were prepared by the action of elemental fluorine on the corresponding 1,3, 5-triazines (Scheme 70) [108]. Substituted 1-fluoro-1,3,5-triazinium salts **121a** are potent electrophilic fluorinating agents capable of fluorinating nonactivated aromatic compounds [109]. The crystal structure of 1-fluoro-2,4,6-trimethoxy-1,3, 5-triazinium cation revealed very short N^+-F bond typical of these types of compounds [110].

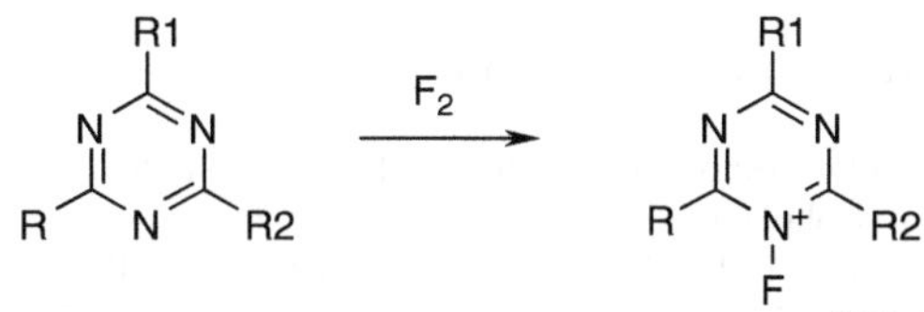

121a

Scheme 70 Synthesis of 1-fluoro-1,3,5-triazinium salts **121a**

6 Conclusions: The "Wish List"

Several years ago, we noted several "white spots" on the map of fluorinated heterocycles – compounds whose convenient synthesis, elucidation of physical–chemical properties, and subsequent utilization might be of academic and practical interest [1]. Many of these compounds were monofluorinated heterocycles, including 3-fluorothiazole (**49**), the facile synthesis of which was reported in 2008 [51]. Parent monofluorinated heterocycles with several heteroatoms, and N-fluorinated heterocycles deserve particular attention since only a few of them are currently available. Some heterocycles from our current "wish list" are presented below (Fig. 6).

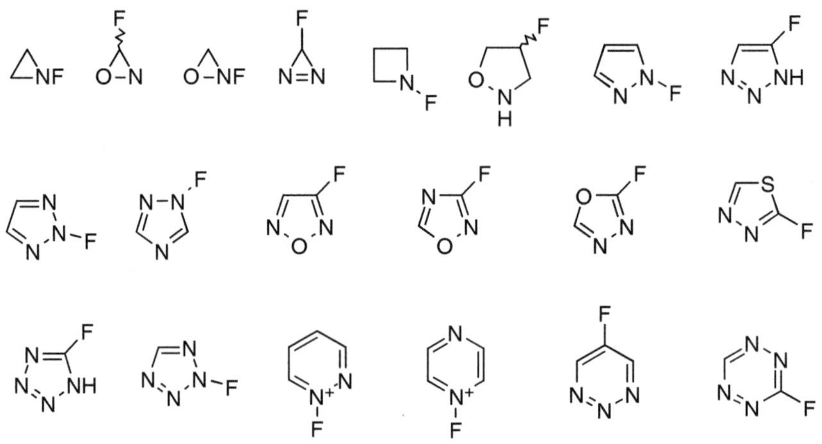

Fig. 6 Monofluoroheterocycles – the "wish list"

Acknowledgments This chapter – a contribution from the Discovery Chemistry Project – is dedicated to the memory of Kirill Gennadievich Nikishin (1967–2010).

References

1. Gakh AA, Kirk KL (eds) (2009) Fluorinated heterocycles. American Chemical Society, Washington DC
2. Petrov VA (ed) (2009) Fluorinated heterocyclic compounds: synthesis, chemistry, and applications. Wiley, Hoboken
3. Chicibabin AE, Rjazancev MD (1915) J Russ Phys Chem Soc 47:1571–1589
4. Boldyrev AI, Schleyer PvR, Higgins D, Thomson C, Kramarenko SS (1992) J Comput Chem 13:1066–1078
5. Hollenstein H, Luckhaus D, Pochert J, Quack M, Seyfang G (1997) Angew Chem Int Ed Engl 36:140–143
6. Verniest G, Colpaert F, Van Hende E, De Kimpe N (2007) J Org Chem 72:8569–8572

7. Tarui A, Kawashima N, Sato K, Omote M, Ando A (2010) Tetrahedron Lett 51:4246–4249
8. Konev AS, Novikov MS, Khlebnikov AF (2005) Tetrahedron Lett 46:8337–8340
9. Rozhkov VV, Makarov KN, Kostyanovsky RG (1998) Mendeleev Commun 8:66–67
10. Firth WC (1968) J Org Chem 33:3489–3492
11. Cox DP, Moss RA, Terpinski J (1983) J Am Chem Soc 105:6513–6514
12. Moss RA, Terpinski J, Cox DP, Denny DZ, Krogh-Jespersen K (1985) J Am Chem Soc 107:2743–2748
13. Moss RA (2010) J Org Chem 75:5773–5783
14. Moss RA, Lawrynowicz W, Turro NJ, Gould IR, Cha Y (1986) J Am Chem Soc 108:7028–7032
15. Van Hende E, Verniest G, Deroose F, Thuring J-W, Macdonald G, De Kimpe N (2009) J Org Chem 74:2250–2253
16. Van Brabandt W, Verniest G, De Smaele D, Duvey G, De Kimpe N (2006) J Org Chem 71:7100–7102
17. Kollonitsch J, Barash L, Doldouras GA (1970) J Am Chem Soc 92:7494–7495
18. Alvernhe G, Laurent A, Touhami K, Bartnik R, Mloston G (1985) J Fluor Chem 29:363–384
19. Genêt J-P, Durand J-O, Roland S, Savignac M, Jung F (1997) Tetrahedron Lett 38:69–72
20. Wojtowicz JA, Polak RJ (1973) J Org Chem 38:2061–2066
21. Lucas SD, Rauter AP, Schneider J, Wessel HP (2009) J Carbohydr Chem 28:431–446
22. Ocampo R, Dolbier WR, Zuluaga F (2002) Collect Czechoslov Chem Commun 67:1325–1334
23. Nishizono N, Sugo M, Machida M, Oda K (2007) Tetrahedron 63:11622–11625
24. Giardina G, Dondio G, Grugni M (1995) Synlett 1995:55–57
25. Caldwell CG, Chen P, He J, Parmee ER, Leiting B, Marsilio F, Patel RA, Wu JK, Eiermann GJ, Petrov A, He H, Lyons KA, Thornberry NA, Weber AE (2004) Bioorg Med Chem Lett 14:1265–1268
26. Shoulders MD, Kamer KJ, Raines RT (2009) Bioorg Med Chem Lett 19:3859–3862
27. Chorghade MS, Mohapatra DK, Sahoo G, Gurjar MK, Mandlecha MV, Bhoite N, Moghe S, Raines RT (2008) J Fluor Chem 129:781–784
28. Hasegawa M, Ishii H, Fuchigami T (2002) Tetrahedron Lett 43:1503–1505
29. Sasaki Y, Ebara R, Nanbu N, Takehara M, Ue M (2001) J Fluor Chem 108:117–120
30. Davis FA, Han W, Murphy CK (1995) J Org Chem 60:4730–4737
31. Yoo BC, Kim JH, Chung Y-H, Lee KS, Paik SW, Ryu SH, Han BH, Han J-Y, Byun KS, Cho M, Lee H-J, Kim T-H, Cho S-H, Park J-W, Um S-H, Hwang SG, Kim YS, Lee Y-J, Chon CY, Kim B-I, Lee Y-S, Yang J-M, Kim HC, Hwang JS, Choi S-K, Kweon Y-O, Jeong S-H, Lee M-S, Choi J-Y, Kim D-G, Kim YS, Lee HY, Yoo K, Yoo H-W, Lee H-S (2007) Hepatology 45:1172–1178
32. Nabhan C, Davis N, Bitran JD, Galvez A, Fried W, Tolzien K, Foss S, Dewey WM, Venugopal P (2011) Cancer 117:1490–1497
33. Du J, Choi Y, Lee K, Chun BK, Hong JH, Chu CK (1999) Nucleos Nucleot 18:187–195
34. Shi J, Du J, Ma T, Pankiewicz KW, Patterson SE, Tharnish PM, McBrayer TR, Stuyver LJ, Otto MJ, Chu CK, Schinazi RF, Watanabe KA (2005) Bioorg Med Chem 13:1641–1652
35. Leroy J, Porhiel E, Bondon A (2002) Tetrahedron 58:6713–6722
36. Zhang Y, Shibatomi K, Yamamoto H (2005) Synlett 2837–2842
37. Onda H, Toi H, Aoyama Y, Ogoshi H (1985) Tetrahedron Lett 26:4221–4224
38. Kim B-M, San Q-Z, Bhatt LR, Kim S-K, Chai K-Y (2010) B Kor Chem Soc 31:31–34
39. Surmont R, Verniest G, Kimpe ND (2009) Org Lett 11:2920–2923
40. Pelzer S, Wichmann K, Wesendrup R, Schwerdtfeger P (2002) J Phys Chem A 106:6387–6394
41. Novak I (2001) J Org Chem 66:9041–9042
42. Dvornikova E, Bechcicka M, Kamienska-Trela K, Krówczynski A (2003) J Fluor Chem 124:159–168
43. Forrest AK, O'Hanlon PJ (1995) Tetrahedron Lett 36:2117–2118

44. Arimitsu S, Jacobsen JM, Hammond GB (2008) J Org Chem 73:2886–2889
45. Li Y, Wheeler KA, Dembinski R (2010) Adv Synth Catal 352:2761–2766
46. VanVleck RT (1949) J Am Chem Soc 71:3256–3257
47. Schuetz RD, Taft DD, O'Brien JP, Shea JL, Mork HM (1963) J Org Chem 28:1420–1422
48. Banks RE, Du Boisson RA, Tsiliopoulos E (1986) J Fluor Chem 32:461–466
49. Onys'ko PP, Kim TV, Kiseleva OI, Rassukana YV, Gakh AA (2009) J Fluor Chem 130:501–504
50. Crestoni ME (1990) J Labelled Comp Radiopharm 28:1109–1112
51. Pomerantz M, Turkman N (2008) Synthesis 2008:2333–2336
52. Cohen LA, Kirk KL (1971) J Am Chem Soc 93:3060–3061
53. Heredia-Moya J, Kirk KL (2007) J Fluor Chem 128:674–678
54. Kirk KL, Nagai W, Cohen LA (1973) J Am Chem Soc 95:8389–8392
55. Schubert DM, Natan DT, Wilson DC, Hardcastle KI (2011) Cryst Growth Des 11:843–850
56. Coad EC, Liu H, Rasmussen PG (1999) Tetrahedron 55:2811–2826
57. Bastos MM, Barbosa JP, Pinto AC, Kover WB, Takeuchi Y, Boechat N (2001) J Braz Chem Soc 12:417–421
58. Sun H, DiMagno SG (2007) Chem Commun 528–529
59. Laali KK, Tanaka M, Forohar F, Cheng M, Fetzer JC (1998) J Fluor Chem 91:185–190
60. Arnold Z (1961) Collect Czechoslov Chem Commun 26:3051–3059
61. Reichardt C, Halbritter K (1975) Liebigs Ann Chem 1975:470–483
62. England K, Mason H, Osborne R, Roberts L (2010) Tetrahedron Lett 51:2849–2851
63. Hanamoto T, Suetake T, Koga Y, Kawanami T, Furuno H, Inanaga J (2007) Tetrahedron 63:5062–5070
64. Surmont R, Verniest G, De Kimpe N (2010) Org Lett 12:4648–4651
65. Caja J, Gakh AA, Dunstan TDJ, Adcock JL, Mamantov G (1994) Proc Electrochem Soc 94–13:752–756
66. Sloop JC, Jackson JL, Schmidt RD (2009) Heteroatom Chem 20:341–345
67. Makino K, Yoshioka H (1988) J Fluor Chem 39:435–440
68. Dalinger IL, Shkineva TK, Vatsadze IA, Popova GP, Shevelev SA (2011) Mendeleev Commun 21:48–49
69. Burger K, Helmreich B, Jendrewski O, Hecht R, Maier G, Nuyken O (1994) Macromol Symp 82:143–160
70. Campbell TF, Stephens CE (2006) J Fluor Chem 127:1591–1594
71. Bumgardner CL, Sloop JC (1992) J Fluor Chem 56:141–146
72. Stephens CE, Blake JA (2004) J Fluor Chem 125:1939–1945
73. Khisamutdinov GK, Okhlobystina LV, Fainzil'berg AA (2010) Russ Chem Bull 58:2182–2184
74. Davis FA, Zhou P, Murphy CK, Sundarababu G, Qi H, Han W, Przeslawski RM, Chen B-C, Carroll PJ (1998) J Org Chem 63:2273–2280
75. Ibrahim H, Togni A (2004) Chem Commun 1147–1155
76. Naik SR, Witkowski JT, Robins RK (1973) J Org Chem 38:4353–4354
77. Zumbrunn A (1998) Synthesis 1998:1357–1361
78. Reck F, Zhou F, Girardot M, Kern G, Eyermann CJ, Hales NJ, Ramsay RR, Gravestock MB (2005) J Med Chem 48:499–506
79. Gakh AA, Romaniko SV, Ugrak BI, Fainzilberg AA (1991) Tetrahedron 47:7447–7458
80. Sheremetev AB, Aleksandrova NS, Dmitriev DE (2006) Mendeleev Commun 16:163–165
81. Schroeder H, Rätz R, Schnabel W, Ulrich H, Kober E, Grundmann C (1962) J Org Chem 27:2589–2592
82. Del Bene JE, Alkorta I, Elguero J (2010) Mol Phys 108:1367–1373
83. Del Bene JE, Alkorta I, Elguero J (2010) J Phys Chem A 114:2637–2643
84. Boudakian MM (1981) J Fluor Chem 18:497–506
85. Kuduk SD, DiPardo RM, Bock MG (2005) Org Lett 7:577–579
86. Darabantu M, Lequeux T, Pommelet J-C, Plé N, Turck A (2001) Tetrahedron 57:739–750

87. Shestopalov AM, Rodinovskaya LA, Fedorov AE, Kalugin VE, Nikishin KG, Shestopalov AA, Gakh AA (2009) J Fluor Chem 130:236–240
88. Shestopalov AM, Fedorov AE, Rodinovskaya LA, Shestopalov AA, Gakh AA (2009) Tetrahedron Lett 50:5257–5259
89. Kahn MGC, Konde E, Dossou F, Labaree DC, Hochberg RB, Hoyte RM (2006) Bioorg Med Chem Lett 16:3454–3458
90. Gakh AA, Tuinman AA, Adcock JL, Compton RN (1993) Tetrahedron Lett 34:7167–7170
91. Gakh AA, Nikishin KG, Kagramanov ND, Semenov VV (1991) Russ Chem Bull 40:2109–2110
92. Umemoto T (2009) In: Gakh AA, Kirk KL (eds) Fluorinated heterocycles. American Chemical Society, Washington DC, pp 37–58
93. Gakh AA, Kiselyov AS, Semenov VV (1990) Tetrahedron Lett 31:7379–7382
94. Bew SP, Fairhurst SA, Hughes DL, Legentil L, Liddle J, Pesce P, Nigudkar S, Wilson MA (2009) Org Lett 11:4552–4555
95. Longley DB, Harkin DP, Johnston PG (2003) Nat Rev Cancer 3:330–338
96. Rangwala HS, Giraldes JW, Gurvich VJ (2011) J Labelled Compd Radiopharm 54:340–34
97. Shkurko OP, Baram SG, Mamaev VP (1972) Chem Heterocycl Compd 8:1159–1162
98. Sun L, Pei W (2007) Chinese J Chem 25:1005–1007
99. Klauke E, Oehlmann L, Baasner B (1983) J Fluor Chem 23:301–308
100. Marque S, Snoussi H, Loupy A, Plé N, Turck A (2004) J Fluor Chem 125:1847–1851
101. Darabantu M, Lequeux T, Pommelet J-C, Ple N, Turck A, Toupet L (2000) Tetrahedron Lett 41:6763–6767
102. Pei W, Yang W (2010) Synth Commun 40:535–539
103. Plé N, Turck A, Heynderickx A, Quéguiner G (1998) Tetrahedron 54:4899–4912
104. Cheeseman GWH, Godwin RA (1971) J Chem Soc C 2973–2976
105. Chambers RD, Shepherd T, Tamura M (1988) Tetrahedron 44:2583–2590
106. Barlow MG, Haszeldine RN, Simon C, Simpkin DJ, Ziervogel G (1982) J Chem Soc Perkin Trans 1:1251–1254
107. Chambers RD, Tamura M, Shepherd T, Ludman CJ (1987) J Chem Soc Chem Commun 1699–1700
108. Schleyer PvR, Buzek P, Klapoetke TM, Tornieporth-Oetting IC, Broschag M, Pickardt J (1993) Inorg Chem 32:1523–1524
109. Banks RE, Besheesh MK, Fraenk W, Klapötke TM (2003) J Fluor Chem 124:229–232
110. Banks RE, Besheesh MK, Pritchard RG (2003) Acta Crystallogr C 59:m141–m143

Top Heterocycl Chem (2012) 27: 65–100
DOI: 10.1007/7081_2011_57

Published online: 3 November 2011

Synthesis of β-Halofurans

Roman Dembinski, Yan Li, Deepthi Gundapuneni, and Alicia Decker

Abstract Furans substituted with halogens at the *beta* carbon, which is less prone to electrophilic reactions compared to an *alpha* position, are important intermediates for accessing highly substituted furans. The regioselectivity of the introduction of a halogen plays an important role in their preparation. This review summarizes efforts for the synthesis of β-halofurans (but not benzofurans) as sorted by halogen (iodo, bromo, chloro, and fluoro). This article provides general reaction schemes that were confirmed with multiple examples and are sometimes applicable to other halogens, and selected reactions specific to a particular substrate and a halogen.

Keywords Bromofurans · Chlorofurans · Fluorofurans · Furan(s) · Iodofurans

Contents

R. Dembinski (✉), Y. Li, D. Gundapuneni and A. Decker
Department of Chemistry, Oakland University, 2200 N. Squirrel Rd, Rochester, MI 48309-4477, USA
e-mail: dembinsk@oakland.edu

Abbreviations

Ac	Acetyl
aq	Aqueous
Ar	Aryl
Bn	Benzyl
bpy	2,2′-Bipyridyl
Bu	Butyl
cat	Catalyst
DCE	1,2-Dichloroethane
DDQ	2,3-Dichloro-5,6-dicyano-1,4-benzoquinone
DME	1,2-Dimethoxyethane
DMF	Dimethylformamide
DMSO	Dimethyl sulfoxide
equiv	Equivalent
Et	Ethyl
i-Pr	*Iso*-propyl
KHMDS	Potassium hexamethyldisilazide
LDA	Lithium diisopropylamide
Me	Methyl
mol	Mole(s)
NBS	*N*-bromosuccinimide
NCS	*N*-chlorosuccinimide
NIS	*N*-iodosuccinimide
NMR	Nuclear magnetic resonance
Nu	Nucleophile
NXS	*N*-halosuccinimide
Ph	Phenyl
PPTS	Pyridinium *p*-toluensulfonate
rt	Room temperature
s-Bu	*Sec*-butyl
TBDMS	*Tert*-butyldimethylsilyl
TBDPS	*Tert*-butyldiphenylsilyl
t-Bu	*Tert*-butyl
THF	Tetrahydrofuran
THP	Tetrahydropyran-2-yl
TMEDA	N,N,N′,N′-tetramethyl-1,2-ethylenediamine
TMP	Tetramethylpiperidyl
Tol	4-Methylphenyl

1 Introduction

Furan motif is found in biologically active natural and artificial compounds [1, 2]. Extension of practical applications of furans include synthetic organic chemistry or material science. Although a variety of methodologies and protocols have been reported, synthesis of highly substituted furans remains the subject of intensive development in recent years [3]. Since furan undergoes electrophilic and metalation reactions more readily at the α-position (C-2 and C-5) rather than at the *β*-position (C-3 or C-4), and is also sensitive to acidic and oxidative environments, the synthesis of a diverse collection of substituted furans remains a challenge. In general, substituted furans are accessed via ring derivatization or via cyclization of acyclic precursors.

Halofurans are important synthetic derivatives, extensively utilized for the preparation of acyclic, carbocylic, and heterocyclic compounds. The presence of a halogen provides an opportunity for further functionalization. Iodo-, bromo-, and chlorofurans are all established substrates for a variety of bond-forming reactions ([4], see, for example, [5], [6–8]) and also serve as building blocks for combinatorial chemistry [9]. Fluorine-containing derivatives are sought after due to their potential in medicinal chemistry. In addition, the Diels–Alder reaction of halofurans leads to oxabicyclo compounds (recent representative examples [10]; [11–13]); some can be converted to halophenols regioselectivity, as referred to the position of a halogen [14]. Chemical routes to nakadomarin A (**1**) and rosefuran (**2**) exemplify practical utility of *beta*-halofurans in organic synthesis (Fig. 1) [15, 16]. The tetracyclic core of **1** is obtained by a three-component coupling using 3-bromofurancarbaldehyde **3** as a key starting material. Rosefuran **2** is obtained via a cross-coupling reaction and decarboxylation of the 3-bromofuran **4** (Fig. 1).

Although numerous authoritative reviews and handbooks describe the synthesis of furans in a systematic way, reactions specific to halofurans are usually scattered throughout the chapters or not included. Only few reviews address halofurans in greater detail [17, 18]. Thus far, no overview detailing the methods of their synthesis across the halogen group exists. This review, focusing (but not limited to) on recent advances, sorts the major approaches to the synthesis of β-halofurans into substitution reactions on the furan core and the construction of the furan ring starting from acyclic precursors. The last one, which can constitute cycloisomerization or cyclocondensation reactions, is divided into two types of reactions: halogenation/cyclizations, and cyclizations of precursors that contain already introduced halogens. Electrophilic cyclization reactions are particularly attractive since they provide versatility to access different halofurans, when treating the same starting material with different halogens. Most commonly, an electrophile serves simultaneously as a cyclization catalyst and a halogen donor, thus improving material economy. Some reactions of this type allow for introductions of two, even different, halogens at the same step. However, fluorine, due to its limited

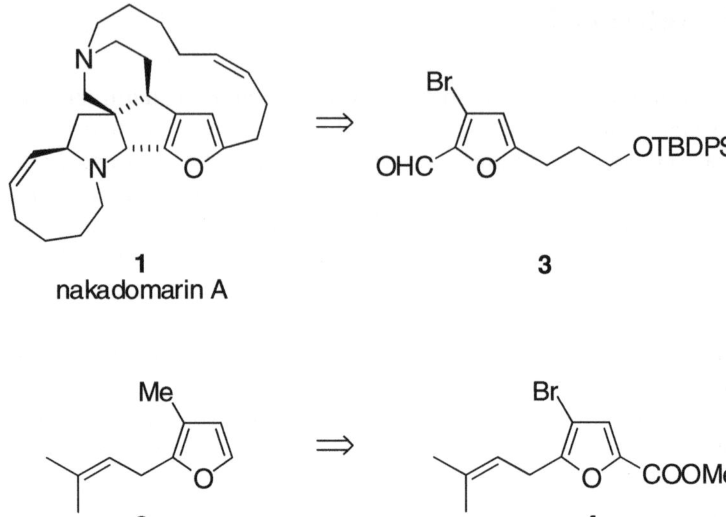

1
nakadomarin A

3

2
rosefuran

4

Fig. 1 Representative compounds synthesized from halofurans

electrophilic character, is not effective in electrophilic cyclizations; synthesis of fluorofurans often requires the use of specific methods.

2 Synthesis of β-Iodofurans

2.1 Iodination of Furans

This group of reactions includes the introduction of iodine onto the already established furan ring. Since the furan does not react directly in a substitution reaction with iodine, the available methods include deprotonation/iodination, and replacement of the silyl group with iodine. As mentioned earlier, deprotonation of the furan ring proceeds preferentially at the *alpha* (C-2- and C-5) positions; thus, the first method often needs the presence of a directing group as a guarantor of the regioselectivity of the reaction. The replacement of the silyl group is limited by available substrates; its introduction into the furan ring based on deprotonation is governed by the same regioselectivity as mentioned above; other methods of synthesis of β-silyl furans are scarce.

Deprotonation of 2-substituted furans takes place at 3- and 5-position. For example, treatment of oxazoline-substituted furan **5** with *s*-BuLi (3.3 equiv) and TMEDA ensures the formation of dilithio-intermediate at −78°C in a procedure optimized by NMR monitoring. Treatment of the intermediate with iodine gives the 3,5-diiodofuran derivative **6** in 94% yield (1) [19].

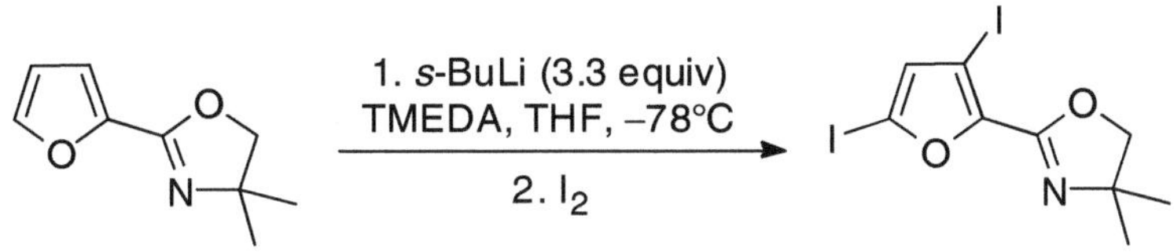

5 **6**, 94% (1)

Selective lithiation at C-4 of a 2,3-substituted furan such as **7** using BuLi has been reported by Keay et al. [20]. The resultant dianion is quenched with iodine in the presence of LiCl at 0°C to provide the iodofuran **8** in high yield (Scheme 1). Lithiation takes place at C-4 due to preferential coordination of the BuLi with the hydroxymethyl or carboxylate group rather than with the furan oxygen atom. Also, the presence of bulky groups at C-2 enhances the *ortho*-directing effect of the hydroxymethyl group by enhancing the presence of a conformation, whereby the hydroxyl is in close proximity to the C-4 hydrogen atom. The silyl group at C-2 can be removed, leading to a solely 3,4-substituted furan **9**. Conversion of the hydroxymethyl group into its *tert*-butyldimethylsilyl derivative leads to the lithiation at C-5.

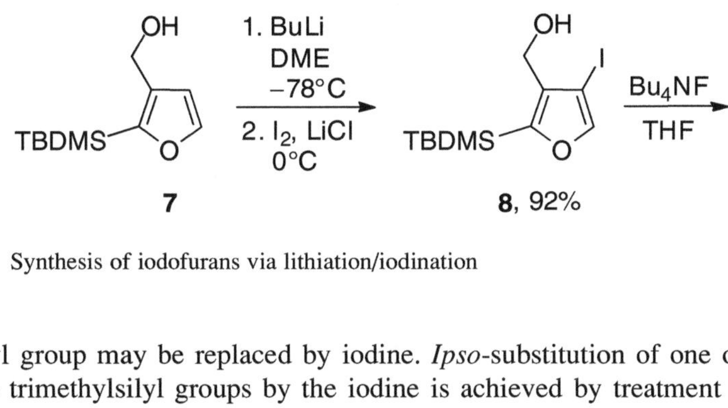

7 **8**, 92% **9**, 91%

Scheme 1 Synthesis of iodofurans via lithiation/iodination

A silyl group may be replaced by iodine. *Ipso*-substitution of one of the two available trimethylsilyl groups by the iodine is achieved by treatment of 3,4-bis (trimethylsilyl)furan **10a** with iodine at −78°C, leading to a substituted furan **11a** (2) [21]. Similarly, iodofuran **11b** can be obtained in good yield through the iodination of furan **10b** with iodine monochloride at low temperature (2) [22].

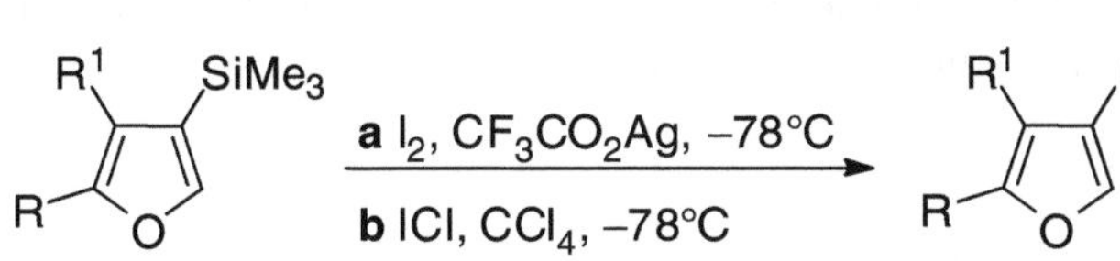

10 **11** (2)

a R = H, R^1 = $SiMe_3$; 80%

b R = PhC(O), R^1 = $(CH_2)_2CH(CO_2Et)_2$; 77%

2.2 Iodination–Cyclization Reactions

5-*Endo-dig* electrophilic cyclization of 1,4-diaryl butynones (propargyl ketones) **12** with *N*-iodosuccinimide (NIS), *N*-bromosuccinimide/acetone, or iodine monochloride/dichloromethane, at room temperature, in the absence of a base, provides 3-halo-2,5-diarylfurans **13** with excellent regiocontrol and high yields (Scheme 2) [23, 24]. This method is effective mainly for aryl substituents, although it facilitates the introduction of substituents such as cyclopropyl [25], which is not easily introduced by other means. The reaction presumably proceeds via an enol intermediate **14**. One of the effective methods to prepare starting material **12** includes ring opening of an oxirane with a lithium acetylide [24]. Analogous iodocyclization and chlorocyclization reactions, leading to 3-fluoro-4-iodofurans [26] and 3-chlorofurans [8], are described in Sects. 5.2 (20) and 4.2 (17), respectively.

Scheme 2 Synthesis of iodofurans via electrophilic cyclization

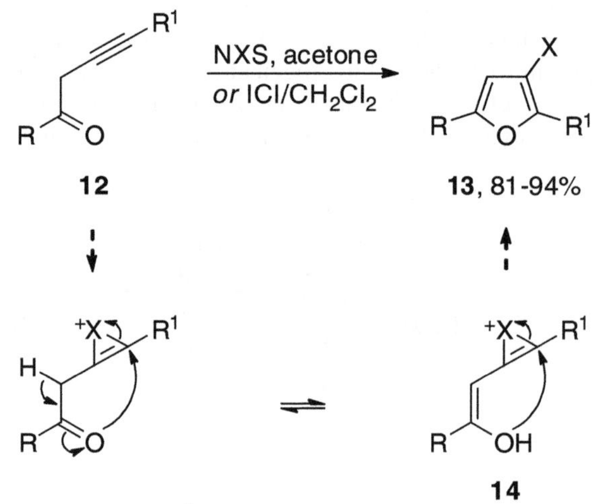

X = I, Br; R, R^1 = aryl

Analogously, an iodocyclization reaction of conjugated enyne acetates **15** leads to 2,5-disubstituted 3-iodofurans **16** under mild conditions, as reported by the Jiang group (3) [27]. Enyne acetates **15** are prepared by the palladium/copper-catalyzed coupling of (*Z*)-*β*-bromoenol acetates and terminal alkynes. Aryl- and alkyl-substituted alkynes undergo iodocyclization in high yields.

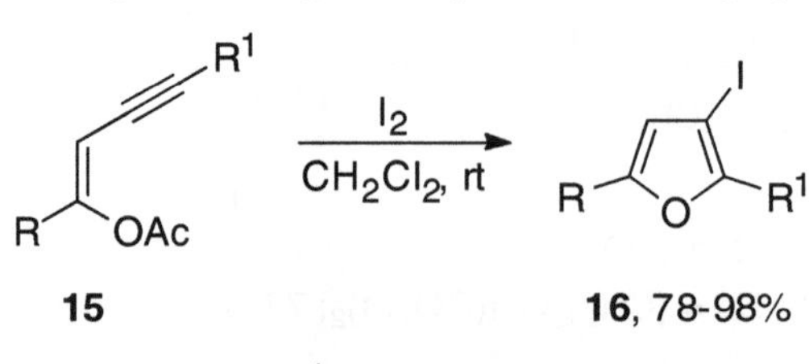

(3)

R = alkyl, aryl; R^1 = alkyl, vinyl, aryl, $SiMe_3$

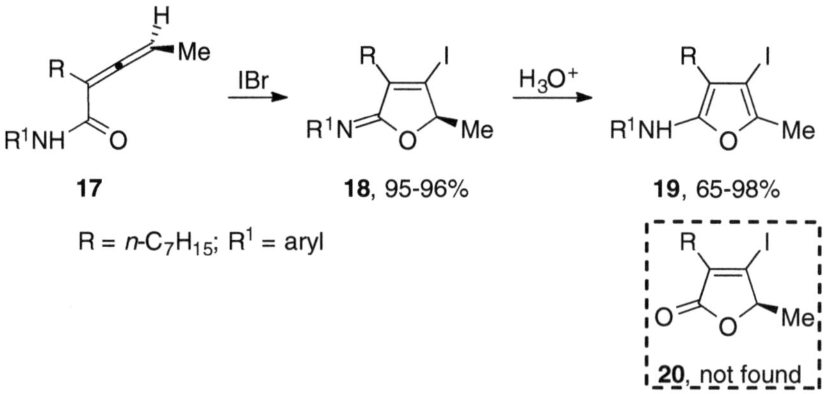

Scheme 3 Synthesis of amino-iodofurans via iodocyclization

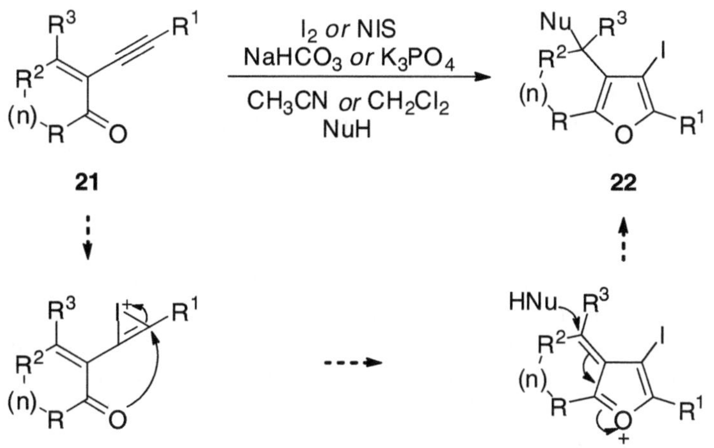

NuH = H_2O, ROH, RCO_2H, $ArNR_2$, $Ar(Me_3SiO)C{=}CH_2$; $n = 0,1$
R = H, Me; R^1 = aryl, 1-cyclohexenyl, $SiMe_3$
R^2 = H, Ph; R^3 = H, Me, Ph

Scheme 4 Electrophilic cyclization/nucleophilic addition of α-alkynyl alkenones

Cyclization of the allenic carboxamides **17** into imino lactone **18** is effected with iodine monobromide; subsequent hydrolysis gives the aminofurans **19**, instead of the anticipated furanone product **20** (Scheme 3) [28].

The electrophilic cyclization of α-alkynyl alkenones **21** using iodine or NIS as an electrophile, in the presence of various nucleophiles including functionally substituted alcohols, water, carboxylic acids, and electron-rich arenes, proceeds under mild reaction conditions, as reported by Larock's group (Scheme 4) [29]. The reaction is applicable to both cyclic and acyclic alkynyl alkenones. Introduction of the nucleophile, which attaches in the vicinity of the furan ring at the

Scheme 5 Electrophilic cyclization/nucleophilic addition of alkynyl-cyclopropyl ketones

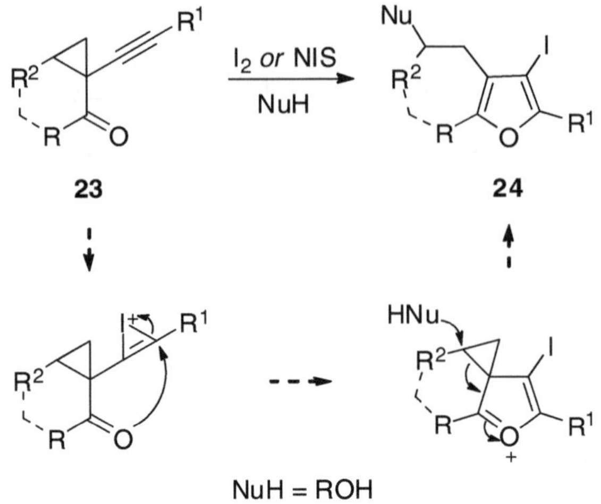

β-substituent, proceeds via Michael-type addition, leading to highly substituted iodofurans **22**.

Concurrently, Liu and Zhou reported parallel chemistry [30]. The electrophilic cyclization of α-alkynyl alkenones **21** using iodine/K_3PO_4 in the presence of various alcohols (including phenol and bulky L-borneol) also leads to an introduction of a nucleophile and proceeds with satisfactory yields to cyclohexyl-fused and other furans **22** (Scheme 4). Carbon-based nucleophiles: silyl enol ether or an electron-rich arene, *N*,*N*-dimethylaniline, were also confirmed as a nucleophile, complementing the iodocyclization process with an option for the formation of a carbon–carbon bond.

The alkynyl-cyclopropyl ketones **23** may also be subjected to electrophilic cyclization, as reported by Huang's group (Scheme 5) [31]. The process is analogous to the reaction of alkynyl alkenones **21** (Scheme 4); cyclopropane serves as a homo-Michael acceptor. Accordingly, the nucleophile is separated from the furan unit by one more carbon, compared to furan **22**. Upon electrophilic activation of the triple bond, the bridged cationic intermediate forms presumably. Next, cyclization facilitates the regioselective ring opening of cyclopropane and is accompanied by nucleophilic addition. When the substrates containing bicyclo[4.1.0]heptan-2-one core are combined with nucleophilic alcohols, as well as iodine or NIS as electrophiles, furans **24** [R-R^2=$(CH_2)_3$] are obtained with good to excellent yields. For monocyclic 1-(1-alkynyl)-cyclopropyl ketones, iodine may serve as both an electrophile and a nucleophile (while alcohols are unreactive).

When iodine is combined in the presence of the base ($NaHCO_3$) with propargyl oxiranes **25**, substituted iodofurans **26** are effectively obtained under mild reaction conditions (room temperature), via the iodonium intermediate **27**, as reported by Liang et al. (Scheme 6, iodonium pathway) [32]. This methodology explores the cyclohexane template, but can be extended to the acyclic as well as the 7-, 8-, and 12-membered ring oxiranes with excellent to moderate yields. For example, 3-iodotetrahydrobenzofuran **26** (R=Ph, R^1=R^2=H) was obtained with 72% yield.

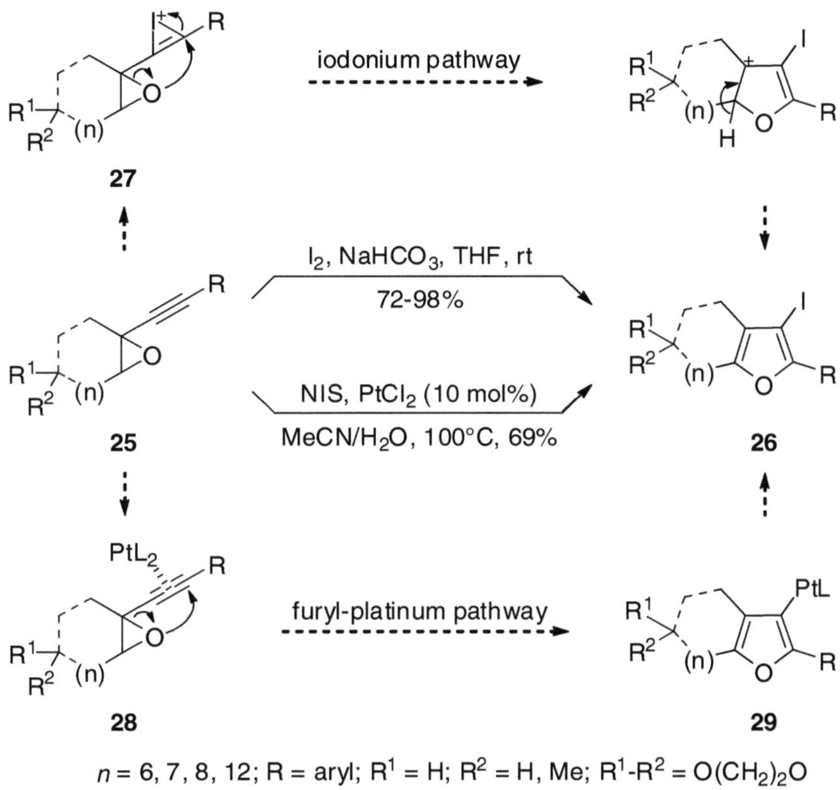

Scheme 6 Base- or platinum-catalyzed iodination/cyclization of oxiranes

A very similar reaction of phenyl-substituted cyclohexenyl oxirane **25** (R=Ph, R^1=R^2=H), this time with the aid of a platinum catalyst but without a base, in acetonitrile/water at 100°C was described by Shishido's group (Scheme 6, furyl-platinum pathway) [33, 34]. The plausible mechanism for the reaction involves the activation of a carbon–carbon triple bond by the platinum catalyst to facilitate the addition of a nucleophile, as illustrated by structure **28**. The oxirane–oxygen attack takes place at the *endo*-position of the alkyne; aromatization via deprotonation leads to the cyclic intermediate **29**. The furanyl-platinum species **29** could be trapped with electrophilic iodine supplied by NIS to afford the 3-iodotetrahydro-benzofuran **26** (R=Ph, R^1=R^2=H) with 69% yield. The increased reactivity under aqueous conditions was reasoned as forming a platinum hydroxide complex, thus enhancing the reactivity of the electrophile-demetalation from the furyl-platinum intermediate. A much lower yield of **26** (22%, R=Ph, R^1=R^2=H) is reported when NIS is used in the absence of a platinum catalyst and a base.

The electrophilic cyclization and 1,2-migration of the thio group of thiobutynols **30** have been developed as a method for the synthesis of 2,4-dihalo-3-thio-furans **31** (Scheme 7) [35]. The reaction proceeds presumably via a thiiranium intermediate

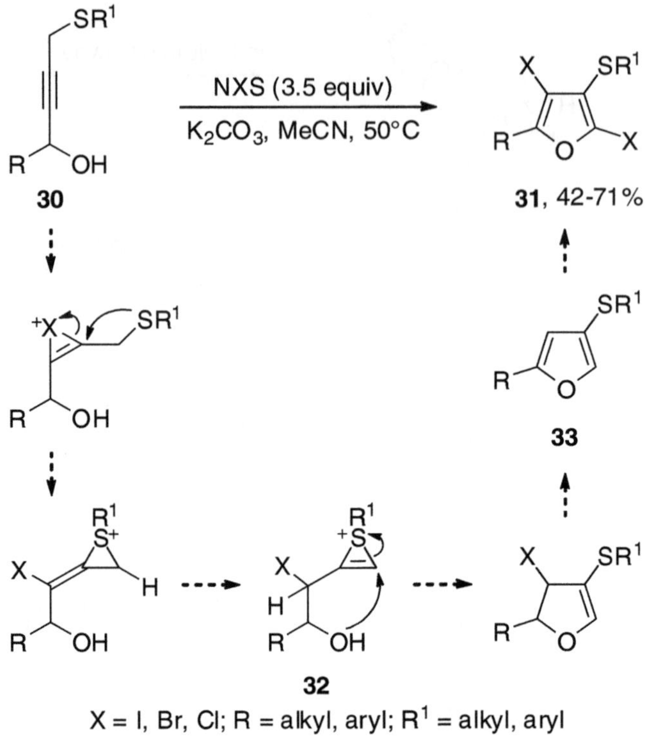

Scheme 7 Synthesis of 2,4-dihalothiofurans from thiobutynols

32 after initial iodine activation, intramolecular nucleophilic attack, and hydrogen transfer. Cyclization and elimination gives 3-thiofuran **33**, which undergoes dihalogenation to give a 2,4-dihalo-3-thio-substituted furan **31**. In this sense, an excessive amount of *N*-halosuccinimide (NXS, X=Cl, Br, I; 3.5 equiv) and a base (K_2CO_3) have to be employed to achieve good to moderate yields.

γ-Hydroxy alkynones **34** are versatile precursors for the synthesis of various furans by acid-catalyzed cyclization reactions. Using the mentioned starting materials, substituted 3-halofurans can be synthesized in good to excellent yields via halogenation/cyclization reactions. The alkynones **34** can be prepared by the addition of the acetylide of THP-protected propargyl alcohols to the aldehyde, followed by oxidation and deprotection. An alternative method consists of double deprotonation of the propargyl alcohols with excess of BuLi and thereafter addition to Weinreb's amide.

The 3-halofurans are obtained by treatment of the γ–hydroxy alkynone **34** with aqueous HX (X=I, Br, Cl), as reported by Obrecht [36]. This strategy allows for the synthesis of the 2,5-disubstituted 3-halofurans **13**, as shown in (4). The mechanism involves the regioselective addition of HX to alkynone **34**, giving rise to a mixture of *E*- and *Z*-isomers, which could undergo rapid interconversion. Subsequent cyclization and dehydration lead to the formation of the 3-halofurans **13**. The methodology allows for the preparation of unsubstituted at C-4 furans.

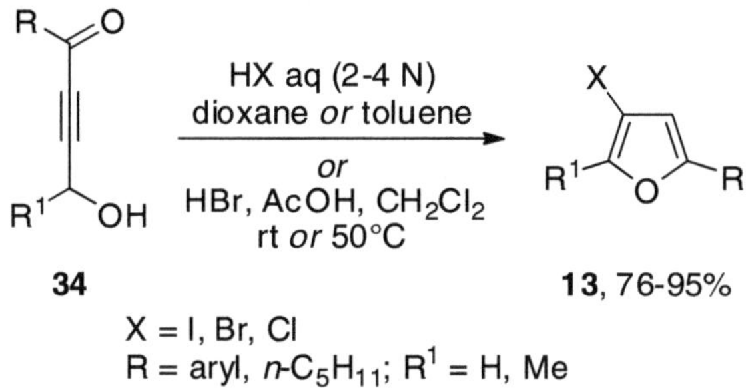

(4)

A sequence of palladium-catalyzed coupling of THP-protected propargyl alcohol **35** with acyl chloride **36** leads to a protected hydroxy alkynone **37** (Scheme 8) [37]. Subsequent electrophilic addition with concomitant deprotection and cyclization creates a one-pot synthesis of 3-halofurans **13** that reduces isolation steps, compared to the reaction described above (4). As illustrated in Scheme 8, the reaction pathway includes solvolysis of **37** under acidic conditions. Michael addition of HI or HCl to the γ–hydroxy alkynone **34** and subsequent cyclocondensation conclude the three-component sequence, leading to the β-iodo- or β-chlorofurans **13**. The use of NaI as a halide source leads to milder reaction conditions compared with the use of NaCl.

The cyclization reaction of the same γ–hydroxy alkynones **34** is facilitated by iodine or iodine monobromide, to afford 3,4-dihalofurans **38**, as demonstrated by

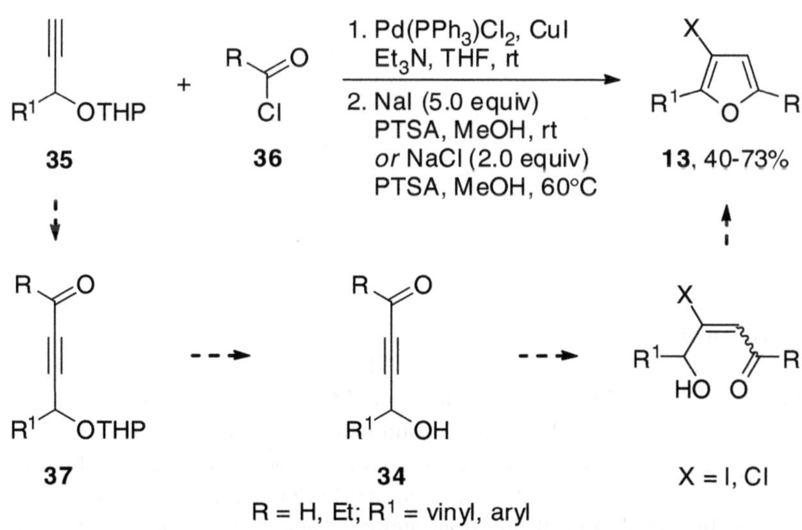

Scheme 8 One-pot synthesis and cyclization of γ–hydroxy alkynones

Scheme 9 Synthesis of 3,4-diiodo- and 3,4-bromoiodofurans

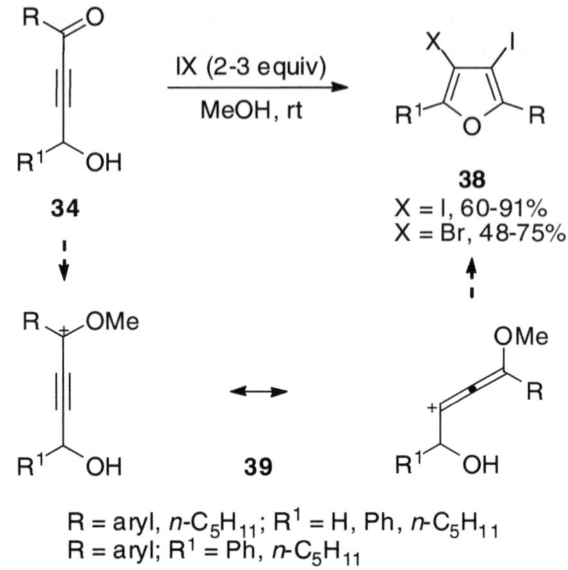

Jin, Yamamoto et al. (Scheme 9) [38]. The mechanistic experiments indicate the importance of methanol, which presumably participates in ketal intermediate formation that leads to the reactive intermediate **39**.

Reported earlier by the Müller group, the same concept is employed for the synthesis of 3,4-chloroiodofurans **40** (5) [7]. Iodine monochloride undergoes Michael addition to the γ–hydroxy alkynone **34** or **37**, obtained in situ from the reaction of alkyne **35** and acyl chloride **36** (as illustrated in more detail in Scheme 8), followed by cyclization to afford dihalofuran **40**, all in a one-pot sequence.

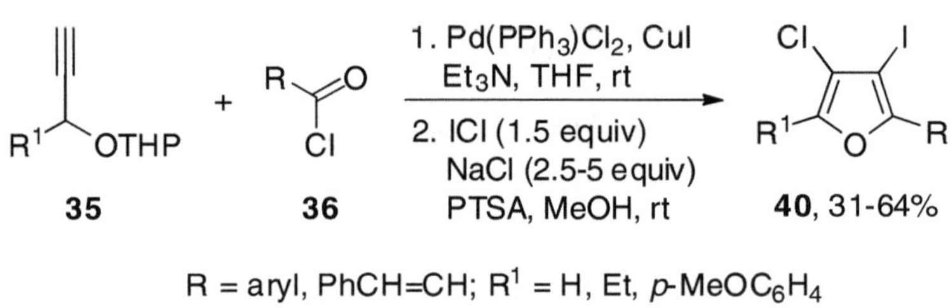

(5)

Butynediol **41** lacks a degree of unsaturation and can be viewed as a reduced homolog of the γ–hydroxy alkynone **34**. The treatment of the diol **41** with iodine leads to the 3,4-diiodo 2,5-dihydrofuran **42** (Scheme 10) [39]. Oxidation/aromatization of the cyclic **42** can be accomplished with the use of DDQ. One specific example, 3,4-diiodo-2,5-diphenylfuran **43**, has been obtained by this method with moderate yield. Without the use of an oxidation agent, methylation of the dianion formed by metalation of the monomethyl ether of 2-butyne-1,2-diol, followed by iodination gives 3-iodo-4-methyl furan in 83% yield [40].

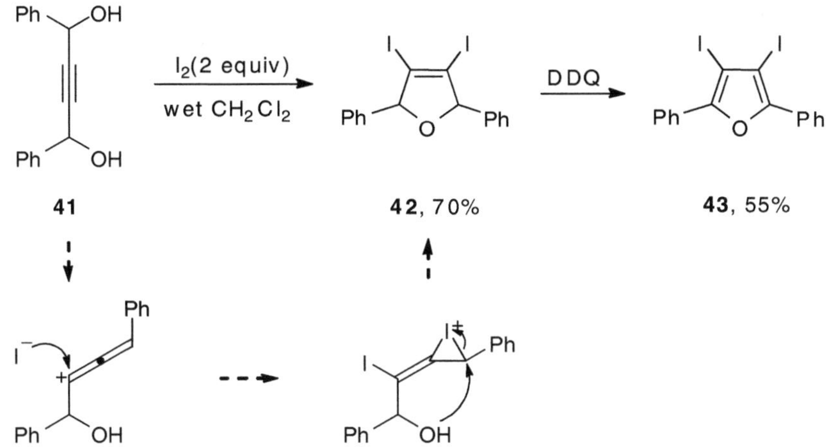

Scheme 10 Synthesis of 3,4-diiodofuran via cyclization/oxidation

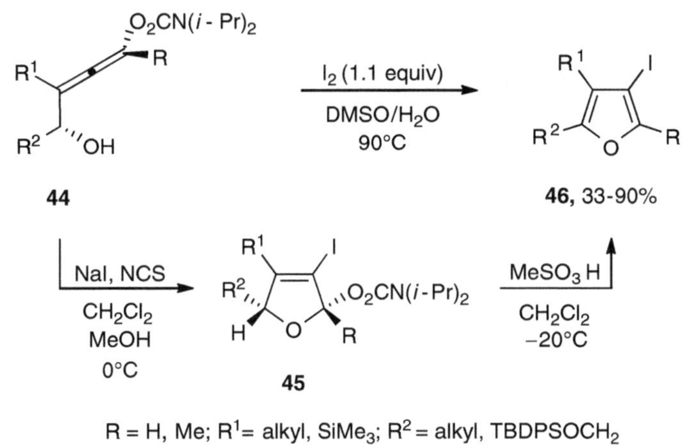

Scheme 11 Synthesis of iodofurans via cyclization of carbamates

The sequential lithiation of alkynyl carbamates, transmetallation with the use of ClTi(O*i*-Pr)$_3$, and the reaction of organotitanium intermediate with aldehydes provide access to the *syn*-4-hydroxy carbamates **44**. The iodine-induced cyclization affords *cis*-2,5-dihydro-3-iodofurans **45** (Scheme 11), which can readily be converted to their corresponding 3-iodofurans **46** via acid-mediated aromatization [41]. Treatment of the carbamates **44** by iodine at an elevated temperature facilitates 1,4-elimination of *N,N*-diisopropylcarbamic acids and leads directly to the di- and trisubstituted 3-iodofurans **46** in good yields.

Iodocyclization of alkyne-1,2-diols **47**, either commercially available or prepared, for example, via bis-hydroxylation of enynes or the addition of acetylides to α-hydroxy carbonyl compounds, delivers corresponding 3-iodofurans **48** generally with high yields (6) [42–45]. This effective reaction developed by Knight's group includes 5-*endo*-dig iodocyclization followed by dehydration using iodine as the electrophile in the presence of a base ($NaHCO_3$).

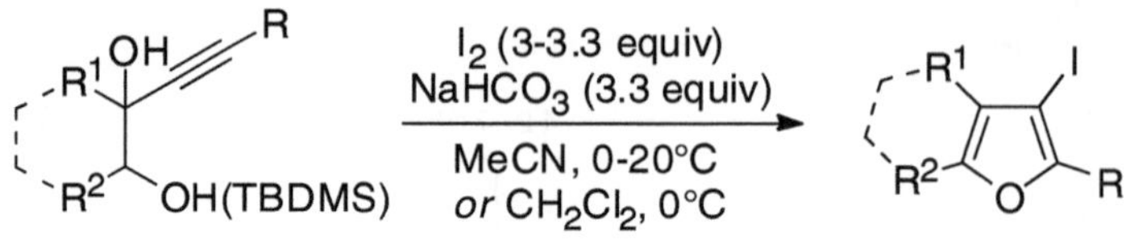

47 **48**, 47-93% (6)

R = aryl, $SiMe_3$, TBDMS
R^1 = Me, aryl, C≡CR' *or* R^1-R^2 = $(CH_2)_4$
R^2 = H, alkyl, aryl, CO_2Me

The tandem ring-opening/cyclization reactions of 1-alkynyl-2,3-epoxy alcohols (glycidol derivatives) **49** was reported by Liang et al. as the synthesis of substituted 3-iodofurans **50** (Scheme 12) [46]. Treatment of oxiranes **49** with an alcohol in the presence of copper(II) triflate as a Lewis acid results in regioselective ring opening. Subsequent iodocyclization/dehydration of the resulting butyne-1,2-diols, such as **51**, with the use of iodine, facilitates access to the 5-alkoxymethyl-substituted 3-iodofurans **50**. Both cyclic and acyclic substrates give good yields under mild reaction conditions. The method constitutes an elegant extension of the method described in (6).

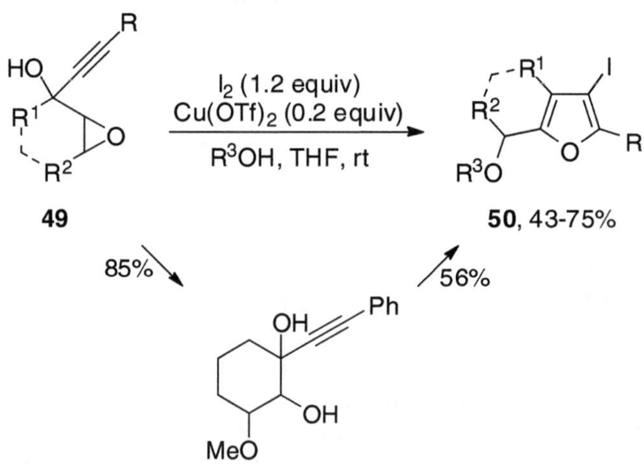

51

R = aryl; R^1 = H; R^2 = Me, Ph; R^1-R^2 = $(CH_2)_3$; R^3 = alkyl

Scheme 12 Synthesis of iodofurans via ring opening of oxiranes and iodocyclization

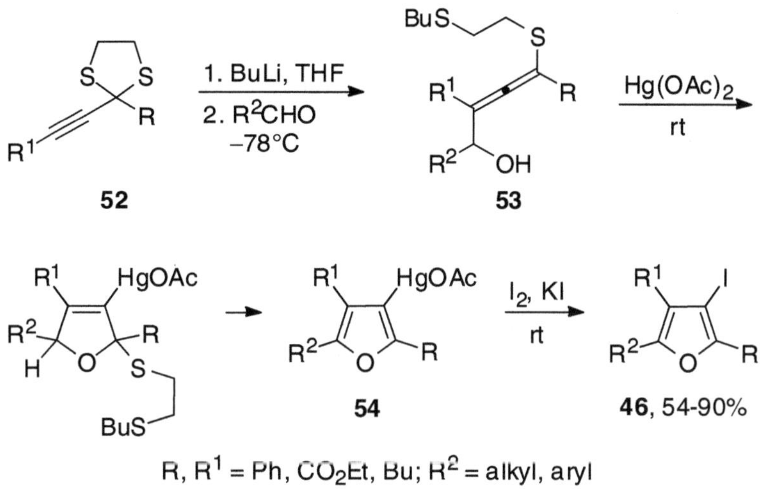

Scheme 13 Synthesis of iodofurans from aryl-substituted dithioacetals

A procedure for the synthesis of tetrasubstituted furans from the corresponding propargyl thioacetals **52** was elaborated by the Luh group (Scheme 13) [47]. Treatment of thioacetals **52** with BuLi and an aldehyde followed by mercuric acetate promotes the cyclization and desulfurization which leads, in the presence of iodine/potassium iodide, to the corresponding trisubstituted iodofurans **46**. This pathway leading via allenic adduct **53** and mercurio-furan **54** is reportedly preferred for aryl-substituted thioacetals **52**.

The same group reported that the addition of BuLi to alkyl-substituted propargyl thioacetals **55** and aldehydes yields ring-opened adducts **56** which react with mercury acetate to give mercurio-furans **57** (Scheme 14) [48]. Demetalation with iodine/potassium iodide offers a route to highly substituted 3-iodofurans **58**, which are regioisomeric to furans **46** obtained via the reaction illustrated in scheme 13, as referring to the position of substituents originating from thioacetal. Information from the authors indicates experimental evidence beyond that communicated, including X-ray structures of relevant derivatives, which documents the pathways (Luh, personal communication).

2.3 Cyclization Reactions of Iodine-Containing Precursors

Gevorgyan's group reported that haloallenyl ketones **59** could undergo cycloisomerization, affording β-halofurans **60** (7) [49, 50]. Synthesis of the halofurans **60** proceeds via halogen 1,2-migration, relevant to the sulfur migration illustrated in Scheme 7. Mechanistic description of the regioselectivity is discussed in Sect. 3.3.

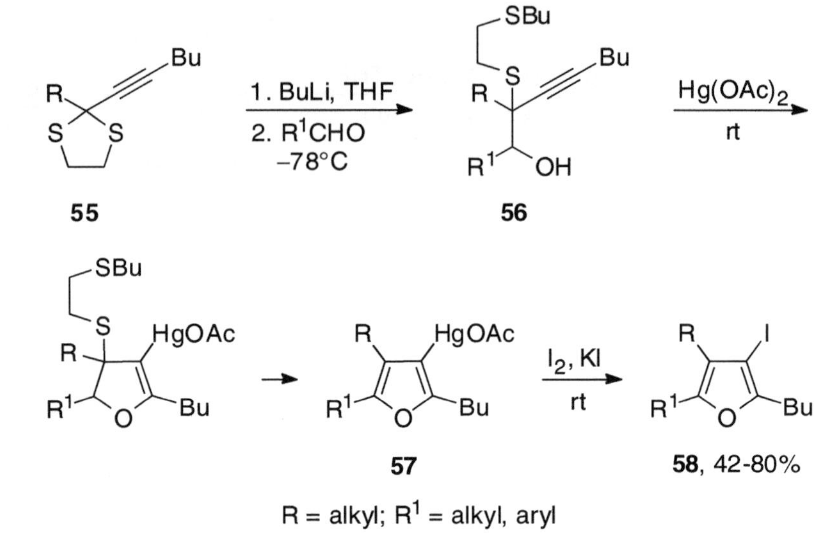

Scheme 14 Synthesis of iodofurans from alkyl-substituted dithioacetals

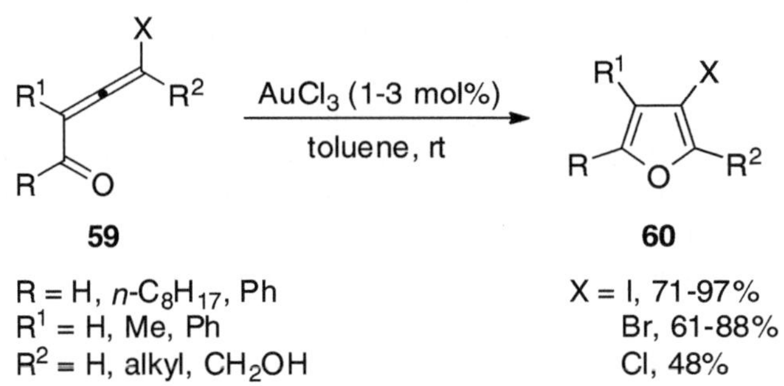

(7)

3 Synthesis of β-Bromofurans

3.1 *Bromination of Furans*

Furan reacts vigorously with bromine (and chlorine) at room temperature leading to polybrominated products. In effect, bromination reactions can generally be applied to trisubstituted furans that specifically offer the only vacant position for the introduction of bromine. Consequently, special methods have been developed for the synthesis of β-bromo-substituted furans with an unsubstituted β′-position.

Regioselective bromination reactions of the type of electrophilic aromatic substitution may be accomplished in specific cases. The selectivity of C-3 vs C-4 position can be achieved in the presence of electronically distinctive substituents. The other challenge includes the sensitivity of the substituents toward bromine.

Bromination of 2,5-disubstituted furan **61** with bromine in the presence of a Lewis acid leads to a bromofuran **62** (8) [51]. The halogen is introduced in the vicinity to the more electrodonating group. However, at the same time, the methyl substituent is affected, yielding a bromomethyl byproduct **63**, which is removed by treating the reaction mixture with triethylamine.

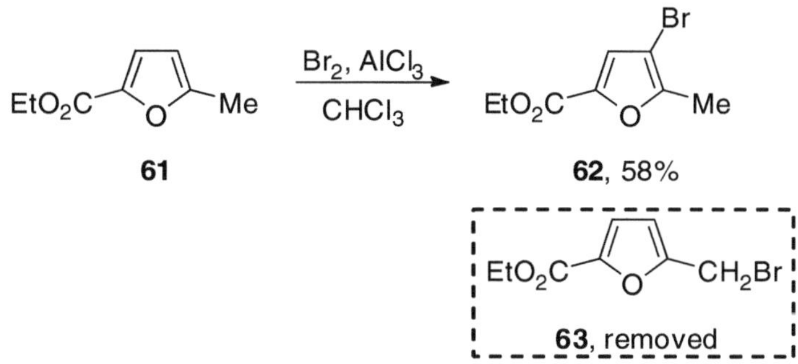

(8)

2-Methoxyphenyl-5-nitrophenylfuran **64a** when treated with NBS in benzene gives 3-bromo-2-(methoxyphenyl)-5-(4-nitrophenyl)furan **65a** (9) [52]. In the same way, phenyl-furan-carbaldehyde **64b** when subjected to NBS in acetonitrile facilitates the formation of 3-bromo-phenylfuran-carbaldehyde **65b** (9) [53]. Electronic effect of substituents, even those attached to the phenyl rings, effects pronounced regioselectivity at the *β*-position. Compounds **65** are isolated in respectable yields, when considering potential bromination of the *β*′-carbon or of the substitutents, and accompanied isolation challenges.

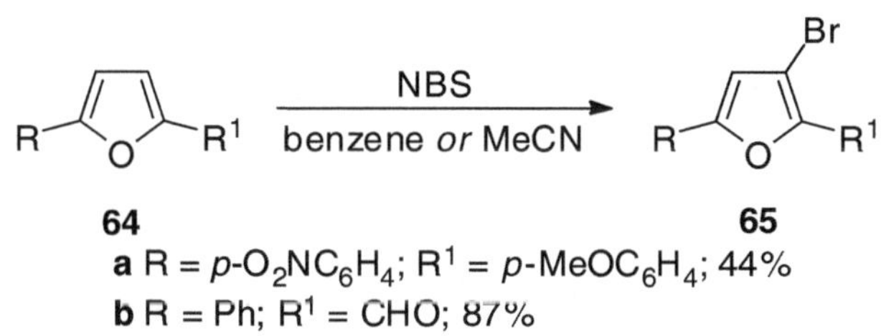

(9)

High yields are achieved when trisubstituted furans **66** are treated with bromine at 0°C, yielding bromo-methylthiofurans **67** (Scheme 15) [54].

3.2 Bromination–Cyclization Reactions

The halogenation/cyclization reactions introducing bromine in neutral conditions with the use of NBS are described in Sects. 2.2 (bromofurans, Scheme 2) and 5.2 (bromofluorofurans, 20) [23, 26]. The bromocyclization of butynones **12** has been

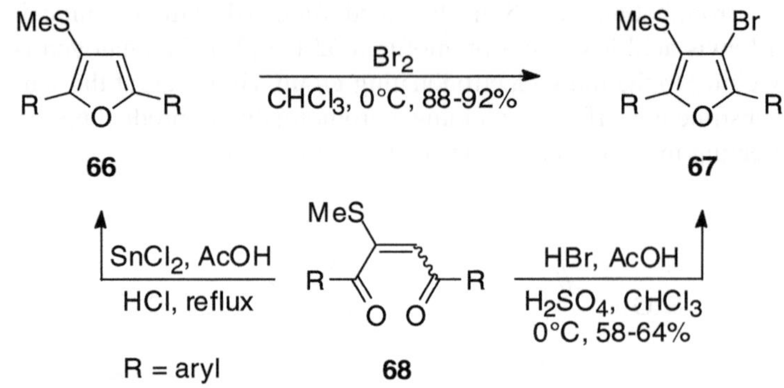

Scheme 15 Synthesis of bromo(methylthio)furans

extended to other examples (53–96%) [55], and also carried out in a one-pot procedure starting from homopropargyl alcohols [25].

1,4-Diketones **68** when treated with hydrogen bromide at 0°C yield bromo (methylthio)furans **67** via a cyclization/dehydration reaction (Scheme 15) [54].

Acetals **69** can similarly be treated with hydrogen bromide (which facilitates the cyclization and is a donor of bromine) or hydrogen chloride. Monosubstituted halofurans **70** can be synthesized this way with high yields (10) [36]. The acetylenic acetals are prepared by addition of the lithium acetylide of 3,3-diethoxyprop-l-yne to an aldehyde.

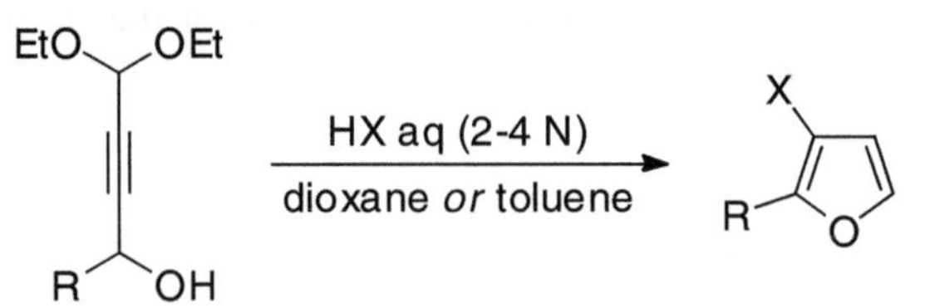

69 **70**, 60-93%

X = Br, Cl; R = alkyl, Ph

(10)

γ-Hydroxy alkynones **34** are effective cyclization precursors for the synthesis of substituted bromofurans **13** and 3,4-bromoiodofurans **38** using strong acids, as described in Sect. 2.2 (4) and Scheme 8 [36, 38]. Additional examples include cyclization with the use of hydrogen bromide in chloroform/acetic acid that leads to the formation of 3-bromo-2,5-dialkylfurans in high yields [9].

Relevant one-pot procedure starts from the reaction of the lithium salt of protected propargylic alcohol **71** with the aldehyde **72** [56]. Obtained diol is oxidized with the use of MnO_2 and tetrahydropyranyl ether is selectively deprotected. Follow-up reaction with HBr leads to the monosubstituted 3-bromofuran **73** (11).

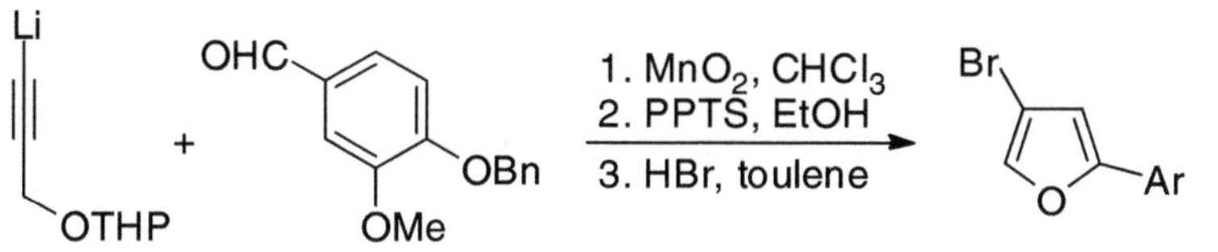

71 **72** (ArCHO) **73**, 45% (11)

By comparison, one-pot reaction of Weinreb's amide with dilithio propargyl alcohol leads to similar γ-hydroxy alkynones **34** and does not require the use of a protecting group. Follow-up HBr cyclization leads to the formation of 5-alkyl-3-bromofuran [57].

Bromination/cyclization reaction, leading for 2,4-dibromofurans **31** (X=Br) with the use of NBS, can be carried out in basic conditions as well (Sect. 2.2, Scheme 7) [33].

3.3 Reactions of Bromine-Containing Compounds

Multiple steps and sacrifice of additional halogen are required to introduce bromine into β-position of unsubstituted furan. Cyclization and dehydration of 2,3-dibromo-2-buten-l,4-diol **74** lead to 3,4-dibromofuran **75** (Scheme 16) [58]. Protonation of 3-bromo-4-lithiofuran, prepared via treatment of **75** with *t*-butylithium (*t*-BuLi), gives the unsubstituted 3-bromofuran **76** [59]. At the occasion, 4,4′-dibromo-3,3′-bifuran **77** is prepared by treating dibromofuran **75** with *n*-butylithium and $ZnCl_2$ via a zinc intermediate, which without isolation is combined with $CuCl_2$ to yield the furan dimer **77** [60].

An interesting halogen migration is observed when 2-bromo-5-methylfuran **78** is subjected to metalation by a combination of LDA and potassium *tert*-butoxide.

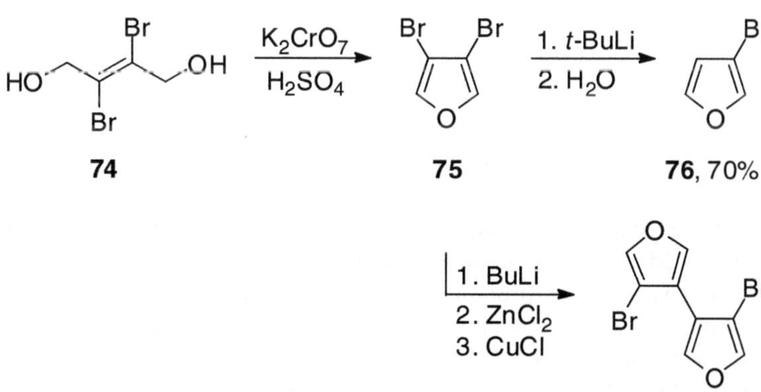

77, 78%

Scheme 16 Synthesis of unsubstituted 3-bromofuran

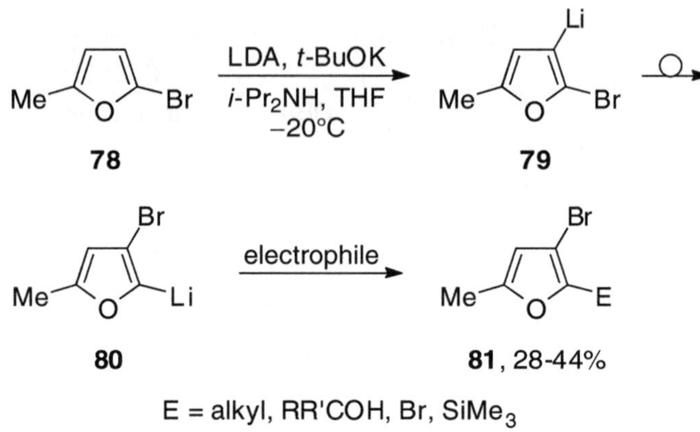

Scheme 17 Synthesis of 3-bromo-5-methyl-2-substituted furans

The reaction conditions allow for the formation of 2-bromo-3-lithio-5-methylfuran intermediate **79** that rearranges to 3-bromo-2-lithio-5-methylfuran **80** via halogen migration, presumably proceeding via an intermolecular transhalogenation process involving 2,3-dibromo-5-methylfuran. Lithiobromofuran **80** when quenched with electrophiles gives rise to 2-substituted-3-bromo-5-methylfurans **81** in moderate yields (Scheme 17) [61].

Deprotonation of furanone **82** with potassium bis(trimethylsilyl)amide (KHMDS) in THF and subsequent reaction with triisopropylsilyl chloride (TIPSCl) leads to the formation of furan **83** (12), which is further used as a diene without isolation [62].

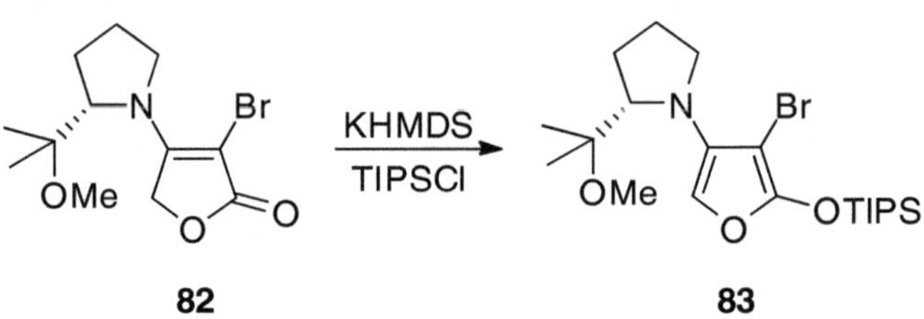

(12)

Cycloisomerization of bromoallenyl ketones **84**, catalyzed by $AuCl_3$, facilitates 1,2-halogen migration and consequently formation of 3-bromofurans **85** (Sect. 2.3, (7)). The mechanism of cycloisomerization of bromoallenes **84** is versatile, and according to the gold catalyst used, may also lead to α-regioisomer **86** (Scheme 18) [49, 50]. More oxophilic Au(III) facilitates the 1,2-halogen migration in haloallenyl ketone, favoring the formation of 3-bromofuran **85**. By comparison, the use of more π-philic Au(I) species switches the selectivity exclusively to 2-bromofuran **86**. The proposed pathway for 1,2-halogen migration involves bromoirenium zwitterion intermediate **87** promoted by coordination of Au(III). Competitive formation of 2-bromofuran is enabled via the activation of the allene double bond by Au(I) to form gold carbenoid species and succeeding 1,2-hydride shift [63].

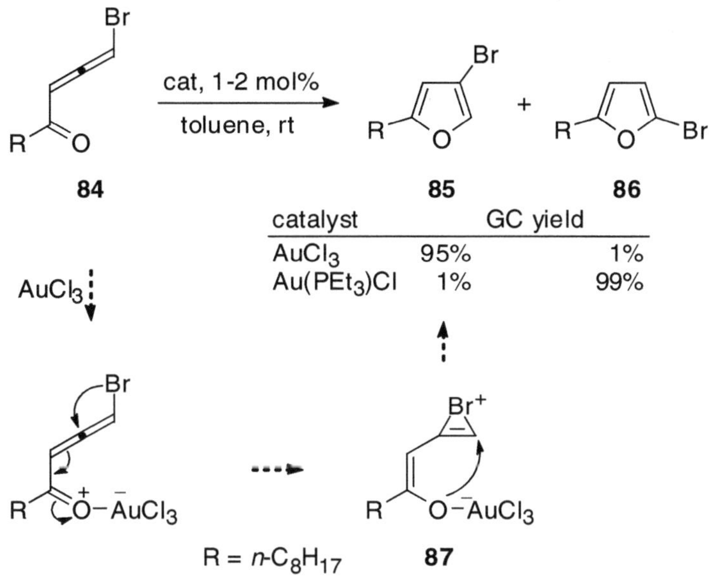

Scheme 18 Gold-catalyzed synthesis of 3-bromo-5-octylfuran

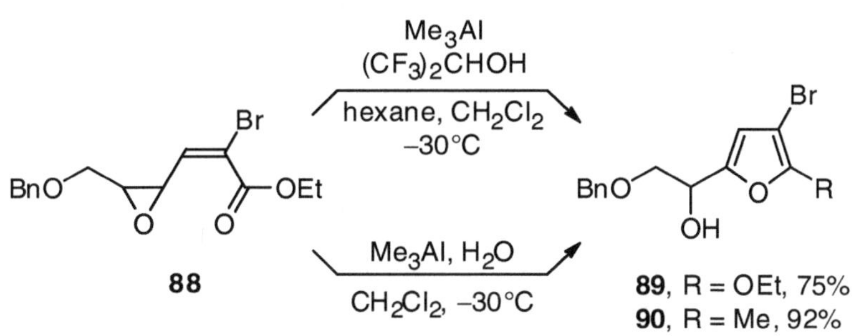

Scheme 19 Synthesis of 3-bromo-2-ethoxy- and 3-bromo-2-methylfurans

α-Bromoacrylate **88** when treated with $Me_3Al/(CF_3)_2CHOH$ or Me_3Al/H_2O reagent at lower temperatures gives, depending on the reagent used, 3-bromo-2-ethoxyfuran **89** or 3-bromo-2-methylfuran **90**, both in high yields (Scheme 19) [64]. It is believed that in the ethoxyfuran **89** case, $Me_2AlOCH(CF_3)_2$ is the acting reagent. Trimethylaluminum is presumably the donor of the methyl group found in the methylfuran **90**.

Diaryl dihalocyclopropyl ketones **91** can be converted to 2,5-diaryl-3-halofurans (**92**, X=Br, Cl) in the presence of aluminum chloride in dichloroethane via a regioselective *gem*-dihalocyclopropane ring opening, followed by cyclization and elimination of hydrogen halide (Scheme 20) [65]. The reaction is extended to the Friedel–Crafts acylation of substituted benzenes with 3-aryl-2,2-dihalocyclopropanecarbonyl chlorides **93** that proceeds in a one-pot manner. Specifically,

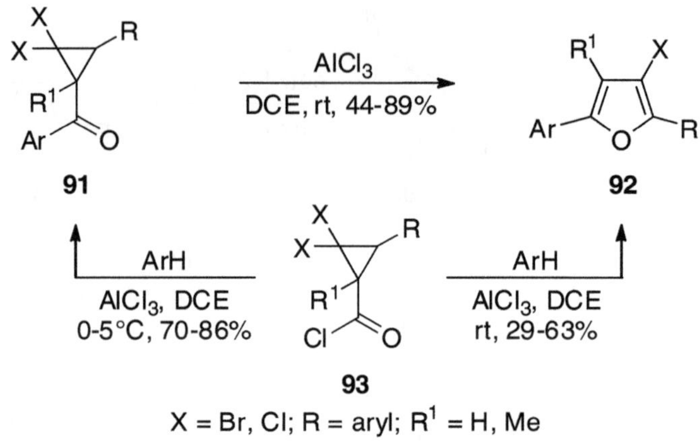

Scheme 20 Synthesis of halofurans from dibromocyclopropyl ketones and carbonyl chlorides

dibromocyclopropyl acyl chloride (**93**, Ar=*p*-MeOC_6H_4, R=*p*-Tol, R^1=Me) when treated with anisole in $AlCl_3$ at rt yields corresponding diarylbromofuran **92** in 63% yield, as reported by Tanabe et al. This reaction combines ring opening/ring closure; mechanistic outline is illustrated in Sect. 5.2 (Scheme 27).

Reduction of 2-bromo-2-cyanoacetophenone **94** at a mercury cathode using DMF/$LiClO_4$ as solvent/supporting electrolyte allows minor amount of 4-bromo-5-amino-2-phenylfuran-3-carbonitrile **95** to be isolated and characterized, although 5-amino-4-benzoyl-3-phenylfuran-2-carbonitrile **96** is the main product (13) [66].

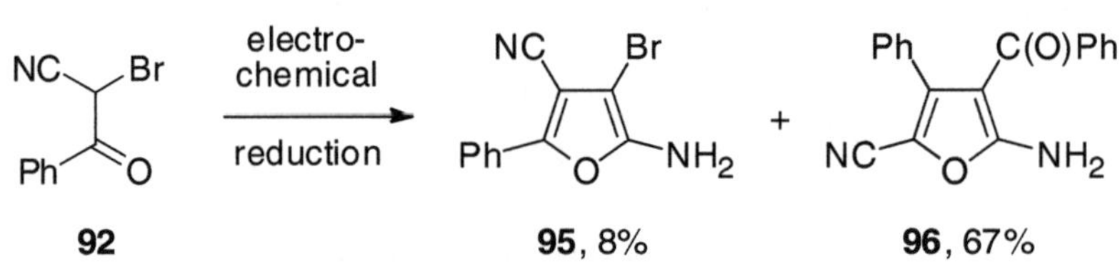

(13)

4 Synthesis of β-Chlorofurans

4.1 Chlorination of Furans

Since chlorine is a less selective halogenation agent than bromine, the issue of potential side reactions exists. General, mild, and selective forms of introducing chlorine into a *beta*-position on the furan ring, by its direct electrophilic chlorination reactions, have been sought to no avail. The use of chloroamine T for chlorination of trimethylsilyl furan **97** does not bring encouraging results, delivering a mixture of β- and α-chlorofurans **98** and **99** in meager yield (14) [67].

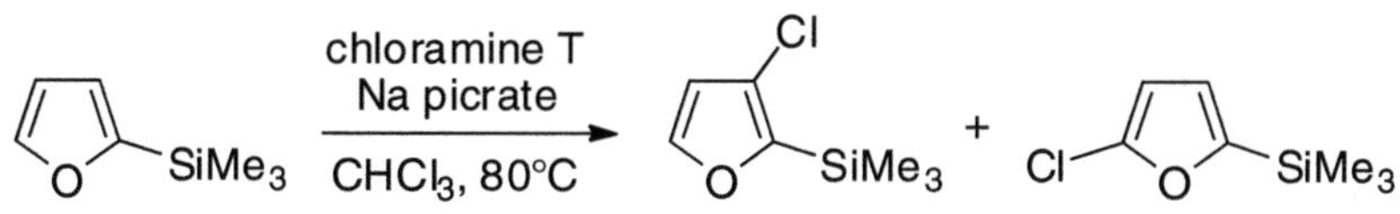

97 98 7%, ratio 7:1 99 (14)

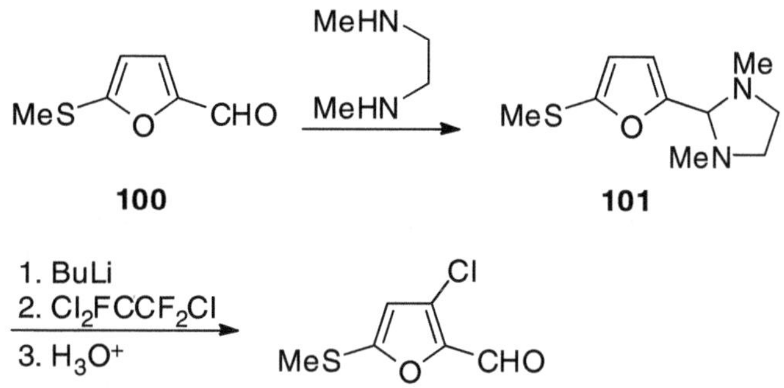

100 101

102, 30%

Scheme 21 Synthesis of chlorofuran carbaldehyde via lithiation/chlorination

The furan can be chlorinated via lithiation. The reactive carbonyl group of furan **100** is protected with the dimethylethane-1,2-diamine (Scheme 21) [68]. This approach allows for regioselective chlorination of intermediate **101** next to the masked aldehyde group, and the chlorofuran **102** is isolated after deprotection. The chlorine atom is supplied from trichlorotrifluoroethane.

Regioselective chlorination can be achieved by treatment of furans **103** with (tetramethylpiperidyl)magnesium chloride–lithium chloride, as reported by Piller and Knochel [69]. The furan-magnesate formed is trapped with chlorine-donating electrophiles, leading to chlorofurans **104** (15). The reaction tolerates the presence of Br, $SiMe_3$, and CO_2Et groups.

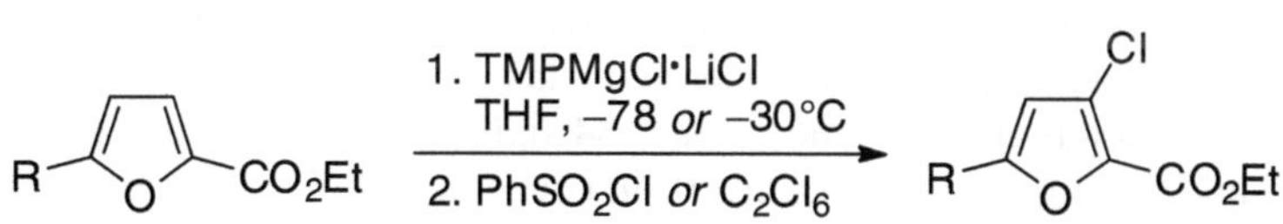

(15)

103 104, 48-59%

R = Br, Me_3Si; TMP - tetramethylpiperidyl

A chlorofuran **105** can be prepared via halogen-lithium exchange when bromofuran **76** is treated with BuLi; chloroalkane similarly serves as a source of chlorine (16) [70].

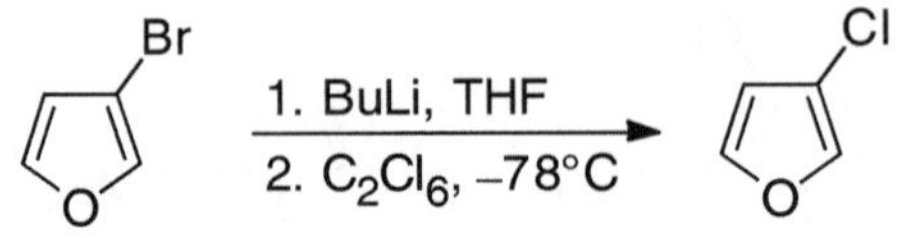

76 **105**, 45%

(16)

4.2 Chlorination–Cyclization Reaction

The 5-*endo-dig* chlorocyclization of butynones **12** with the use of trichloro-*s*-triazinetrione (trichloroisocyanuric acid, TCCA; 0.4 equiv) in toluene at room temperature, in the absence of a base, provides disubstituted 3-chlorofurans **106** in high yields (17) [8]. The mechanistic rationale is illustrated in Scheme 2. The reaction can also be accomplished by using a commercially available swimming pool sanitizer.

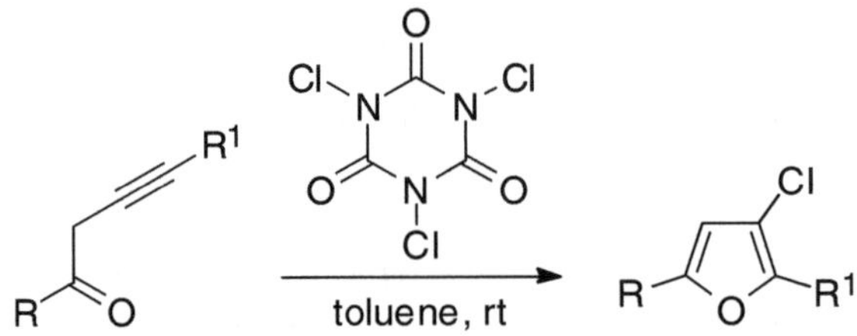

12 **106**, 79-96%

R = aryl, R^1 = aryl, c-C_3H_5

(17)

An interesting application of the Vilsmeier reaction begins with 1,2 diketone **107** and leads to the synthesis of furans **108** (Scheme 22) [71]. Enolizable carbonyl group is transformed into chlorovinyliminium salt **109** by the mixture of DMF and phosphorus oxychloride. Cyclization of salt **109** leads to furan **110** that undergoes second Vilsmeier reaction affording, after hydrolysis, dimethylamino-substituted chlorofuran carbaldehyde **108** in moderate yields.

Chlorocyclization of acetylenic ketones **34** and acetals **69** has been presented in Sects. 2.2 and 3.2 (Equations 4, 5, and 10, and Scheme 8) [36, 37]. Another example of chlorocyclization of acetylenic acetal **69** that is illustrated in (10) has been reported for R=C≡CCH(OH)C≡CH [72].

Synthesis of dichlorothiofurans **31** is described in Sect. 2.2, Scheme 7 [35].

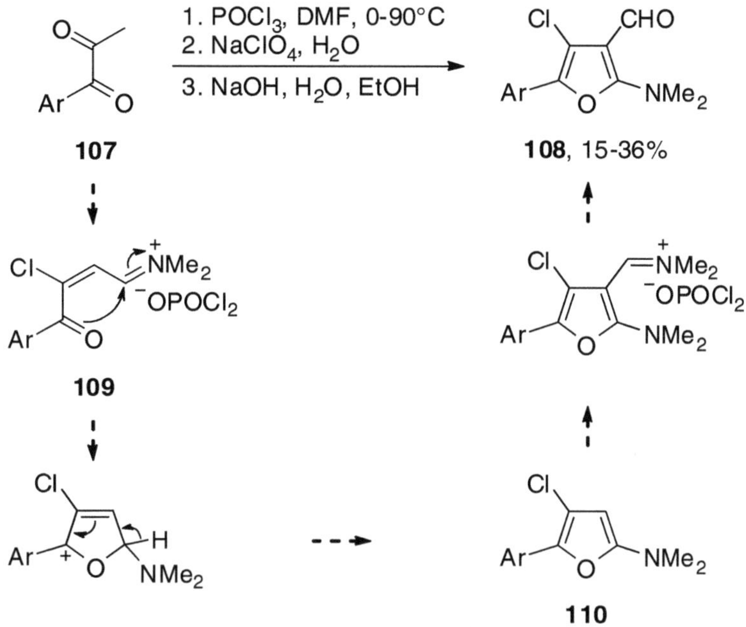

Scheme 22 Synthesis of chlorofurans via Vilsmeier reaction

4.3 Cyclization Reactions of Chlorine-Containing Compounds

In a palladium-catalyzed reaction, 2-halobutynones **111** undergo cycloisomerization to halofurans **112** (Scheme 23), as described by Lie Ken Jie et al. [73]. The reaction can be extended by the addition of allyl bromides to the reaction mixture, which leads to completely substituted furans **113**, which are regioisomeric to furan **106**. Reaction of oxiranes with LiCl forms a chlorohydrin, and its subsequent chromic acid oxidation leads to the starting material **111**.

Treatment of 3,3-dichlorotetrahydrofurans with a base such as KOH or *t*-BuOK has led to an equimolar mixture of derived chlorofuran and dihydrofuran [74]. However, this concept has been brought to practicality by the Ram group, using trichloroethyl allyl ethers **114** as starting materials [75]. The synthesis of 4-alkylchlorofurans **115** is presented in Scheme 24. The ethers **114** are converted to 3,3-dichloro-4-(1-chloroalkyl)tetrahydrofuran **116** in high yields via a halogen atom transfer/radical cyclization reaction mediated by a copper complex. Double dehydrochlorination and isomerization of tetrahydrofurans **116** with DBU afford disubstituted chlorofurans **115**. A one-pot procedure has been elaborated that affords chlorofurans **115** in good overall yields from the allyl ethers **114**, which can be prepared via reaction of trichloromethyl carbinols with allylic bromides.

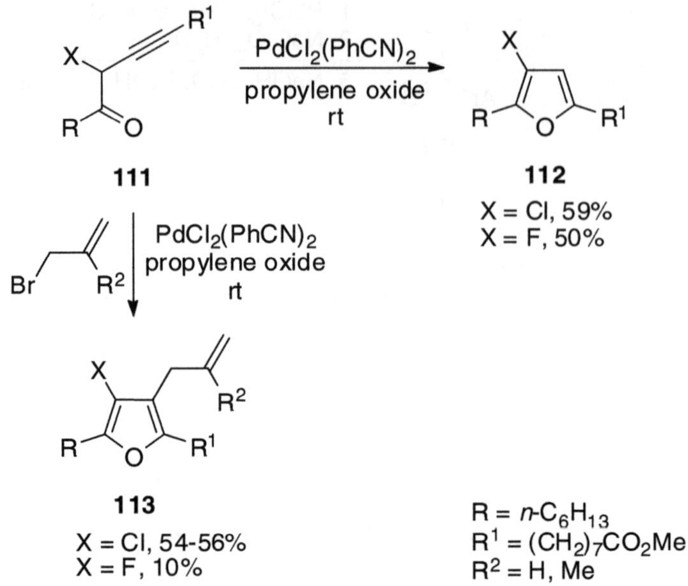

Scheme 23 Synthesis of halofurans via cycloisomerization of α-halobutynones

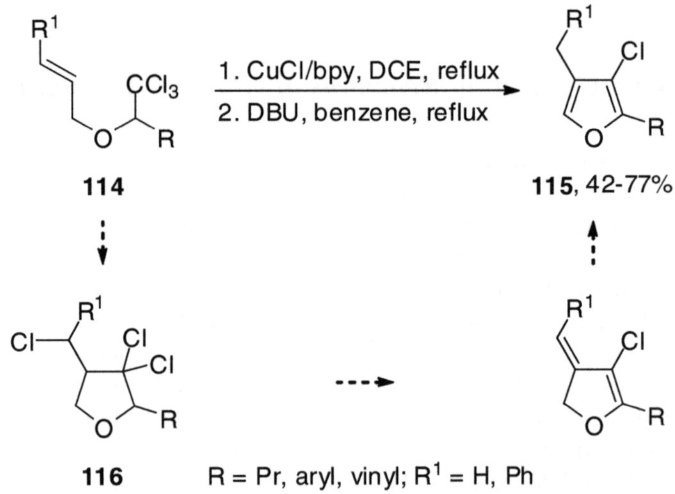

Scheme 24 Synthesis of chlorofurans via cyclization/dehydrohalogenation of trichloroethyl ethers

When, instead of tetrahydrofuran **116**, an acetoxy derivative **117** is investigated, the reaction leads via dechloroacetoxylation to disubstituted chlorofurans **118**. Acetoxy tetrahydrofuran **117** can be aromatized with the use of Zn and a base (*t*-BuOK) in THF [76], or sodium iodide in DMF (18) [77].

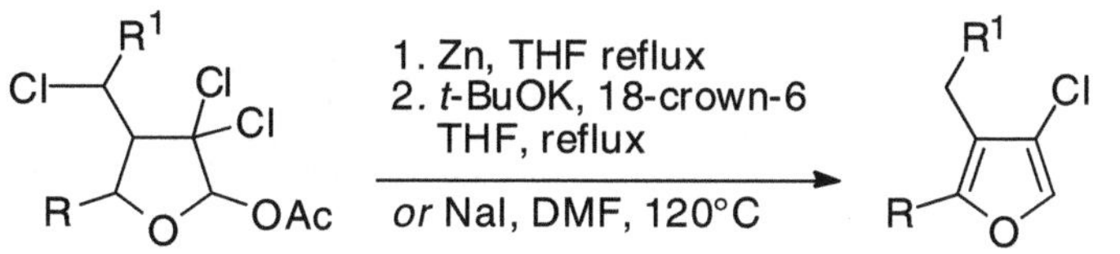

117 118, 23-46% (18)

R, R^1 = H, alkyl, Ph

Synthesis of chlorofurans from chloroallenyl ketones **59** [X=Cl, (7)] [49, 50] and cyclopropyl ketones **91** (Scheme 20) [65] is described in Sects. 2.3 and 3.3, respectively.

5 Synthesis of β-Fluorofurans

5.1 Fluorination of Furans

Synthetic methods leading to β-fluorofurans include limited options. The lithiation of 3-bromo-2-octylfuran **119** and its subsequent reaction with *N*-fluoro-*N*-(phenylsulfonyl)benzenesulfonamide (NFSI) gives 3-fluoro-2-octylfuran **120** with poor yield (19) [78].

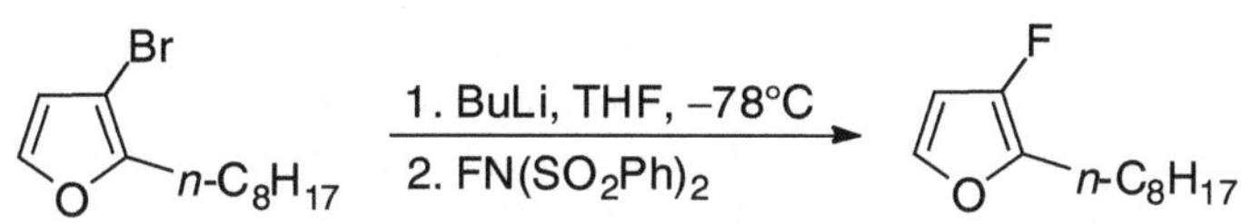

119 120 (19)

5.2 Cyclization Reactions of Fluorine-Containing Compounds

Reaction of fluoroalkynone **111** (X=F, Scheme 23, Sect. 4.3) with $PdCl_2(PhCN)_2$ (50 mol%) yields fluorofuran **112** (X=F) [73]. When the same substrate was treated with a mixture of $PdCl_2(PhCN)_2$, allyl bromide, or methyl-allyl bromide, tetrasubstituted C_{18} furan derivatives **113** were obtained with low yield. Synthesis of fluoroalkynone **111** has been reported via a low-yielding nucleophilic fluoride ring opening of an epoxide precursor and subsequent oxidation.

Limitations of the above method are avoided applying fluorination of 1,4-disubstituted silyl enol ethers **121** with Selectfluor that gives mono-fluorobutynones **122** in practically quantitative yields (Scheme 25). Subsequent 5-*endo-dig* cyclization in the presence of chlorotriphenylphosphine gold(I)/silver trifluoromethanesulfonate (both 5 mol%) under ambient conditions provides a facile method for the generation of mainly 2,5-diaryl-fluorofurans **123** in high yields (89–96%) [79]. The silyl enol

Scheme 25 Synthesis of fluorofurans from silyl enol ethers

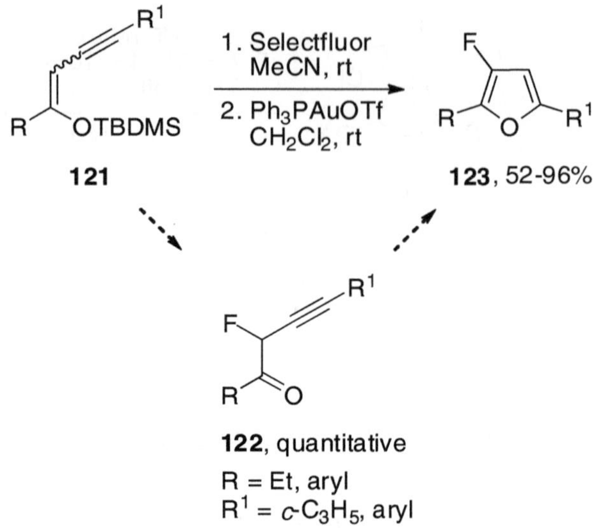

ethers **112** can be prepared by silylation of butynones **12** [80]. Furans such as **123** can be further functionalized [81, 82].

5-*Endo-dig* halocyclization of 2-fluoroalk-3-yn-1-ones **121** with the use of *N*-iodo- and *N*-bromosuccinimide in the presence of gold chloride/zinc bromide (5:20 mol%), under ambient conditions, provides a facile method for the synthesis of 2,5-disubstituted 3-bromo-4-fluoro- and 3-fluoro-4-iodofurans **124** (20). The sequential procedure starts, as above, at monofluorination of the silyl enol ether **121** with Selectfluor and proceeds with good overall yields [26]. One other example of 3-bromo-4-fluorofuran **124** (X=Br; R, R^1=Ph), with undisclosed preparative yield, has been prepared by a sequential lithiation/bromination reaction of 3-fluoro-2,5-diphenylfuran **123** (R, R^1=Ph) [83].

(20)

A base-promoted cyclization of the electrodeficient *gem*-difluorohomopropargyl alcohols (2,2-difluoroalk-3-yn-1-ols) **125** with potassium *tert*-butoxide or DBU, at an elevated temperature, provides the corresponding fluorofurans **123** (Scheme 26) [83, 84]. A series of fluorinated 3,3′-bifurans and tetrasubstituted furans has been prepared by Zhu et al. via a functionalization process [83]. The preparation of

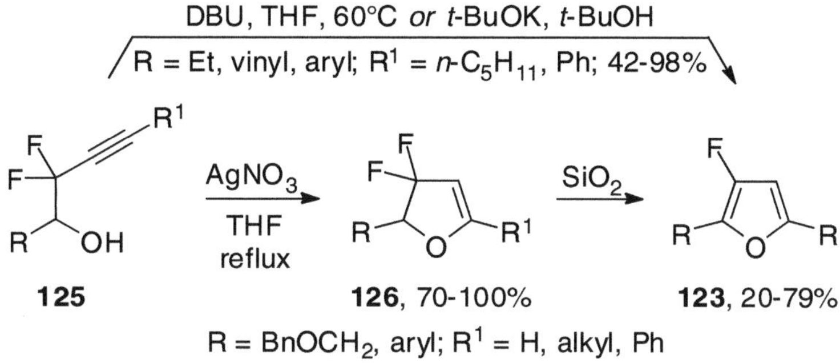

Scheme 26 Cyclization of *gem*-difluorohomopropargyl alcohols

alcohols **125** involves bromine substitution of bromochlorodifluoromethane with lithium acetylide, and the follow-up reaction with aldehydes.

Corresponding transition metal-catalyzed cyclization of *gem*-difluorohomopropargyl alcohols **125** to 3,3-difluoro-dihydrofurans **126** has been reported by the Hammond group (Scheme 26) [85]. The electron-deficient triple bond is effectively activated by silver nitrate, and the formation of dihydrofurans **126** is observed in excellent NMR yields. Subsequent aromatization by elimination of hydrogen fluoride via silica gel chromatography yields the corresponding 3-fluorofurans **123**. However, non-aromatic difluoro-2,3-dihydrofurans may remain as the final product, depending on the substituents. Interestingly, attempts to induce aromatization of **126** using basic (NaOH, *t*-BuOK, NaH) or acidic (BF_3·Et_2O, BCl_3) conditions failed.

The iodocyclization of difluorohomopropargyl alcohols **125** can also be induced by iodine monochloride (21) [4]. The unreactive nature of alcohols **125** requires a combination of strong electrophile (ICl), base (Na_2CO_3), and microwave irradiation, to access 3,3-difluoro-4-iodo-2,3-dihydrofurans **127**. Subsequent silica gel aromatization leads to the desired fluoroiodofurans **128**. The iodo substituent in furans **128** is utilized to prepare multisubstituted β-fluorofurans using a cross-coupling reaction [4].

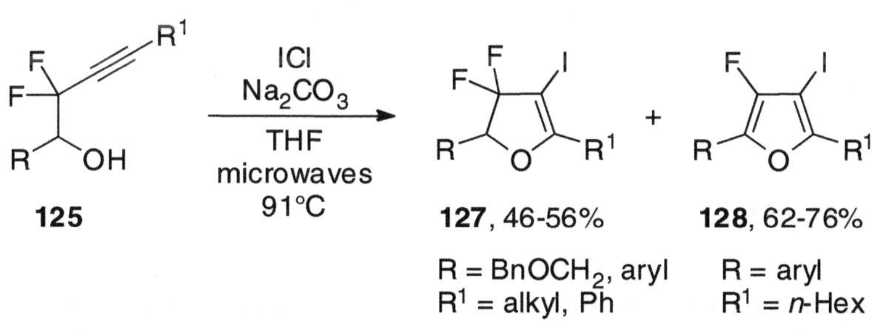

(21)

Rearrangement of *gem*-difluorocyclopropyl aryl ketones, analogous to the reaction of dihalocyclopropyl ketones **91** (Scheme 20, Sect. 3.3), offers a method for the preparation of fluoro-2,5-diarylfurans with good yield [86]. However, a high

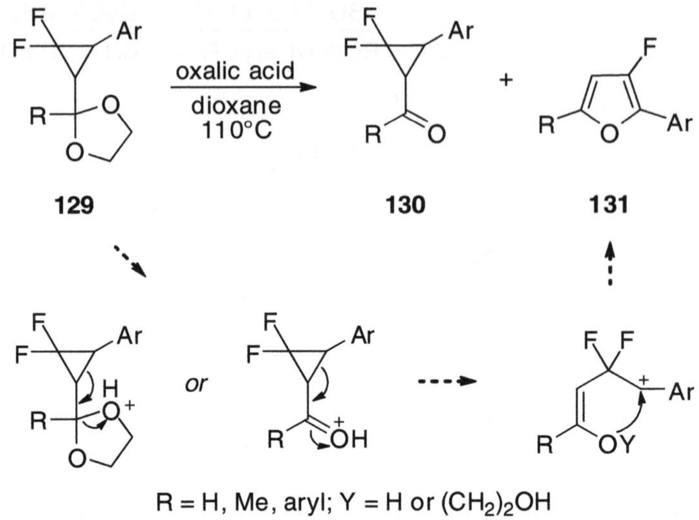

Scheme 27 Acid hydrolysis of *gem*-difluorocyclopropyl ketals

reaction temperature (216°C) is required to accomplish this process effectively. Analogously, acid-catalyzed hydrolysis of *gem*-difluorocyclopropyl acetals and ketals **129** proceeds at 110°C via the intramolecular carbonium rearrangement with simultaneous ring cleavage (Scheme 27) [87]. Acetals with electron-withdrawing groups in the *p*-position of the phenyl ring give a mixture of *gem*-difluorocyclopropyl ketones **130** and fluorofurans **131**, with a predominance of the latter. However, when the electron-donating group in the *p*-position of the aromatic ring is present, fluorofurans **131** are the dominant product. The hydrolysis of ketals containing electron-withdrawing groups yields ketones **130** as the sole or major products. Difluorocyclopropyl starting materials **129** can be obtained via the cycloaddition of difluorocarbene to α,β-unsaturated carbonyl compounds or their derivatives. Similar to the reaction of *gem*-difluorohomopropargyl alcohols **125** [(21) and Scheme 26], this method includes elimination of hydrogen fluoride.

Acid catalyzed cyclization of 2-fluoroalk-2-enoates **132** leads to 3-fluorofuran-2(5*H*)-ones **133**. Silyloxy substituted β-fluorofurans **134** are obtained as a result of metallation-silylation of the furanones **133** (Scheme 28) [88a, b].

6 Concluding Remarks

Although halofurans have been known for over a century, the challenge of selective introduction of a halogen into the β-position still remains. Significant progress has been made in recent years. However, general methods, especially those that would offer versatility toward the introduction of halogen as the last step, thus allowing the use of a single substrate to deliver a library of diverse halofurans, are not so

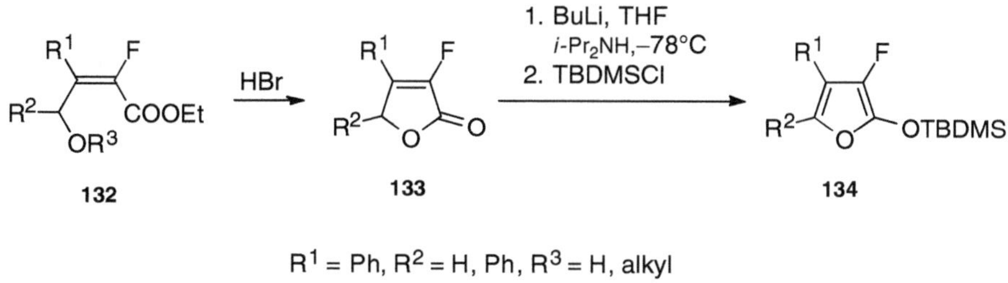

Scheme 28 Synthesis of silyloxy substituted β-fluorofurans

common. The development of halocyclization reactions in recent years has brought the possibility of mild reaction conditions to access iodo-, bromo-, and even chlorofurans. However, the fluorocyclization reactions still constitute the holy grail to synthetic organic chemists. Accordingly, synthesis of fluorofurans requires insertion of fluorine into a substrate. It should be noted that the already synthesized halofurans in numerous cases can be further functionalized by introducing additional substituents, including C-H functionalization. These reactions have not been emphasized in this article.

References

1. Graening T, Thrun F (2008) Furans and their benzo derivatives: synthesis. In: Katritzky AR (ed) Comprehensive heterocyclic chemistry III, vol 3. Elsevier, New York, pp 497–569
2. König B (2001) Product class 9: furans. In: Maas G (ed) Hetarenes and related ring systems; science of synthesis, vol 9. George Thieme Verlag, Germany, pp 183–286, Chap. 9.9
3. Yeung K-S, Yang Z, Peng X-S, Hou X-L (2011) Five-membered ring systems: furans and benzofurans. In: Gribble G, Joule JA (eds) Progress in heterocyclic chemistry, vol 22. Elsevier, New York, pp 181–216, Chap. 5.3
4. Arimitsu S, Jacobsen JM, Hammond GB (2008) Synthesis of 2,4,5-trisubstituted 3-fluorofurans via sequential iodocyclization and cross-coupling of *gem*-difluorohomopropargyl alcohols. J Org Chem 73:2886–2889. doi:10.1021/jo800088y
5. Hussain M, Khera RA, Hung NT, Langer P (2011) Site-selective Suzuki–Miyaura cross-coupling reactions of 2,3,4,5-tetrabromofuran. Org Biomol Chem 9:370–373. doi:10.1039/C0OB00695E
6. Schweizer SA, Bach T (2010) Regioselective Pd(0)-Catalyzed Hiyama Cross-Coupling Reactions at Dihalo-Substituted Heterocycles. Synlett 81–84. doi:10.1055/s-0029-1218528
7. Karpov AS, Merkul E, Oeser T, Müller TJJ (2006) One-Pot three-component synthesis of 3-halofurans and 3-chloro-4-iodofurans. Eur J Org Chem 2991–3000. doi:10.1002/ejoc.200600225
8. Sniady A, Morreale MS, Wheeler KA, Dembinski R (2008) Room-temperature electrophilic 5-*endo-dig* chlorocyclization of Alk-3-yn-1-ones with the use of pool sanitizer: synthesis of 3-chlorofurans and 5-chlorofuropyrimidine nucleosides. Eur J Org Chem 3449–3452. doi:10.1002/ejoc.200890051
9. a) Cho C-H, Larock RC (2011) Highly Substituted Lactone/Ester-Containing Furan Library by the Palladium-Catalyzed Carbonylation of Hydroxyl-Substituted 3-Iodofurans. ACS Comb Sci 13:272–279. doi:10.1021/co100088q. b) Sauers RR, Van Arnum SD (2004) γ-Hydroxyalkynyl

ketones as useful scaffolds for the preparation of combinatorial libraries of furans, isoxazoles and pyrazoles. J Comb Chem 6:350–355. doi:10.1021/cc030029
10. Webster R, Lautens M (2009) Conformational effects in diastereoselective aryne Diels-alder reactions: synthesis of benzo-fused [2.2.1] Heterobicycles. Org Lett 11:4688–4691. doi:10.1021/ol9019869
11. Pieniazek SN, Houk KN (2006) The origin of the halogen effect on reactivity and reversibility of Diels-alder cycloadditions involving furan. Angew Chem Int Ed 45:1442–1445. doi:10.1002/anie.200502677
12. Morton GE, Barrett AGM (2005) 1,4-Difluoro-2,5-dimethoxybenzene as a precursor for iterative double benzyne-furan Diels–Alder reactions. J Org Chem 70:3525–3529. doi:10.1021/jo050112e
13. Milkiewicz KL, Neagu IB, Parks DJ, Lu T (2003) Synthesis of a novel series of 10-oxa-3-aza-tricyclo[5.2.1.01,5]dec-8-en-4-ones through an intramolecular Diels-Alder reaction. Tetrahedron Lett 44:7341–7343. doi:10.1016/S0040-4039(03)01865-3
14. Ram NR, Kumar N (2008) Synthesis of highly substituted *m*-chlorophenolic biaryls by chlorine substituent-promoted Diels–Alder reaction of 2-aryl-3-chlorofurans. Tetrahedron 64:10267–10271. doi:10.1016/j.tet.2008.07.050
15. Young IS, Kerr MA (2007) Total synthesis of (+)-nakadomarin a. J Am Chem Soc 129: 1465–1469. doi:10.1021/ja068047t
16. Bach T, Krüger L (1999) The preparation of 2,3,5-Tri- and 2,3-disubstituted furans by regioselective palladium(0)-catalyzed coupling reactions: application to the syntheses of rosefuran and the F5 furan fatty acid. Eur J Org Chem 2045–2057. doi:10.1002/(SICI)1099-0690(199909)1999:9<2045::AID-EJOC2045>3.0.CO;2-6
17. König, B. Product Class 9: Furans. In *Hetarenes and Related Ring Systems*; *Science of Synthesis*; Maas, G., Ed.; George Thieme Verlag: Stuttgart, Germany, 2001; Vol. 9, Ch. 9.9.4.1.12. p 249–252.
18. Serdyuk O, Butin A, Abaev V (2010) Synthesis of fluorofurans and perfluoroalkylfurans. J Fluorine Chem 131:296–319. doi:10.1016/j.jfluchem.2009.11.009
19. Carpenter AJ, Chadwick DJ (1985) Dimetallations of 2-substituted furans and thiophenes: approaches to the 2, 3, 5-trisubstituted heterocycles. Tetrahedron Lett 26:5335–5338. doi:10.1016/S0040-4039(00)95032-9
20. Bures E, Nieman JA, Yu S, Spinazzé PG, Bontront J-LJ, Hunt IR, Rauk A, Keay BA (1997) Regioselective Preparation of 2,4-, 3,4-, and 2,3,4-Substituted Furan Rings. 2. Regioselective Lithiation of 2-Silylated-3-substituted Furan Rings. J Org Chem 62:8750–8759 and references cited therein. doi:10.1021/jo971098b
21. Song ZZ, Wong HNC (1994) 3,4-Disubstituted Furans, 5. Regiospecific Mono-ipso-Iodination of 3,4-Bis(trimethylsilyl)furan and Regiospecific ipso-Iodination of Tris[(4-alkyl- or -aryl)furan-3-yl]boroxines to 4-Substituted 3-(Trimethylsilyl)furans and Unsymmetrical 3,4-Disubstituted Furans. Liebigs Ann Chem:29–34
22. Lopez FJ, Jett M-F, Muchowski JM, Nitzan D, O'Yang C (2002) Synthesis and biological evaluation of ketorolac analogs. Heterocycles 56:91–95. doi:10.3987/COM-01-S(K)44
23. Sniady A, Wheeler KA, Dembinski R (2005) 5-*Endo-Dig* electrophilic cyclization of 1,4-disubstituted but-3-yn-1-ones: regiocontrolled synthesis of 2,5-disubstituted 3-bromo- and 3-iodofurans. Org Lett 7:1769–1772. doi:10.1021/ol050372i
24. Sniady A, Morreale MS, Dembinski R (2007) Electrophilic cyclization with *N*-Iodosuccinimide: Preparation of 5-(4-bromophenyl)-3-iodo-2-(4-methylphenyl)furan. Org Synth 84: 199–208; Coll 11:794-801 (2009). doi:10.1002/0471264229.os084.20
25. Li Y, Gundapuneni D, Sniady A, Morreale MS, Marcinek A, Wheeler KA, Dembinski R, manuscript in preparation
26. Li Y, Wheeler KA, Dembinski R (2011) Electrophilic cyclizations of 2-fluoroalk-3-yn-1-ones: room-temperature synthesis of diversely 2,5-disubstituted 3,4-fluorohalofurans. Eur J Org Chem 2767–2771. doi:10.1002/ejoc.201100344
27. Chen Z, Huang G, Jiang H, Huang H, Pan X (2011) Synthesis of 2,5-disubstituted 3-iodofurans via palladium-catalyzed coupling and iodocyclization of terminal alkynes. J Org Chem 76: 1134–1139. doi:10.1021/jo1023987

28. Marshall JA, Wolf MA (1996) Amination, aminocarbonylation, and alkoxycarbonylation of allenic/propargylic Pd intermediates derived from nonracemic propargylic mesylates: synthesis of nonracemic propargyl amines, allenic amides, and butenolides. J Org Chem 61: 3238–3239. doi:10.1021/jo960442m
29. Yao T, Zhang X, Larock RC (2005) Synthesis of highly substituted furans by the electrophile-induced coupling of 2-(1-Alkynyl)-2-alken-1-ones and nucleophiles. J Org Chem 70:7679–7685. doi:10.1021/jo0510585
30. Liu Y, Zhou S (2005) Electrophilic cylization of 2-(1-alkynyl)-2-alken-1-ones using the I_2/K_3PO_4 system: an efficient synthesis of highly substituted iodofurans. Org Lett 7: 4609–4611. doi:10.1021/ol051659i
31. Huang X, Fu W, Miao M (2008) An efficient synthesis of highly substituted furans via the electrophilic cyclization of 1-(1-alkynyl)-cyclopropyl ketones. Tetrahedron Lett 49:2359–2362. doi:10.1016/j.tetlet.2008.02.081
32. Xie Y-X, Liu X-Y, Wu L-Y, Han Y, Zhao L-B, Fan M-J, Liang Y-M (2008) Efficient synthesis of substituted 3-iodofurans by electrophilic cyclization of propargylic oxirane derivatives. Eur J Org Chem 1013–1018. doi:10.1002/ejoc.200700963
33. Yoshida M, Al-Amin M, Matsuda K, Shishido K (2008) Synthesis of substituted furans by platinum-catalyzed cyclization of propargylic oxiranes in aqueous media Tetrahedron Lett 49:5021–5023. doi:10.1016/j.tetlet.2008.06.043
34. Yoshida M, Al-Amin M, Shishido K (2009) Syntheses of substituted furans and pyrroles by platinum-catalyzed cyclizations of propargylic oxiranes and aziridines in aqueous media. Synthesis:2454–2466. doi:10.1055/s-0029-1216867
35. Zhou H, Yao J, Liu G (2008) Electrophilic cyclization of 4-thio-but-2-yn-1-ols via 1,2-migration of the thio group: efficient synthesis of 2,4-dihalo-3-thio-substituted furans. Tetrahedron Lett 49:226–228. doi:10.1016/j.tetlet.2007.11.092
36. Obrecht D (1989) Acid-catalyzed cyclization reactions of substituted acetylenic ketones: a new approach for the synthesis of 3-halofurans, flavones, and styrylchromones. Helv Chim Acta 72:447–456. doi:10.1002/hlca.19890720305
37. Karpov AS, Merkul E, Oeser T, Müller TJJ (2005) A novel one-pot three-component synthesis of 3-halofurans and sequential Suzuki coupling. Chem Commun 2581–2583. doi:10.1039/B502324F
38. Yang F, Jin T, Bao M, Yamamoto Y (2011) Facile synthesis of 3,4-dihalofurans *via* electrophilic iodocyclization. Chem Commun 47:4541–4543. doi:10.1039/C1CC10584A
39. Ji K-G, Zhu H-T, Yang F, Shaukat A, Xia X-F, Yang Y-F, Liu X-Y, Liang Y-M (2010) Synthesis of five- and six-membered dihalogenated heterocyclic compounds by electrophile-triggered cyclization. J Org Chem 75:5670–5678. doi:10.1021/jo101085f
40. Reich HJ, Olson RE (1987) Preparation of 3-Substituted 4-Methylfurans: 3-Iodo-4-methyl- and 3-Formyl-4-methylfuran. J Org Chem 52:2315–2317. doi:10.1021/jo00387a043
41. Schultz-Fademrecht C, Zimmermann M, Fröhlich R, Hoppe D (2003) Synthesis of enantiopure 2,5-dihydro-3-iodofurans and substituted 3-iodofurans by iodocyclization of 4-hydroxy-1,2-alkadienyl carbamates. Synlett:1969–1972. doi:10.1055/s-2003-42046
42. Bew SP, Knight DW (1996) A brief synthesis of β-iodofurans. Chem Commun 1007–1008. doi:10.1039/CC9960001007
43. El-Taeb GMM, Evans AB, Jones S, Knight DW (2001) Practical alternatives for the synthesis of β-iodofurans by 5-*endo*-dig cyclisations of 3-alkyne-1,2-diols. Tetrahedron Lett 42: 5945–5948. doi:10.1002/ejoc.200700681
44. Bew SP, El-Taeb GMM, Jones S, Knight DW, Tan W-F (2007) Expedient syntheses of β-iodofurans by 5-*endo-dig* cyclisations. Eur J Org Chem 5759–5770. doi:10.1002/ejoc.200700681
45. Evans AB, Flügge S, Jones S, Knight DW, Tan W-F (2008) A new synthesis of the F_5 furan fatty acid and a first synthesis of the F_6 furan fatty acid. Arkivoc 10:95–102

46. Wen S, Liu W, Liang Y (2007) Convenient synthesis of polysubstituted 3-iodofurans through the tandem ring-opening/cyclization reaction of 1-Alkynyl-2,3-epoxy Alcohols. Synthesis: 3295–3300. doi:10.1055/s-2007-990830
47. Tseng J-C, Chen J-H, Luh T-Y (2006) A Convenient Synthesis of Tetrasubstituted Furans from Propargylic Dithioacetals. Synlett:1209–1212. doi:10.1055/s-2006-939686
48. Chen C-W, Luh T-Y (2009) Mercuric acetate-mediated annulation of homopropargylic alcohols having thioether substituent. A general route for the synthesis of tetrasubstituted furans from propargylic dithioacetals Tetrahedron Lett 50:3263–3265. doi:10.1016/j.tetlet.2009.02.043
49. Sromek AW, Rubina M, Gevorgyan V (2005) 1,2-halogen migration in haloallenyl ketones: regiodivergent synthesis of halofurans. J Am Chem Soc 127:10500–10501. doi:10.1021/ja053290y
50. Dudnik AS, Sromek AW, Rubina M, Kim JT, Kel'in AV, Gevorgyan V (2008) Metal-catalyzed 1,2-shift of diverse migrating groups in allenyl systems as a new paradigm toward densely functionalized heterocycles. J Am Chem Soc 130:1440–1452. doi:10.1021/ja0773507
51. Pevzner LM (2009) Synthesis of isomeric bromo(diethoxyphosphorylmethyl)furans. Russ J Gen Chem 79:362–372. doi:10.1134/S1070363209030050 (Zh Obshch Khim 2009, 79: 372–382
52. Rees CW, Yue T-Y (1997) Further reactions of furans with trithiazyl trichloride; mechanistic considerations. J Chem Soc Perkin Trans 1:2247–2252. doi:10.1039/A700488E
53. Milkiewicz KL, Parks DJ, Lu T (2003) Synthesis of a novel series of tetra-substituted furan [3,2-*b*]pyrroles. Tetrahedron Lett 44:4257–4260. doi:10.1016/S0040-4039(03)00895-5
54. Yin G, Wang Z, Chen A, Gao M, Wu A, Pan Y (2008) A new facile approach to the synthesis of 3-methylthio-substituted furans, pyrroles, thiophenes, and related derivatives. J Org Chem 73:3377–3383. doi:10.1021/jo702585s
55. Wang X, Liu LY, Chang WX, Li J (2009) Sonogashira coupling reaction of homopropargyl ether with aryl bromides and synthesis of 2,5-disubstituted 3-bromofurans. Sci China Ser B-Chem 52:1314–1320. doi:10.1007/s11426-009-0203-z
56. Fuganti C, Serra S (1998) A new approach to 2-aryl-7-alkoxy-benzofurans: synthesis of ailanthoidol, a natural neolignan. Tetrahedron Lett 39:5609–5610. doi:10.1016/S0040-4039(98)01053-3
57. Toro A, Deslongchamps P (2003) Furanophane transannular Diels–Alder approach to (+)-chatancin: an asymmetric total synthesis of (+)-anhydrochatancin. J Org Chem 68:6847–6852. doi:10.1021/jo034123o
58. Gorzynski M, Rewicki D (1986) Synthese von alkyl-2,3-dihydrothieno[2,3-*c*]furanen. Aromastoffen des Kaffees Liebigs Ann Chem 625–637. doi:10.1002/jlac.198619860404
59. Alvarez-Ibarra C, Quiroga ML, Toledano E (1996) Synthesis of polysubstituted 3-thiofurans by regiospecific mono-*ipso*-substitution and *ortho*-metallation from 3,4-dibromofuran. Tetrahedron 52:4065–4078. doi:10.1016/S0040-4020(96)00069-5
60. Kabir SM, Miura M, Sasaki S, Harada G, Kuwatani Y, Yoshida M, Iyoda M (2000) New syntheses of tricyclic thiophenes and cyclic tetrathiophenes using transition-metal-catalyzed cyclization. Heterocycles 52:761–774. doi:10.3987/COM-99-S86
61. Frohlich J, Hametner C (1996) Synthesis of trisubstituted furans from 2-bromo-5-methylfuran *via* halogen migrations and their selective preventions. Monatsh Chem 127:435–443. doi:10.1007/BF00810886
62. Schlessinger RH, Pettus TRR, Springer JP, Hoogsteen K (1994) Diastereoselective Diels–Alder reactions using furan substituted with a nonracemic amine. J Org Chem 59:3246–3247. doi:10.1021/jo00091a002
63. Xia Y, Dudnik AS, Gevorgyan V, Li Y (2008) Mechanistic insights into the gold-catalyzed cycloisomerization of bromoallenyl ketones: ligand-controlled regioselectivity. J Am Chem Soc 130:6940–6941

64. Yoshimura F, Takahashi M, Tanino K, Miyashita M (2009) An efficient synthetic method for 3-bromofuran derivatives via stereoselective cyclization of γ, δ-epoxy (E)-α-bromoacrylates. Heterocycles 77:201–206. doi:10.3987/COM-08-S(F)25
65. Tanabe Y, Wakimura K, Nishii Y, Muroya Y (1996) Synthesis of 2,5-Diaryl-3-halofurans via Regioselective Ring Cleavage of Aryl 3-Aryl-2,2-dihalocyclopropyl Ketones. Synthesis: 388–392. doi:10.1055/s-1996-4218
66. Batanero B, Vago M, Barba F (2000) Electrochemical synthesis of 5-amino-4-benzoyl-3-phenylfuran-2-carbonitrile. Heterocycles 53:1337–1342. doi:10.3987/COM-99-8843
67. Lukevics E, Ignatovich L, Goldberg Y, Polyak F, Gauhkman A, Rozite S, Popelis J (1988) Reactions of trialkyl(2-furyl)germanes with electrophilic reagents. J Organomet Chem 348: 11–23. doi:10.1016/0022-328X(88)80334-6
68. Basarab GS, Hill PJ, Rastagar A, Webborn PJH (2008) Design of helicobacter *pylori* glutamate racemase inhibitors as selective antibacterial agents: a novel pro-drug approach to increase exposure. Bioorg Med Chem Lett 18:4716–4722. doi:10.1016/j.bmcl.2008.06.092
69. Piller FM, Knochel P (2011) Regio- and chemoselective synthesis of fully substituted furans. Synthesis 1751–1758. doi:10.1055/s-0030-1260035
70. Zhang HZ, Kasibhatla S, Kuemmerle J, Kemnitzer W, Mason KO, Qiu L, Grundy CC, Tseng B, Drewe J, Cai SX (2005) Discovery and structure-activity relationship of 3-aryl-5-aryl-1,2,4-oxadiazoles as a New series of apoptosis inducers and potential anticancer agents. J Med Chem 48:5215–5223. doi:10.1021/jm050292k
71. Jones AM, Simpson AJ, Stanforth SP (1990) Synthesis of tetra-substituted furan derivatives from 1-arylpropane-2,3-diones. J Heterocycl Chem 27:1843–1844. doi:10.1002/jhet.5570270662
72. Chauvin R (1995) "Carbomers". II. En route to $[C,C]_6$Carbo-benzene. Tetrahedron Lett 36:401–404. doi:10.1016/0040-4039(94)02276-H
73. Lie Ken Jie MSF, Lau MML, Lam CNW (2003) Synthesis of novel Tri- and tetrasubstituted C18 furan fatty esters. Lipids 38:1293–1297. doi:10.1007/s11745-003-1192-8
74. Claus H, Schäfer HJ (1985) 3,3-Dichlorotetrahydrofurans by reductive addition of 3-bromo 1,1,1-trichloroalkanes to carbonyl compounds. Tetrahedron Lett 26:4899–4902. doi:10.1016/S0040-4039(00)94980-3
75. Ram RN, Kumar N (2008) Synthesis of 2,4-disubstituted 3-chlorofurans and the effect of the chlorine substituent in furan Diels–Alder reactions. Tetrahedron Lett 49:799–802. doi:10.1016/j.tetlet.2007.11.193
76. Ram NR, Charles I (1999) Synthesis of 3-substituted and 2,3-disubstituted 4-chlorofurans. Chem Commun 2267–2268. doi:10.1039/A907270E
77. Ram RN, Charles I (2008) Unexpected aromatization of 2-acetoxy-3,3-dichloro-4-(*a*-chloroalkyl)tetrahydrofurans to 3-alkyl-4-chlorofurans with NaI/DMF. Synth Commun 38:1946–1951. doi:10.1080/00397910801997678
78. Dvornikova E, Bechcicka M, Kamieńska-Trela K, Krówczyński A (2003) Synthesis and NMR studies of 2- and 3-fluorosubstitued five-membered heterocycles. J Fluorine Chem 124:159–168. doi:10.1016/j.jfluchem.2003.07.001
79. Li Y, Wheeler KA, Dembinski R (2010) Gold(I)-catalyzed cycloisomerization of 2-fluoroalk-3-yn-1-ones: synthesis of 2,5-substituted 3-fluorofurans. Adv Synth Catal 352:2761–2766. doi:10.1002/adsc.201000411
80. Li Y, Wheeler KA, Dembinski R Room Temperature Syntheses of Entirely Diverse Substituted β-Fluorofurans. Org Biomol Chem doi: 10.1039/c1ob06693e
81. Li P, Gu J-W, Ying Y, He Y-M, Zhang H-F, Zhao G, Zhu S-Z (2010) Palladium-catalyzed alkenylation of fluoro-substituted furans via C–H activation to form tetrasubstituted furans. Tetrahedron 66:8387–8391. doi:10.1016/j.tet.2010.08.064
82. Li P, Chai Z, Zhao G, Zhu S-Z (2009) Pd-catalyzed direct arylation of 3-fluorofurans utilizing the neighboring effect of fluorine atom: facile synthesis of tetrasubstituted monofluoro furans. Tetrahedron 65:1673–1678. doi:10.1016/j.tet.2008.12.026

83. Li P, Chai Z, Zhao G, Zhu S-Z (2008) Synthesis of 3-fluoro-2,5-disubstituted furans and further derivative reactions to access fluorine-containing 3,3'-bifurans and tetrasubstituted furans. Synlett:2547–2551. doi:10.1055/s-2008-1078053
84. Sham HL, Betebenner DA (1991) A new and concise synthesis of 3-fluoro-2,5-disubstituted furans. Chem Commun 1134–1135. doi:10.1039/C39910001134
85. Arimitsu S, Hammond GB (2007) Selective synthesis of fluorinated furan derivatives via $AgNO_3$-catalyzed activation of an electronically deficient triple bond. J Org Chem 72: 8559–8561. doi:10.1021/jo701616c
86. Hodge P, Derenberg M (1971) Reaction of difluorocarbene with 2-benzylidenecyclohexanones to afford phenylfluorofurans. J Chem Soc D Chem Commun 233–234. doi:10.1039/C29710000233
87. Xu W, Chen Q-Y (2003) A novel approach of cycloaddition of difluorocarbene to α, β-unsaturated aldehydes and ketones: synthesis of gem-difluorocyclopropyl ketones and 2-fluorofurans. Org Biomol Chem 1:1151–1156. doi:10.1039/B212232D
88 a) Pomeisl K, Kvíčala J, Paleta O (2006) Convenient synthesis of 3-fluoro-4,5-diphenylfuran-2 (*5H*)-one from benzoin ethers. Novel and efficient *Z-E* isomerisation and cyclisation of 2-fluoroalkenoate precursors, substitution of vinylic fluorine. J Fluorine Chem 127:1390–1397. doi:10.1016/j.jfluchem.2006.05.012. b) Pomeisl K, Čejka J, Kvíčala J, Paleta O (2007) Synthesis of 3-Fluorofuran-2(*5H*)-ones Based on *Z*/*E* Photoisomerisation and Cyclisation of 2-Fluoro-4-hydroxybut-2-enoates. Eur J Org Chem:5917–5925. doi:10.1002/ejoc.200700439

Top Heterocycl Chem (2012) 27: 101–138
DOI: 10.1007/7081_2011_65

Published online: 23 September 2011

Synthesis of Halogenated 5- and 6-Membered Sulfur- and Sulfur, Nitrogen-Containing Heterocycles

Yuriy Shermolovich and Sergiy Pazenok

Abstract This review describes the data concerning the chemistry of halogenated 5- and 6-membered sulfur- and sulfur, nitrogen-containing heterocycles published in literature since 2000. The only monocyclic derivatives of thiophene, thiazole, thiopyrane, and thiazine with halogens atom(s) or perhalogenoalkyl group directly bound to heterocycle are analyzed. The main attention is paid to the new approaches to the synthesis of these compounds by halogenations of the heterocycle molecule or by the usage of the acyclic halogenated synthons. The consideration of the chemical properties of these heterocycles concentrates mainly on the examples of their usage for the synthesis of compounds with valuable properties, for example biologically active.

Keywords Halogenated heterocycles · Isothiazole · Thiazine · Thiazole · Thiophene · Thiopyran

Contents

Y. Shermolovich (✉)
Institute of Organic Chemistry NAS of Ukraine, Kiev, Ukraine
e-mail: shermolovich@yandex.ru

S. Pazenok
BayerCropscience, Monheim, Germany

Abbreviations

DAST Dialkylamino sulfur trifluoride
NBS *N*-bromosuccinimide
NCS *N*-chlorosuccinimide
NIC *N*-iodosuccinimide

1 Introduction

The present review updates data on the synthesis and chemical properties of the halogen-containing derivatives of thiophene, thiazole, isothiazole, thiopyrane, 1,2-, 1,3-, and 1,4-thiazines published in literature since 2000. Every section of this chapter begins with discussion on the synthetic methods for unsaturated heterocycles and follows by the synthesis of dihydro- and tetrahydro-derivatives. Some representatives of these types of compounds have been previously described in reviews [1–7].

As far as the publication volume is limited, only monocyclic compounds in which halogen atom(s) or perhalogenoalkyl groups are directly connected with the heterocycle are discussed.

The main attention will be paid to the new approaches toward halogeno compounds using direct halogenations of the heterocyclic moieties or by utilization of the acyclic halogenated synthons.

The consideration of the chemical properties of these heterocycles will concentrate mainly on the examples of their usage for the synthesis of compounds with valuable properties, for example biological activity.

2 Halogenated Thiophenes

2.1 Fluorinated Thiophenes

The chemistry of fluorinated thiophenes has been previously discussed in numerous reviews; see for instance [1–7]. Among new methods for the synthesis of fluorinated thiophenes, it is worth to mention the recently described direct electrophilic fluorination leading to 2- and 3-fluoro-containing thiophenes.

N-fluorodibenzensulfonimide was successfully used as a fluorinating agent for the preparation of 2-fluorothiophene **1** [8, 9].

The treatment of 3-bromothiophenes **2**, **4** with butyllithium and *N*-fluorodibenzenesulfonimide results in formation of 3-fluorothiophenes **3**, **5** in 60–78% yield [9–11] (Scheme 1).

The synthetic approaches toward 2 fluorothiophenes using acyclic synthones are still quite rare. For example, 5-fluoro-2,3-dihydrothiophene **6** was successfully prepared by cyclization of 1,1-difluoro-1-butenes **7** with allylic mercapto group [12] (Scheme 2).

Some recent publications describe preparation of the new α,α-difluoroamides **8** via deoxofluorination of α-ketoacid **9** with Deoxofluor® or DAST [13].

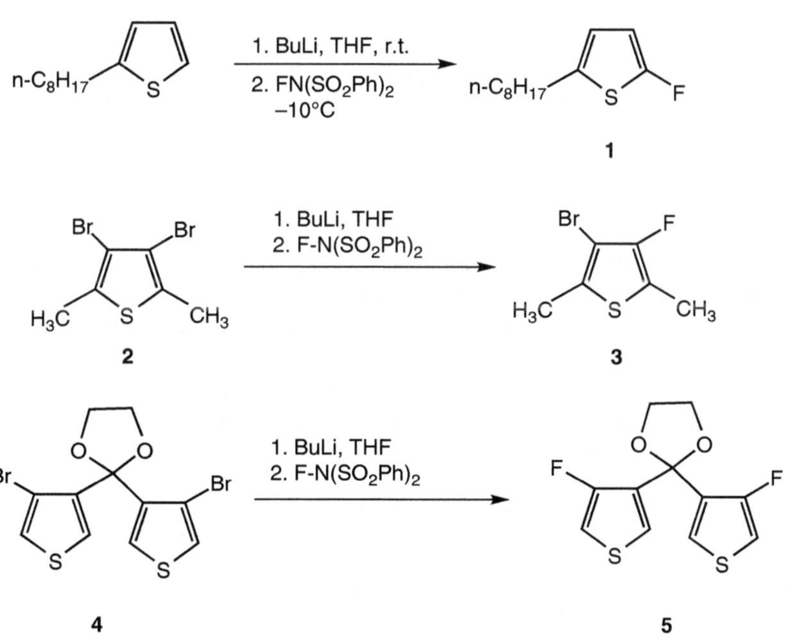

Scheme 1 Synthesis of 2- and 3-fluorothiophenes

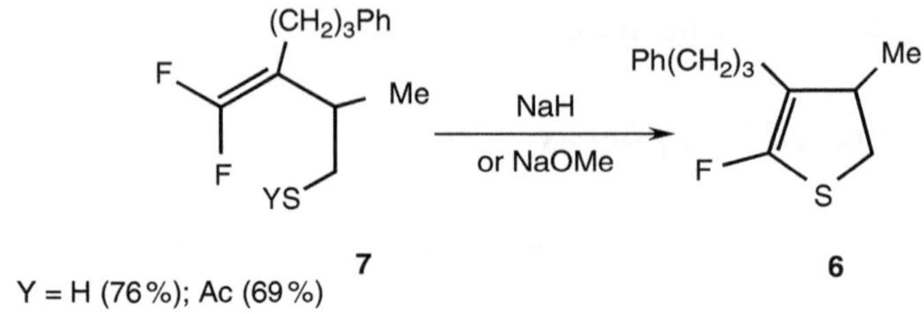

Scheme 2 Cyclization of 1,1-difluoro-4-mercapto-1-btene

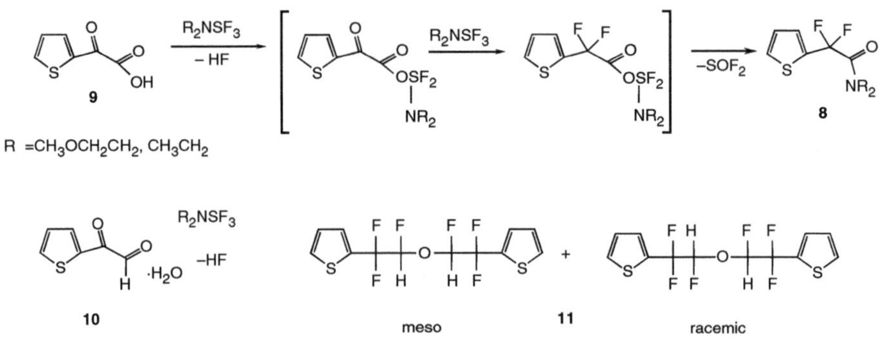

Scheme 3 Fluorination of thiophene-based α-ketoaldehyde and acid

Deoxofluor® smoothly fluorinates glyoxal hydrate **10** to form polyfluoroethers **11** as a mixture of meso and racemic (1:1) compounds in a good yield [14, 15] (Scheme 3).

Fluorine-containing unsaturated sulfides **12**, **13** and nitrile **14** easily undergo base-catalyzed cyclization forming 3-hydroxy(amino)-4-trifluoromethylthiophenes **15–17** [16] (Scheme 4).

Photochemical alkylation of thiophene with selenydes **18**, **19** gave 2-substituted thiophenes **20**, **21** in moderate yield [17, 18] (Scheme 5).

A more general approach to the perfluoroalkylthiophenes comprises a halogen/perfluoroaklyl exchange using perfluoroalkyl copper, generated in situ from perfluoroalkyliodides and copper. Several thiophenes **22–25** could be successfully synthesized using this convenient general method [19, 20].

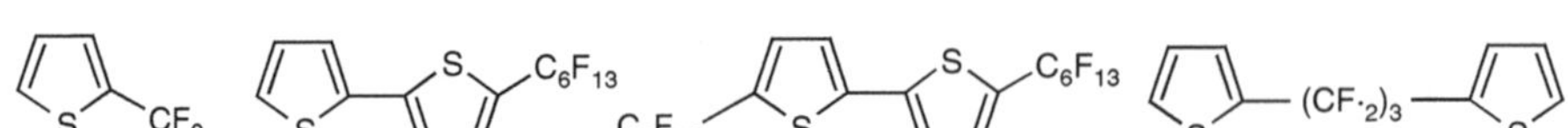

22 **23** **24** **25**

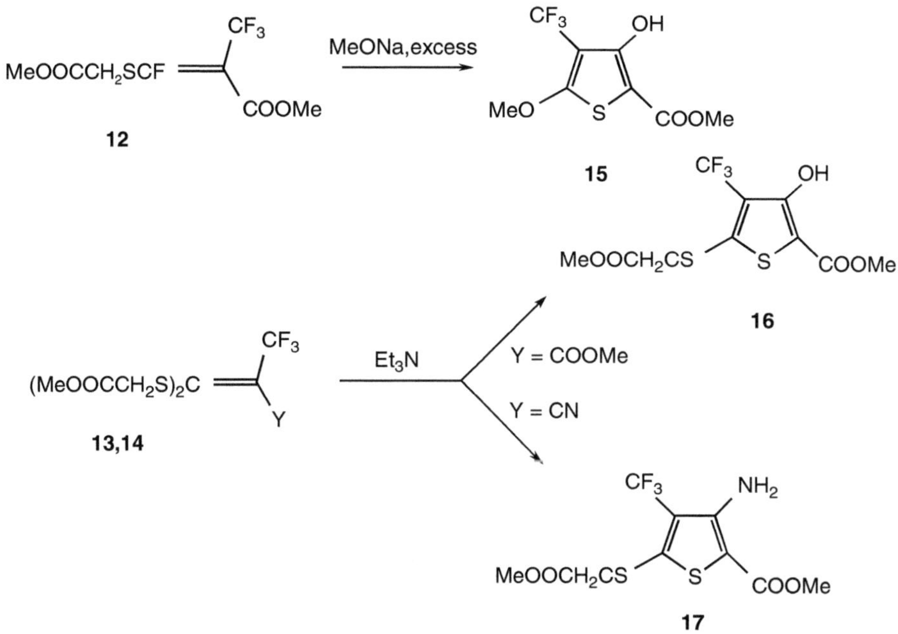

Scheme 4 Synthesis of 3-hydroxy(amino)-4-trifluoromethylthiophenes

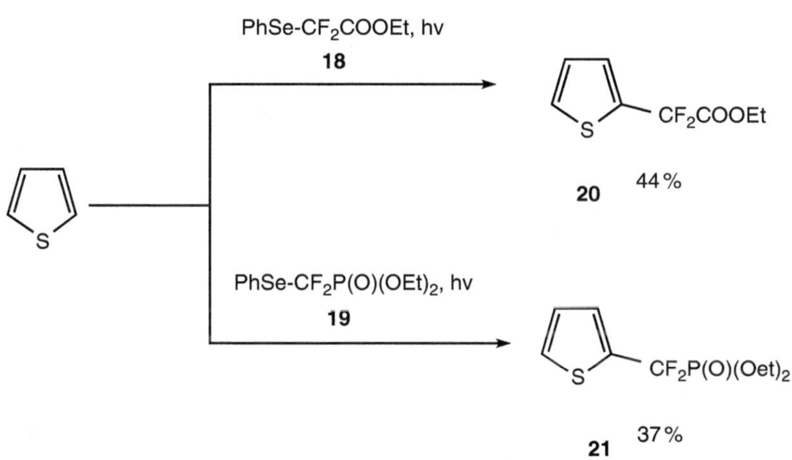

Scheme 5 Photochemical alkylation of thiophene with α,α-difluorodelenides

2.2 *Chlorinated Thiophenes*

Chlorination of the corresponding thiophenes is still a method of choice for the preparation of chlorothiophenes. 2-Chlorothiophene and 2,5-dichlorothiophene

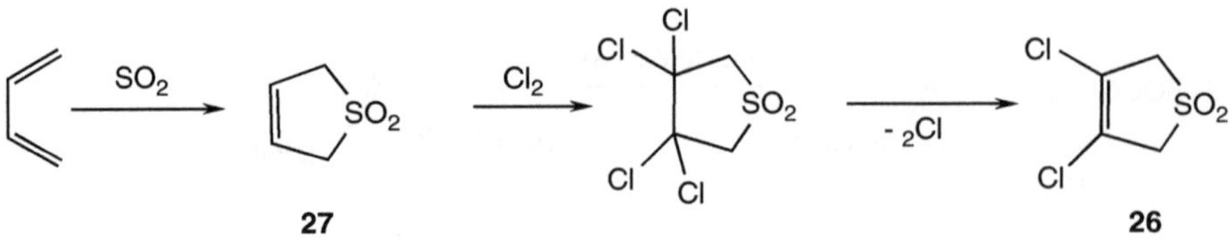

Scheme 6 Chlorination of 2,5-dihydrothiophene-1,1-dioxide

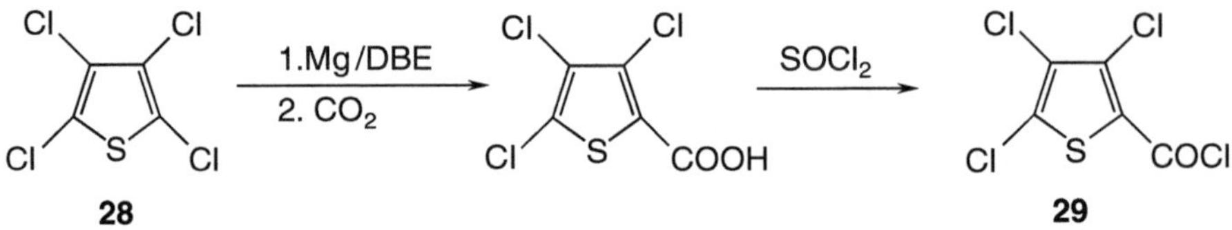

Scheme 7 Synthesis of perchlorinated 2-thiophenecarboxylic acid chloride

were produced by the chlorination of thiophene by applying an excess of chlorine at 50°C. Mono- and dichloroderivatives could be separated by fractional distillation [21]. Highly efficient method for the halogenation of thiophene with *N*-chlorosuccinimide (NCS) was reported by Tanemura et al. [22]. 2-Chlorothiophene (59% yield) and 2,5-dichlorothiophene (10% yield) resulted from the reaction of thiophene with NCS in the presence of 10 equiv of NH_4NO_3 [22]. The same authors reported on a preparation of 3,4-dichlorothiophene 1,1-dioxide **26** starting from 2,5-dihydrothiophene-1,1-dioxide 27 [22] (Scheme 6).

The preparation of tetrachlorothiophene **28** from tetrachloroethylene and H_2S or from hexachlorobutadiene and sulfur has been reported 30 years ago [23]. Based on this compound, the potential manufacturing route to trichlorinated 2-thiophenecarboxylic acid **29** derivatives has been developed [24]. It opened a new way for the development of commercial processes for the new class of polyhaloheteryl-1,2,4-triazole insecticides [24] (Scheme 7).

Chlorination of tetrahydrothiophene with molecular chlorine, sulfuryl chloride [25], or NCS [26] furnish a mixture of 2-chloro- and 2,3-dichlorotetrahydrothiophene. The ratio of isomers is highly solvent dependent and slightly affected by the molar ratio of the reactants [26]. An excess of chlorine during the chlorination leads to the formation of 2,3,4,5-tetrachloroterhydrothiophene [27].

Schmidt et al. demonstrated that chlorination of tetrahydrothiophene with chlorine gives a chlorine complex, which not only undergoes Pummerer reaction to produce the 2-chloro- and 2,3-dichlorotetrahydrothiophene but also forms 1-(2-chlorotetrahydro-3-thienyl)-tetrahydrothiophenium chloride **30** [28].

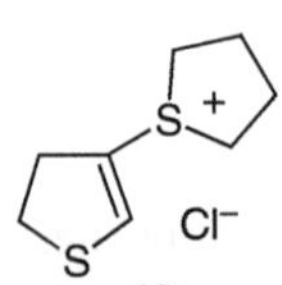

30

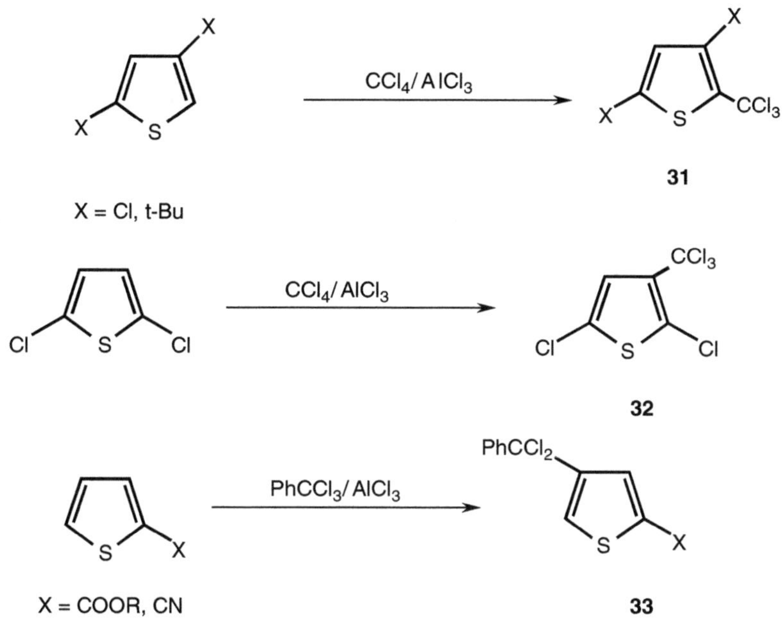

Scheme 8 Synthesis of polychoroalkylthiophenes

Contrary to tetrahydrothiophene, the reaction of tetrahydrothiophene-1-oxide with chlorine afforded only 2-chloro-tetrahydrothiophene-1-oxide as a mixture of diastereomers in a ratio of 4:1 [29].

Belenki and coworkers reported on the synthesis of polychoroalkylthiophenes **31–33** via utilization of CCl_4 or $PhCCl_3$ in the presence of $AlCl_3$ [30–32] (Scheme 8).

2.3 *Brominated and Iodinated Thiophenes*

2-Bromothiophene and 2,5-dibromothiophene were synthesized by heating thiophene with excess of bromine [21]. The same authors converted dibromo compound into monobromo derivative using reduction with magnesium. The corresponding 2-iodothiophene was obtained by the reaction of thiophene with iodine in the presence of mercury oxide. Hull Jr et al. reported very recently on a convenient access to 3,5-dibromo-2-thiophenecarboxylic acid **34** starting from 3-methylthiophene [24] (Scheme 9).

Carbamates of 3-aminothiophene and 3,4-diaminothiophene can be easily transformed to the corresponding 2-bromo- or 2-iodo derivatives **35–37** by the reactions with *N*-bromo- or *N*-iodosuccinimides in mild conditions [33].

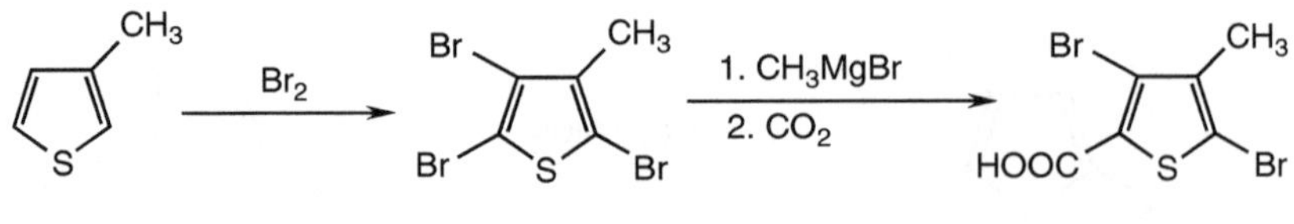

34

Scheme 9 Synthesis of brominated thiophenes

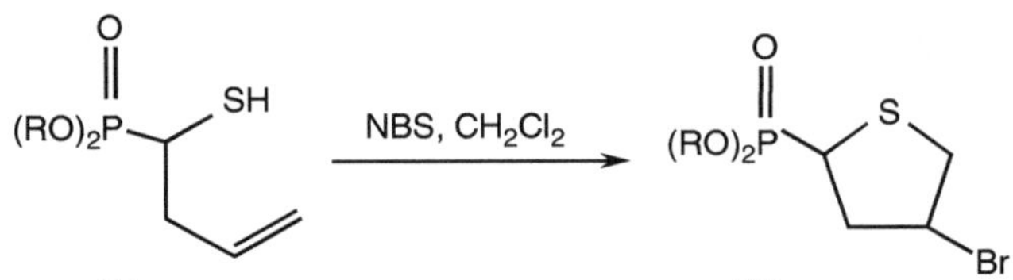

39 38

Scheme 10 Synthesis of 4-bromo-2-phoshono-tetrahydrothiophene

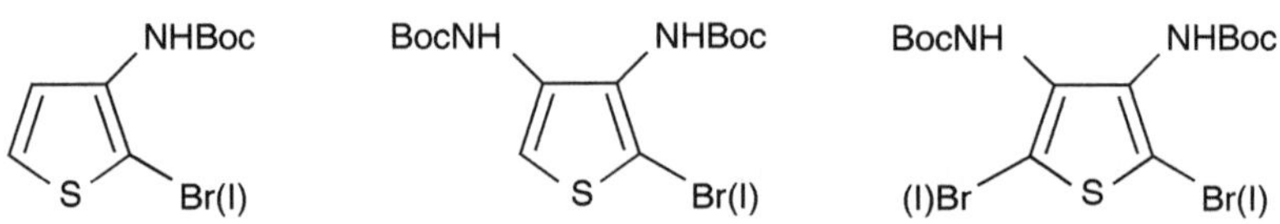

35 36 37

4-Bromo-2-phosphono-tetrahydrothiophene **38** was prepared by the reaction of acyclic phosphonate **39** with *N*-bromosuccinimide [34]. α-Heterosubstituted phosphonates of such type have great interest due to their potential biological activities and synthetic applications [35, 36] (Scheme 10).

3 Chemical Properties and Perspectives for the Utilization of Halogenated Thiophene Derivatives

3.1 Transformation Involving Heterocycle C–H and C–B Fragments

The synthesis of polythiophenes with alkyl and perfluoroalkyl substituents [37, 38] involves a metallation of fluorine-containing thiophene **40** with acidic α-hydrogen followed by C–C-bond formation under Ni catalysis (Scheme 11).

2-Chlorothiophene ring can be easily acylated under Fridel–Crafts conditions with aroylchlorides yielding aglycons **41** which were used for the preparation of C-glycosides possessing high antihyperglycemic effects in high-fat diet KK mice [39] (Scheme 12).

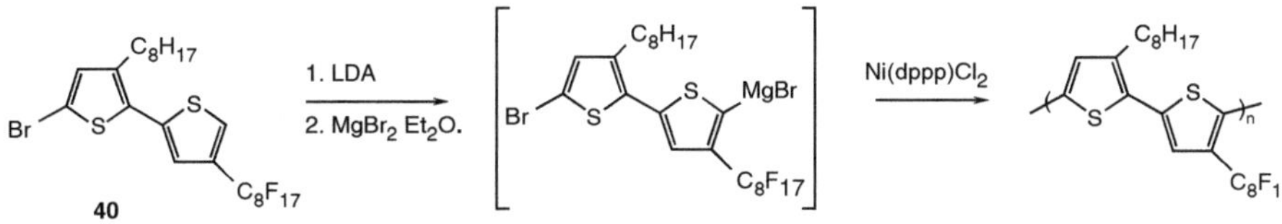

Scheme 11 Synthesis of polyfluorinated thiophenes

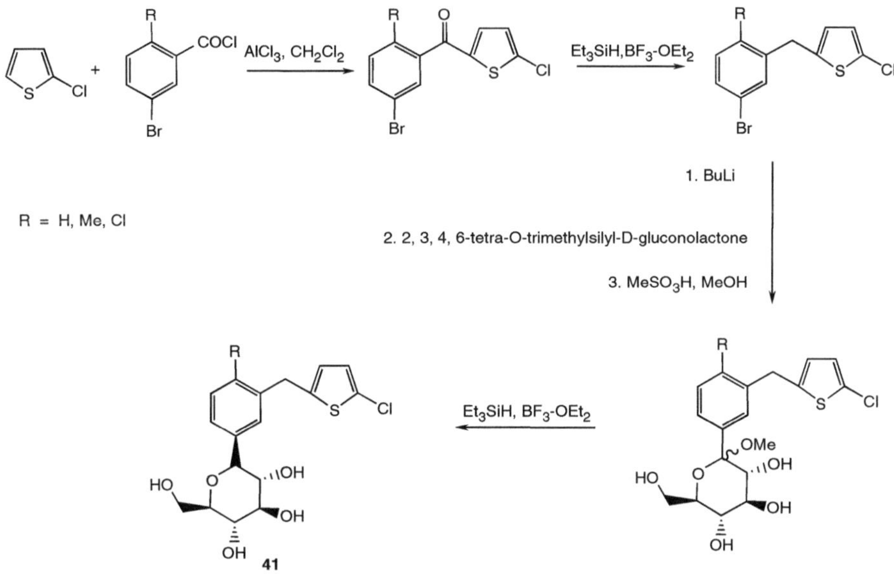

Scheme 12 The use of 2-chlorothiophene for the synthesis of C-glycosides

The new highly potent and selective nonsteroidal classes of inhibitors of estrogen action **42–44** were prepared based on the Suzuki cross-coupling reaction of commercially available 5-chlorothiophen-2-boronic acid and diarylderivatives [40] (Scheme 13).

3.2 Transformation of Heterocyclic C–Halogen Fragment

This type of transformation is very frequently used for the derivatization of halothiophenes via nucleophilic displacement (Scheme 14).

Burger and coworkers described the replacement of an α-fluorine activated additionally by a CF_3 group in β-position upon reaction with different nucleophiles and prepared a series of trifluoromethyl-containing thiophene derivatives **46** [41] and butenolides **47** via thermal Claisen rearrangement [42]. In contrary, thioethers **48**, synthesized from the corresponding thiophenes **45**, and allyl mercaptane reveal

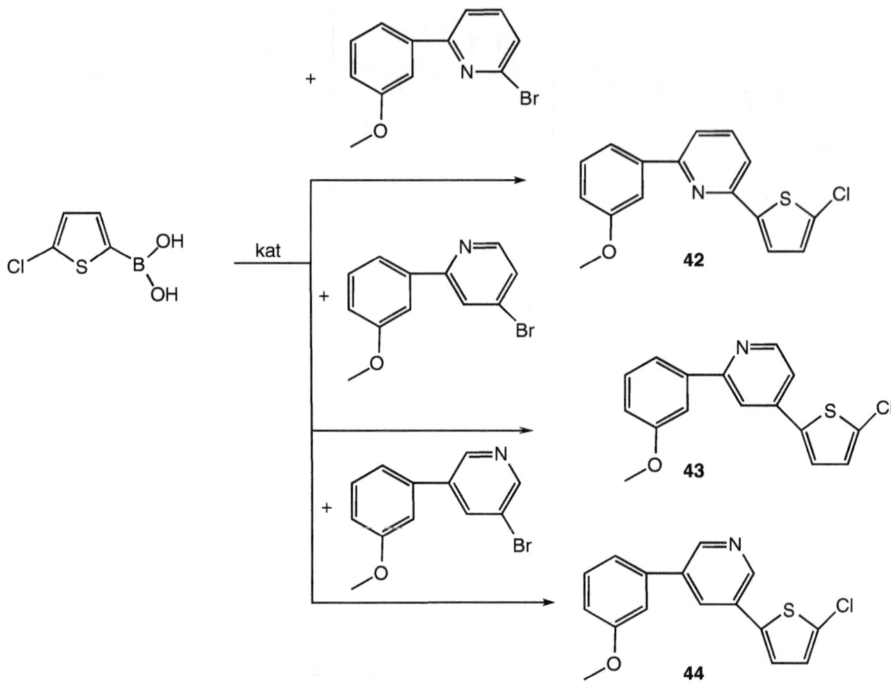

Scheme 13 Suzuki cross-coupling reaction of 5-chloro-thiophen-2-boronic acid

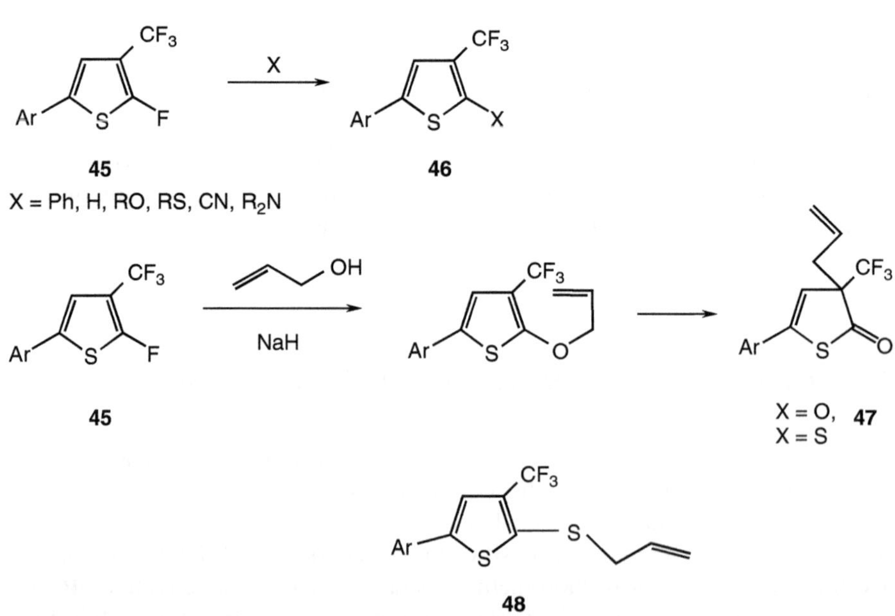

Scheme 14 Synthesis of trifluoromethyl-containing thiophene derivatives

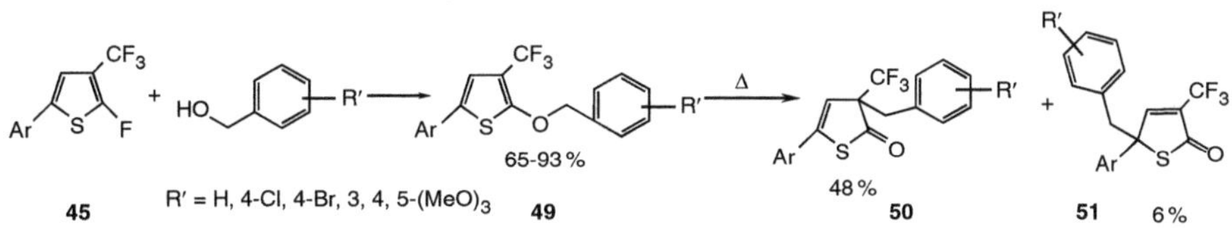

Scheme 15 Benzyl group migration in 3-trifluoromethyl thiophene derivatives

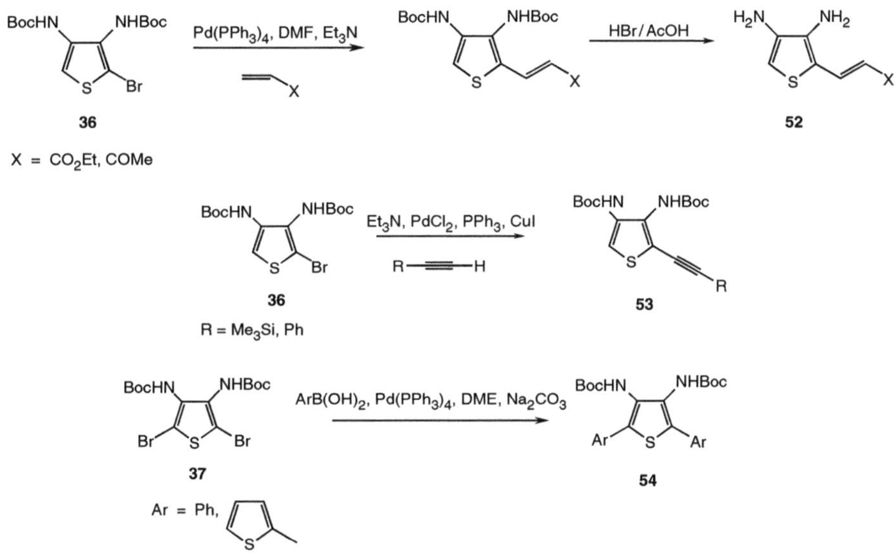

Scheme 16 Coupling reactions of mono- and dibromo(iodo)thiophenes

high thermal stability and show no evidence for a thio-Claisen rearrangement [42]. The fluorine atom in 2-fluoro-3-trifluoromethylthiophenes can be easily replaced by nucleophilic substitution with benzyl alcohols. Primary formed 2-benzyloxyderivatives are susceptible to [1,3]- and [1,5] benzyl group migration [43] (Scheme 15).

The mono- and dibromo(iodo)thiophenes **36, 37** have been subjected to coupling reactions with acetylenic, olefinic compounds, or boronic acids. The formed products **52–54** are useful precursor for the synthesis of biotin analogs [33] (Scheme 16).

The thiophene isosteres of indole derivatives **55, 56** display a wide range of biological activity. The new method of their synthesis was elaborated in [44] (Scheme 17).

A family of annelated 4H-thiopyran-4-ones **57** was readily synthesized in good yields by a microwave-assisted coupling-addition sequence starting from chloride **58**, alkynes, and sodium sulfide [45] (Scheme 18).

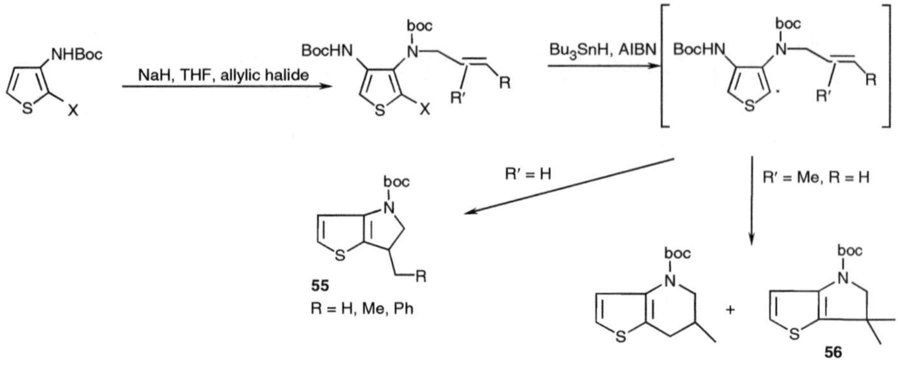

Scheme 17 Synthesis of indole's thiophene isosters

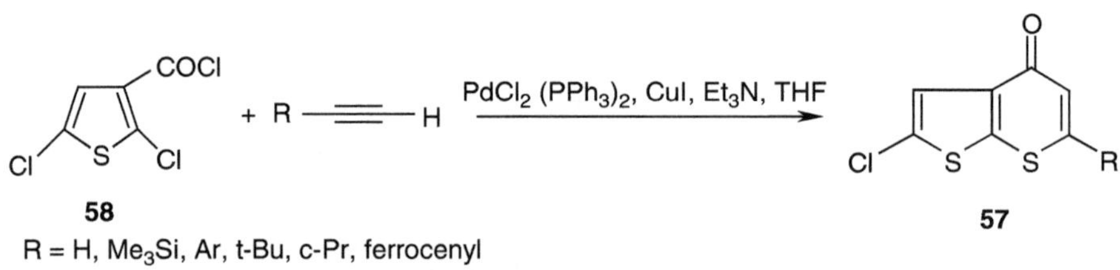

Scheme 18 Synthesis of annelated 4H-thiopyran-4-ones via the coupling reactions

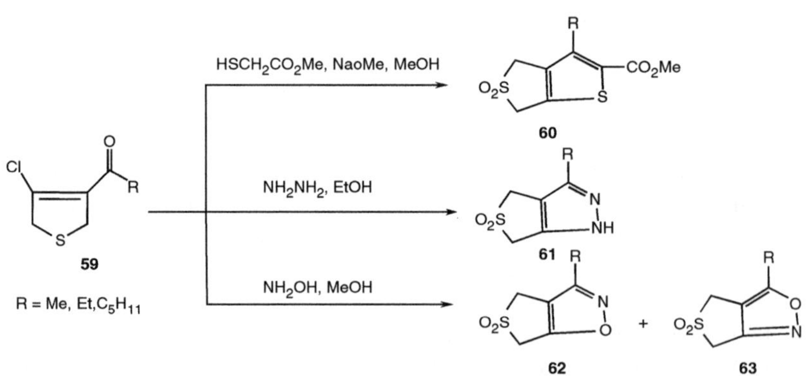

Scheme 19 Synthesis of fused 3-sulfolenes from 3-chloro-4-acyl-3-sulfolenes

The use of 3-chloro-4-acyl-3-sulfolenes **59** for the synthesis of the fused 3-sulfolenes **60–63** as stable precursors to the heterocyclic *o*-quinodimethanes was described in [46] (Scheme 19).

When 4-bromo-2-sulfolene **64** was treated with an excess of ultrasonically dispersed potassium, the fused heterocycle **65** was obtained as a result of elimination–dimerization process followed by extraction of SO_2 [47] (Scheme 20).

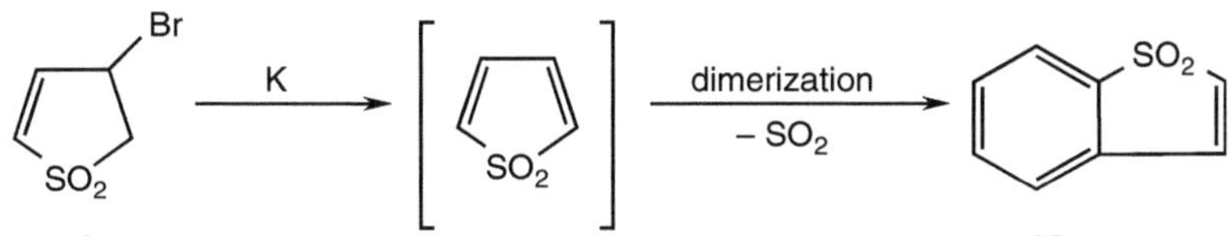

64 65

Scheme 20 Transformation of 4-bromo-2-sulfolene under the action of ultrasonically dispersed potassium

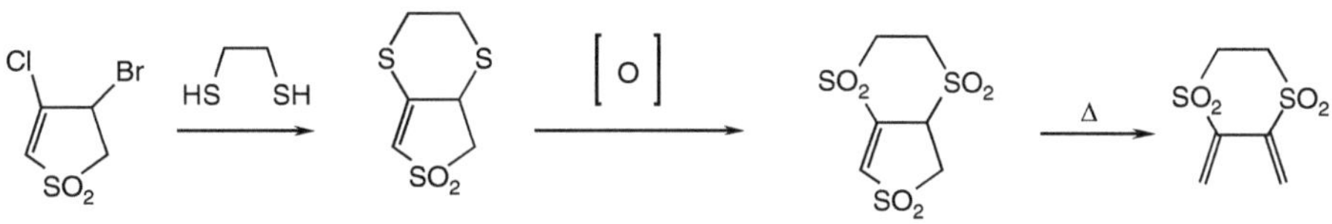

66 67 68

Scheme 21 The substitution reaction of 4-bromo-3-chloro-2-sulpholene

The substitution reaction of 4-bromo-3-chloro-2-sulfolene **66** with sodium salt of ethane-1,2-dithiol gave sulfone **67**. The subsequent transformation of this sulfone leads to the formation of a new rigid electron-deficient diene **68** which can be used in Diels–Alder reactions with a number of dienophiles [48] (Scheme 21).

The usage of the reaction products of 2-chloro-tetrahydrothiophene **69** with different, N-, P-, S-, and C-nucleophiles and their deprotonated derivatives was described earlier [49–51].

Fused *cis*-octahydrothieno[3,2-b]thiepine **71**, *cis*-hexahydrothieno[2,3-b]-1,4-oxathiine **72**, and *cis*-hexahydrothieno[2,3-b]-1,4-dithiine **73** have been synthesized by the reaction of nucleophiles with *trans* 2,3-dichlorotetrahydrothiophene [52].

Acylation of the racemic tetrahydrothiophen-1,1-dioxide **74** with chiral α-methoxyphenylacetic acid chloride produces a mixture of diastereomic esters from which pure diasteromers of benzoyloxy-2,3-dihydrothiophene-1,1-dioxide **75** were isolated [53] (Scheme 22).

The thioanalog of ribostamycin antibiotic **76** was prepared based on the reaction of neamine derivative **77** with 2,3,5-tri-*O*-acetyl-thioribosyl-1-chloride [54] (Scheme 23).

The convenient procedure for the synthesis of new heterocyclic system **78** via S, N-tandem heterocyclization of 3,4-dibromotetrahydrothiophene-1,1-dioxide **79** with thiouracils was proposed [55] (Scheme 24).

3.3 Transformation Involving Endocyclic Double Bond

3-Chloro-4-fluorothiophene-1,1-dioxide **80**, a new synthetically useful fluorodiene, reacts with different types of dienophiles such as acetylenes, alkenes, furans,

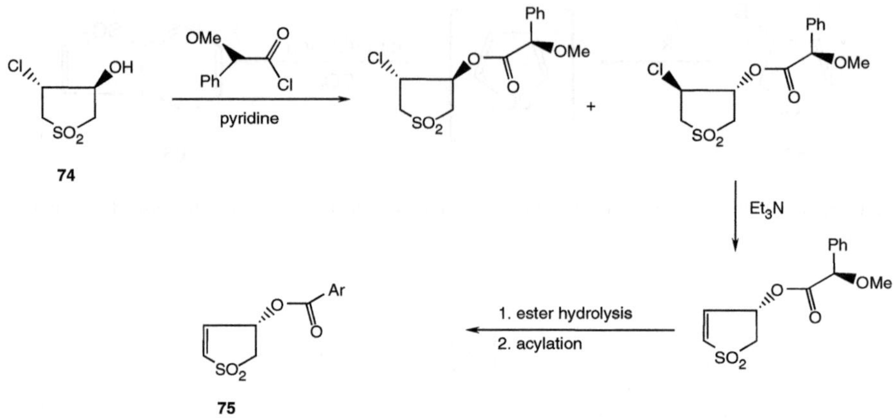

Scheme 22 Synthesis of diasteromer of benzoyloxy-2,3-dihydrothiophene-1,1-dioxide

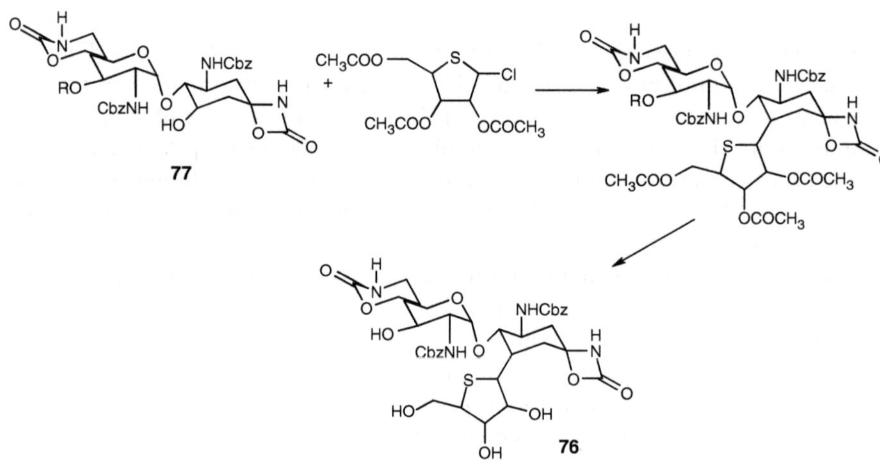

Scheme 23 2,3,5-Tri-*O*-acetyl-thioribosyl-1-chloride in the synthesis of ribostamycin thioanalog

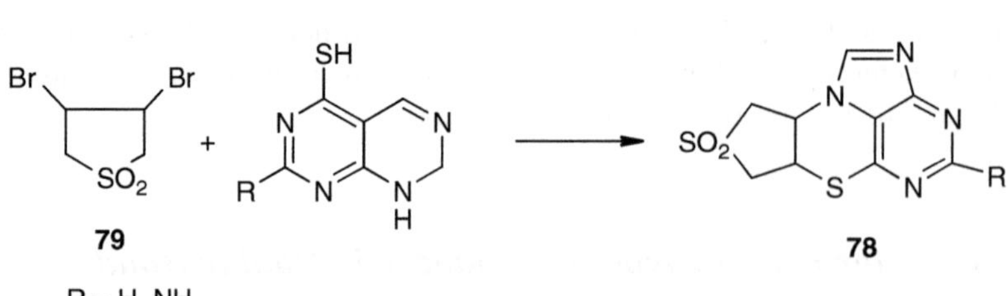

Scheme 24 S,N-tandem heterocyclization of 3,4-dibromotetrahydrothiophene-1,1-dioxide with thiouracils

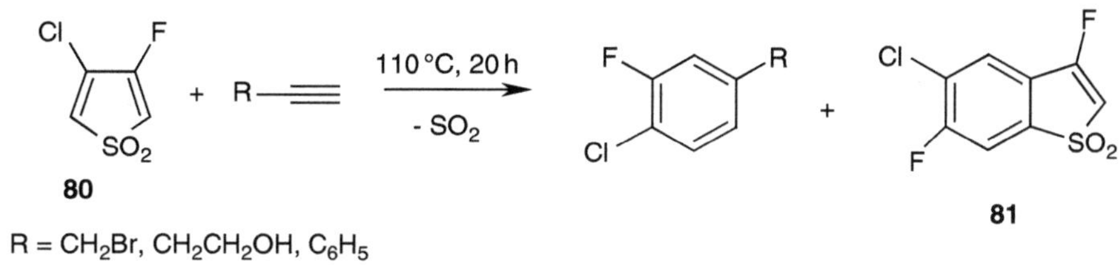

Scheme 25 3-Chloro-4-fluorothiophene-1,1-dioxide as dienophile

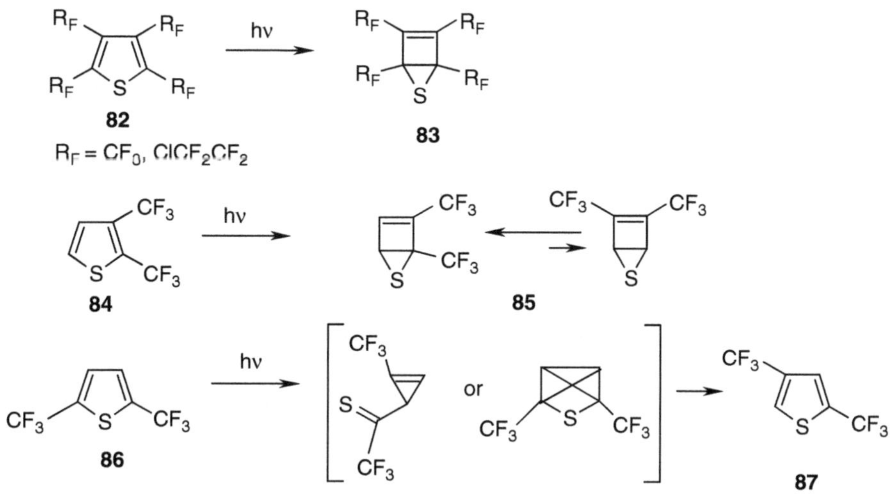

Scheme 26 Photoreactions of fluorinated thiophenes

quinine, and anthracene giving the 3-fluoro-4-chloro substituted aromatics or bicyclic chlorofluorsulfones **81** [56] (Scheme 25).

3.4 Transformation of Thiophene Cycle

Dewar-type isomeres **83**, **85** were obtained by the photoreactions of fluorinated thiophenes **82**, **84** [57–59]. In contrast, photoirradiation of 2,5-bis(trifluoromethyl) thiophene **86** gave no Dewar isomer, but 2,4-bis(trifluoromethyl)thiophene **87** [60] (Scheme 26).

3.5 Miscellaneous Reactions

Iodide salt of 2-heteroaryl 5-chlorothiophene derivative **88**, used as a potential antitumor agent, was prepared by the reaction of aldehyde **89** with 1,2-dimethyl quinolinium iodide **90** [61] (Scheme 27).

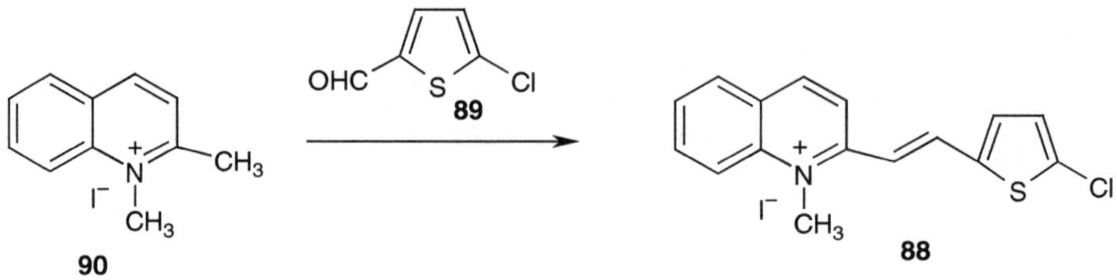

Scheme 27 Condensation reaction of 2-chloro-thiophene-5-aldehyde

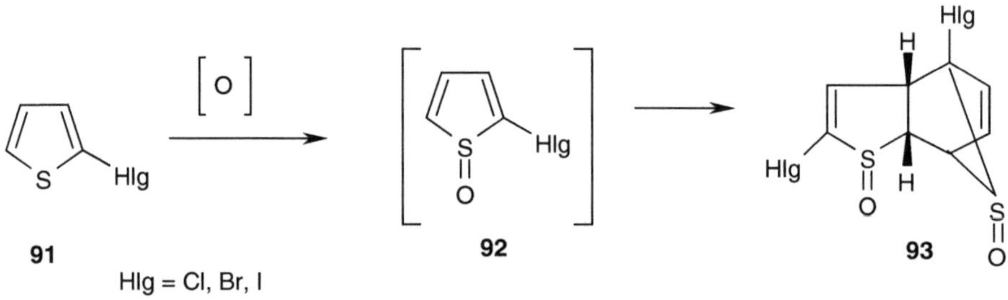

Scheme 28 Transformation of 2-halogeno-thiophene oxides

Attempts to oxidize the halogenoderivatives with peroxyacids leads to the formation of unstable thiophene oxides which are spontaneously dimerized yielding the disulfoxide cycloadducts [62] (Scheme 28).

4 Halogenated Thiazole Derivatives

4.1 Fluorinated Thiazole Derivatives

Introduction of fluorine atom into thiazole moiety is in contrast to the thiophene described very few in the literature. For instance, 2-fluorothiazole was prepared from 2-nitrothiazole by the reaction with KF in *N*-methyl-2-pyrrolidone with 20% yield [63].

The direct fluorination of naphtylthiazole **94** was unsuccessful, but the deprotonation of this naphtylthiazole with *n*-BuLi followed by treatment with $(PhSO_2)_2NF$ yielded the 2-fluorothiazole derivative **95**. Compound **95** was used for the preparation of thiazole amide isoster **96**. This compound exhibited the high activity as antagonists of inhibitor of apoptosis proteins [64] (Scheme 29).

Electrochemical anodic fluorination of (4R)-3-acyl-4-carboxymethoxy-2,2-dimethyl-thiazolidine **97** was carried out in an undivided cell using platinum electrodes [65] (Scheme 30).

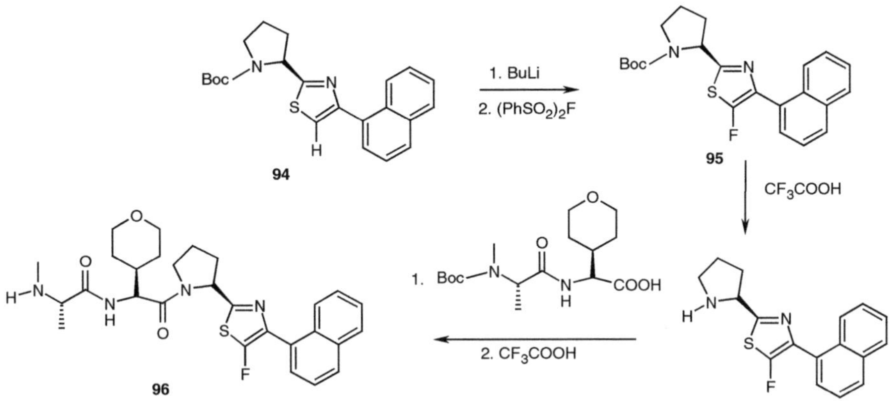

Scheme 29 Fluorination of thiazole derivatives with $(PhSO_2)_2NF$

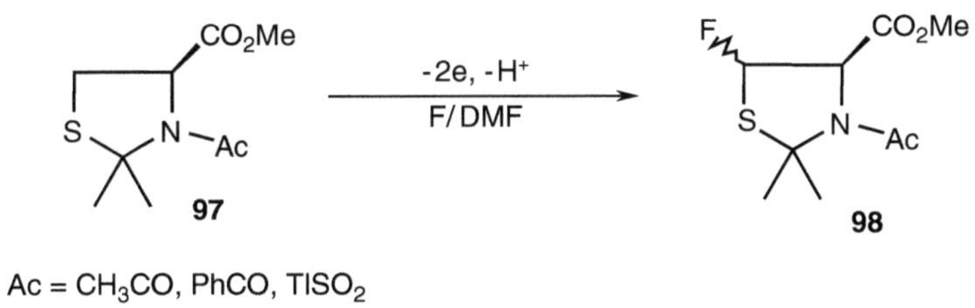

Scheme 30 Electrochemical anodic fluorination of (4R)-3-acyl-4-carboxymethoxy-2,2-dimethyl-thiazolidine

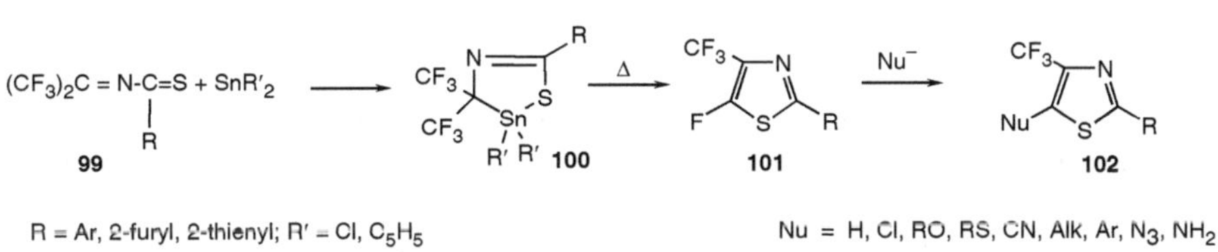

Scheme 31 Synthesis of fluorinated thiazole derivatives from thiaheterodienes

The diastereoselectivity of fluorination was rather high (de = 58–94%) and decreased with the size of the *N*-acyl substituents.

Building blocks approach using fluoro-containing acyclic synthons is still a most important method for the synthesis of fluorinated thiazoles.

Fluorinated thiazole derivatives **101**, **102** were prepared from thiaheterodienes **99** [66–68] (Scheme 31).

Cyclocondensation of thioamides **103** with dimethyl acetylendicarboxylate primary leads to thiazolidinones **104** (Scheme 32).

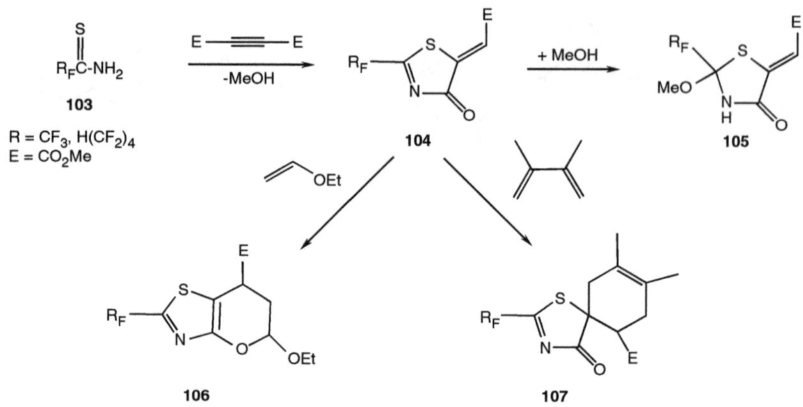

Scheme 32 Cyclocondensation of fluorinated thioamides with dimethyl acetylendicarboxylate

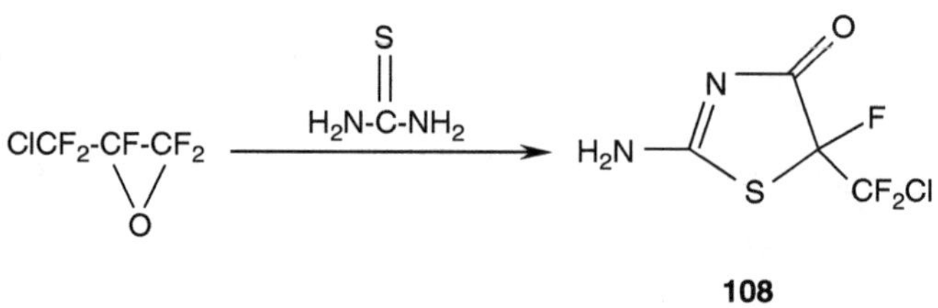

Scheme 33 Reaction of thiourea with 3-chloropentafluoropropene-1,2-oxide

Thiazolin-4-ones **104** participates in [2 + 4] cycloaddition reactions with 1,3-dienes to give spiro-cycloadducts **107** and in [4 + 2] cycloaddition reaction with vinyl ethyl ether as hetero-1,3-dienes giving the dihydropyrano[2,3-d]-thiazole **106** [69].

A novel approach to chlorodifluoromethylated thiazolines comprising a reaction of thiourea with 3-chloropentafluoropropene-1,2-oxide to the 2-amino-5-(chlorodifluoromethyl)-5-fluorothiazolin-4-ones **108** was described in [70] (Scheme 33).

The reactions of internal fluoroolefines with thiourea and thiosemicarbazide result in new type of thiazolines, which are of interest as biologically active compounds [71, 72] (Scheme 34).

The ring opening has been found to occur mainly near the bulkier perfluoroalkyl group [71, 72]. Similar, early unknown fluoro-containing camphor thiazolinyl hydrazones **112** were prepared from oxides of internal perfluoroolefines with (1S, 4S) or racemic camphor thiosemicarbazone [73].

An interesting thiazoline derivative – potassium salt of 4-mercapto-2,2,5,5-tetrakis(trifluoromethyl)-thiazoline **113** – was prepared by the reaction of hexafluorothioacetone dimer with KNCS [74].

The 2-trifluoromethyl-1,3-thiazolidin-4-ones **114** with ester group in the hetero ring was prepared by the reaction of 1-aryl-1-chloro-2,2,2-trifluoroethyl **115** isocyanates with alkyl sulfanylacetates [75] (Scheme 35).

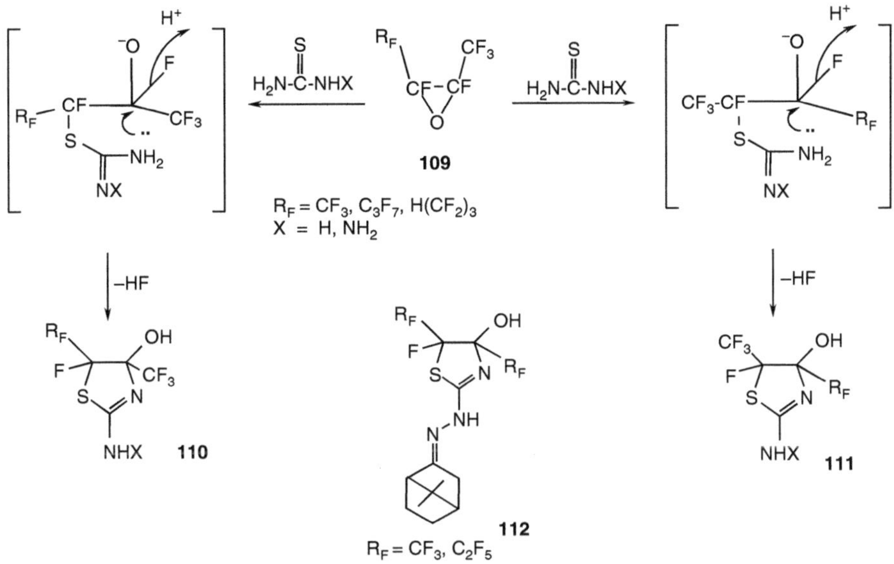

Scheme 34 Reactions of internal fluoroolefines with thiourea and thiosemicarbazide

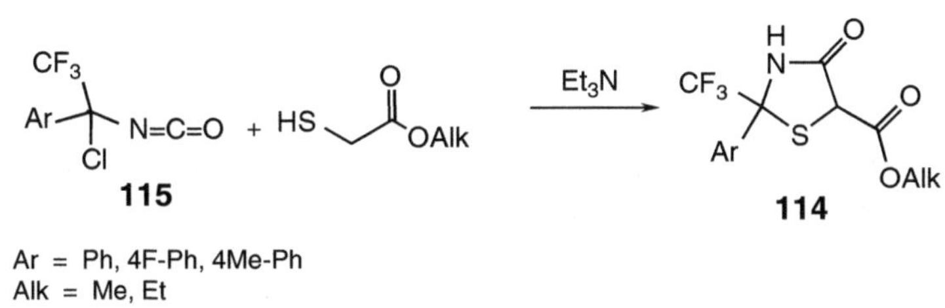

Scheme 35 Reaction of 1-aryl-1-chloro-2,2,2-trifluoroethyl isocyanates with alkyl sulfanylacetates

4.2 Chloro(Bromo)(Iodo) Thiazole Derivatives

In the last decades, the biologically active compounds having halogenated thiazoles have been attracting a growing interest. Some pesticides having thiazole ring with chlorine in positions 2 and 4 possess insecticidal, herbicidal, and antibiotic activity [76–78].

Main methods of synthesis of such compound (in comparison to the F-thiazoles) is the halogenation of the corresponding thiazoles accomplished with exocyclic C–H or C–N bond breaking.

Chlorination of aminothiazolyl(syn-methoxyimino)acetic acid **116** with *N*-chlorosuccinimide in AcOH at 70°C followed by a coupling of the formed chloro acid **117** with 7-amino-3-chloromethyl-3-cephen-4-carboxylic acid *p*-methoxybenzyl

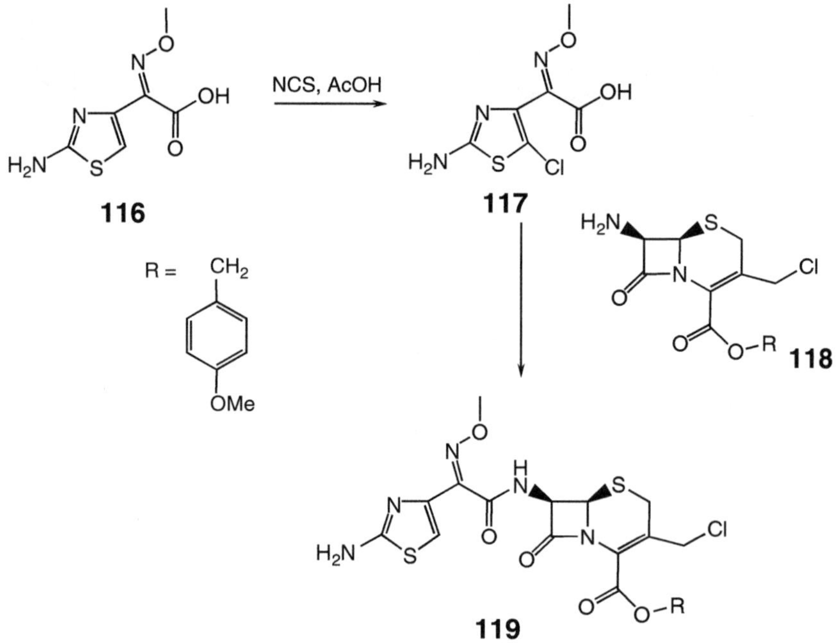

Scheme 36 Key steps for the preparation a new cephalosporin antibiotics

ester **118** are the key steps in the preparation of a new cephalosporin antibiotics **119** [79, 80] (Scheme 36).

These antibiotics inhibit Gram-positive bacteria cell wall biosynthesis. The chloro- or bromo-containing thiazole amide isosteres **120** were prepared through the halogenation of thiazole **121** and were tested as antagonists of inhibitor of apoptosis proteins [64] (Scheme 37).

2,4-Dichlorothiazole-5-carbaldehyde **122** was prepared via chlorination of 1,3-thiazolidine-2,4-dione **123** with $POCl_3$. This compound was used for the synthesis of a new dihydropyridine derivatives **124** having a heterocyclic substituent in position 4 showing a moderate calcium channel antagonist activity [81] (Scheme 38).

Reaction of 5-methoxy-2-(2-pyridyl)-thiazole **125** with *N*-iodosuccinimide produces the iodosubstituted derivative **126** [82] (Scheme 39).

It is worth to mention a new publication describing multistep synthesis of 2-chlorosubstituted thiazoles **127**, **128** via utilization of 2,3-dichloro-1-propene [83] (Scheme 40).

Treatment of 1-tosyl-2,2-dichloroethenyl isothiocyanate **129** with alcohols, thiols, primary, and secondary amines leads to the formation of the corresponding 2-functionally substituted 4-tosyl-5-chloro-1,3-thiazoles **130–132** [84] (Scheme 41).

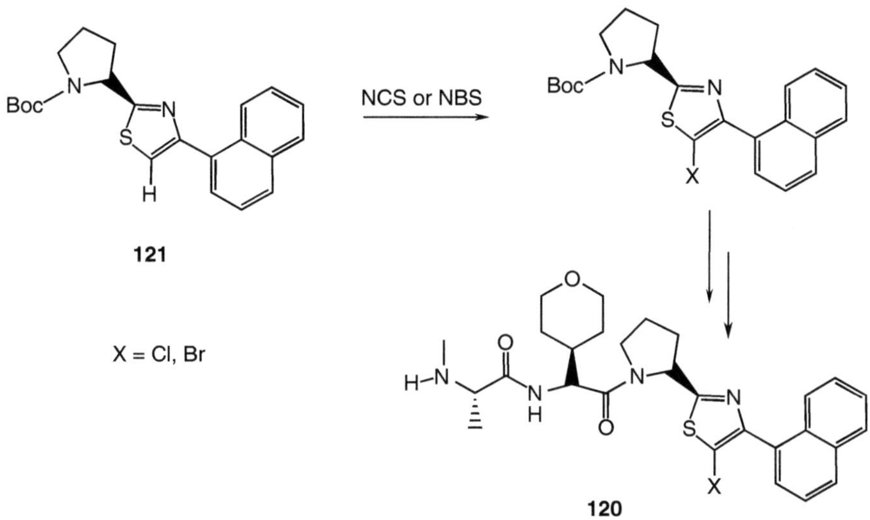

Scheme 37 Synthesis of the antagonists of inhibitor of apoptosis proteins

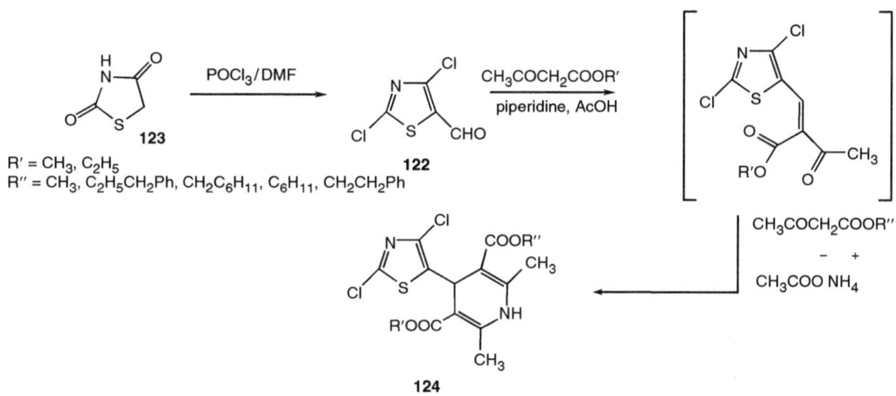

Scheme 38 Chlorination of 1,3-thiazolidine-2,4-dione

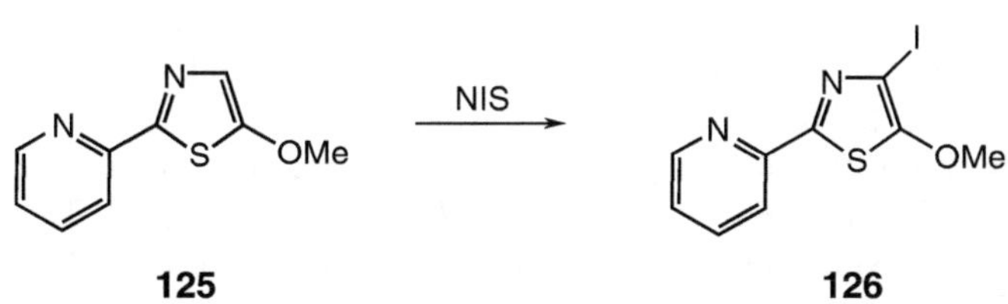

Scheme 39 Iodation reaction of 5-methoxy-2-(2-pyridyl)-thiazole

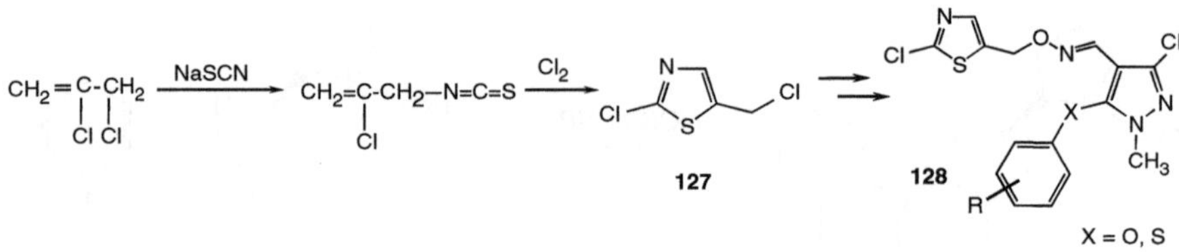

Scheme 40 Synthesis of 2-chlorosubstituted thiazoles from 2,3-dichloro-1-propene

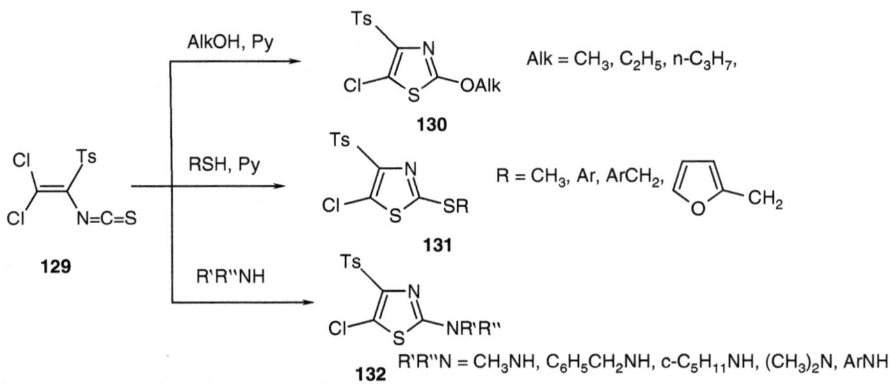

Scheme 41 Synthesis of 2-functionally substituted 4-tosyl-5-chloro-1,3-thiazoles from 1-tosyl-2,2-dichloroethenyl isothiocyanate

Halothiazoles are valuable synthones in organic synthesis. Numerous cyclizations including replacement of the halogen atom have been described in the literature, for instance, for the synthesis of biologically active 2-arylamino-thiazol[4,5-e] diazepines **133** from 4-chloro- **134** or 4-iodo thiazoles **135** [85] (Scheme 42).

The synthesis of simplified analogs of effective antimitotic peptides tribulysins was based on the bromine substitution in 2-bromothiazole **136** with the formation of tubuvaline fragment **137** [86] (Scheme 43).

The Sonogashira coupling of 4-iodosubstituted thiazole **138** with phenylacetylene **139** produces alkyne **140**, which was used for the preparation of ^{18}F-labeled benzonitrile **141** as a high affinity radioligand in positron emission tomography [87] (Scheme 44).

Similarly, via a typical Sonogashira cross-coupling reaction, the desired 1,2-di (5-methoxy-2-(2-pyridyl)thiazolyl) ethyne **142** was prepared [82]. This compound exhibits the interesting fluorescent changes in the presence of different ions and could be used for potential applications in the information storage and processing devices [88] (Scheme 45).

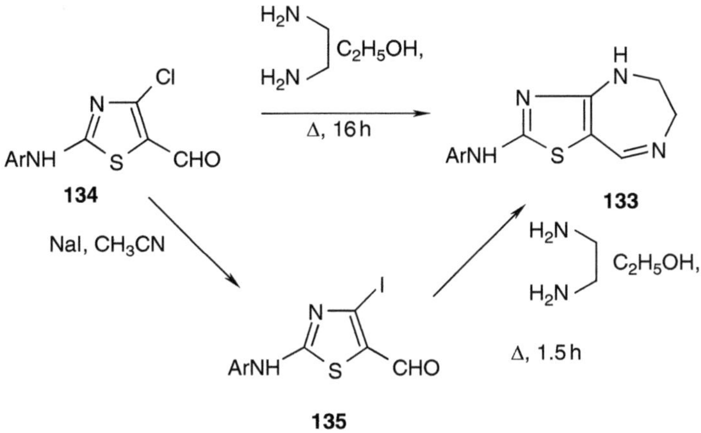

Scheme 42 Synthesis of biologically active 2-arylamino-thiazol[4,5-e] diazepines from 4-chloro- or 4-iodo thiazoles

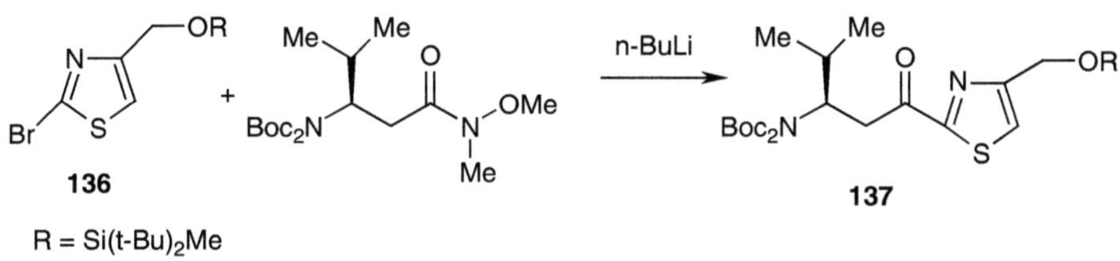

Scheme 43 Synthesis of tubuvaline fragment based on the bromine substitution in 2-bromothiazole

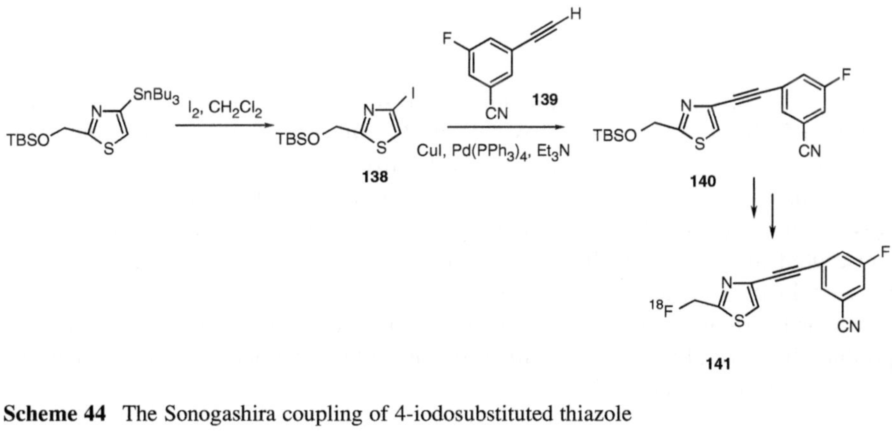

Scheme 44 The Sonogashira coupling of 4-iodosubstituted thiazole

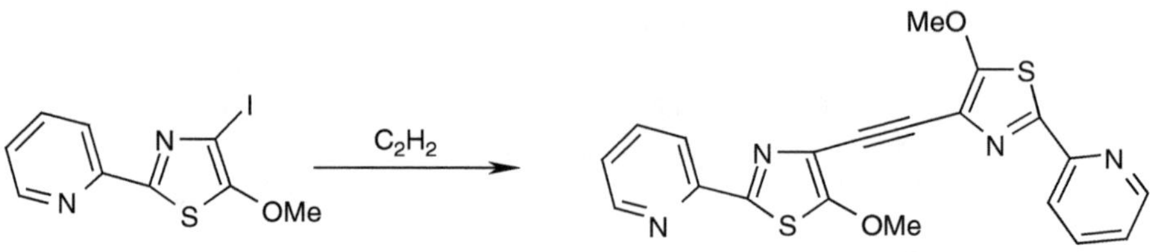

142

Scheme 45 Synthesis of 1,2-di(5-methoxy-2-(2-pyridyl)thiazolyl) ethyne

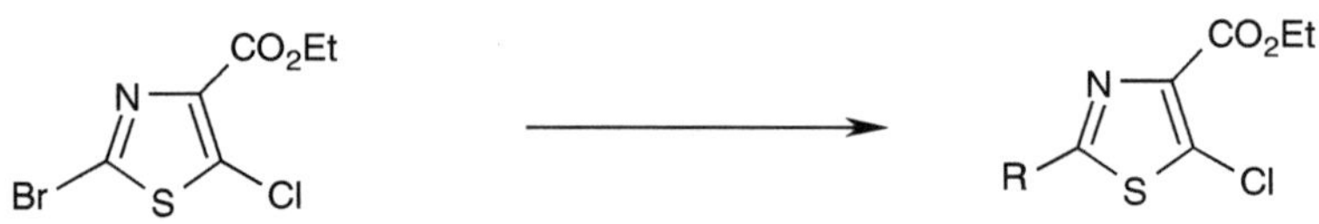

143 **144-147**

R	Conditions	Yield,%	
Ph	$Pd(Ph_3P)_4$, $PhB(OH)_2$, aq.K_2CO_3	81	**144**
$CH_2=CH$	$Pd(Ph_3P)_2Cl_2$, $CH_2=CHSnBu_3$	91	**145**
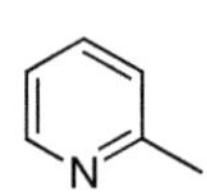	$Pd(Ph_3P)_4$, 2-pyridilzinc bromide	74	**146**
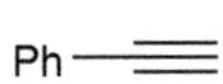	$Pd(Ph_3P)_2Cl_2$, CuI, phenyl acetylene	19	**147**

Scheme 46 The reactions of ethyl 2-bromo-5-chloro-4-thiazolecarboxylate with a variety of organometallic reagents

The regioselectivity of the palladium-catalyzed cross-coupling reactions of ethyl 2-bromo-5-chloro-4-thiazolecarboxylate **143** with a variety of organometallic reagents was examined. Under standard Suzuki conditions with 1 equiv of phenylboronic acid, an exclusive coupling at the more electron-deficient 2-position was observed [89]. The Stile coupling leads to the formation of only 2-vinyl substituted thiazole, as well as the Negishi reaction with 2-pyridylzinc bromide gave exclusive coupling at C-2 position.

Sonogashira reaction with phenylacetylene gave **147** as the major-coupled product; however, the yield was rather low and a large amount of resinous material was formed [89] (Scheme 46).

5 Halogenated Isothiazole Derivatives

5.1 Fluorinated Isothiazole Derivatives

Fluoro-containing isothiazoles are very rare. Typical methods of their preparations comprise utilization of the corresponding acyclic synthons.

β-Cyano-containing sulfenyl chloride **149** underwent intramolecular thermal cyclization at the nitrile group to form substituted isothiazole **150** or reacts with sulfuryl chloride at 10–20°C giving the product of cyclization and simultaneous chlorination **151** [90] (Scheme 47).

3-Trifluoromethyl-isothiazole **152** was obtained by the reaction of 2-cyanothioacetamide **153** with trifluoroacetonitrile with subsequent oxidative cyclization of unsaturated intermediate **154** [91] (Scheme 48).

Compound **152** can be an interesting starting material for the synthesis of new CF_3-substituted calcium-sensing receptor antagonists [91].

New possibilities of synthesizing perfluoroalkyl-containing izothiazolines and isothiazolidines starting from *N*-alkyl(aryl)-C-perfluoroalkyl-C-chlorosulfinimides were investigated [92, 93].

Sulfinimides react with styrenes to give the isothiazolidines as a mixture of diastereomeres (Scheme 49).

From the reactions of sulfinimide **160** and styrene, 2-vinylpyridine or acrylic acid derivatives 5,5-bis(trifluoromethyl)isothiazolidines **161** were obtained (Scheme 50).

Reactions are generally rather regiospecific, and only in the case of acrylamide the formation of the second regioisomer **162** in 25% yield was observed [94].

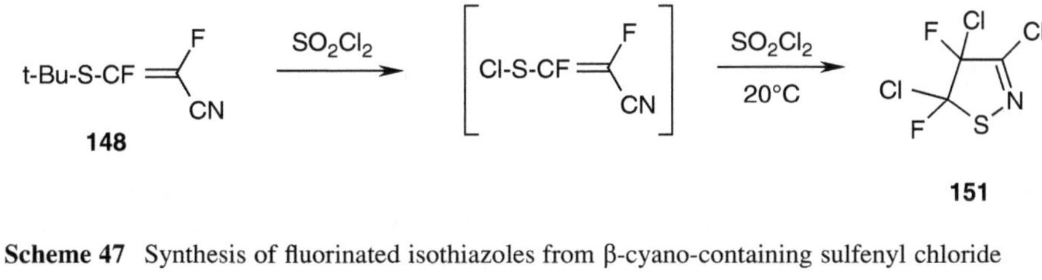

Scheme 47 Synthesis of fluorinated isothiazoles from β-cyano-containing sulfenyl chloride

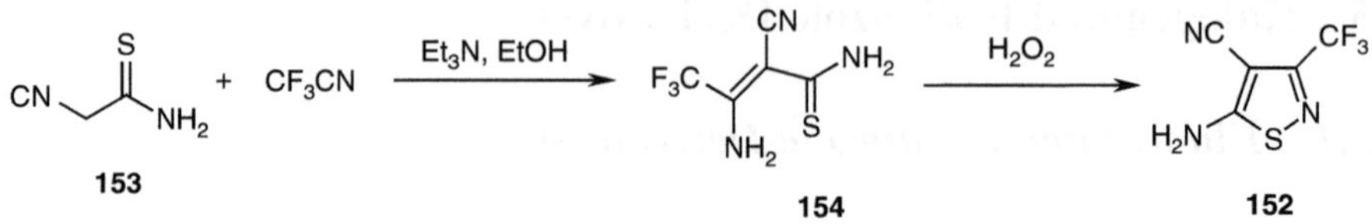

Scheme 48 Reaction of 2-cyanothioacetamide 153 with trifluoroacetonitrile

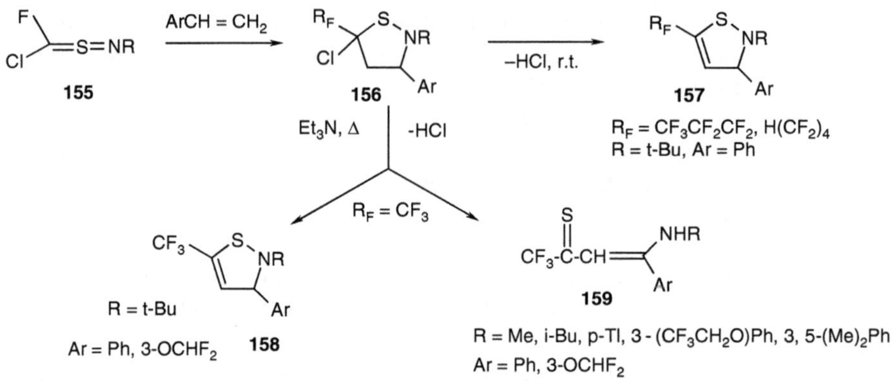

Scheme 49 Synthesis of perfluoroalkyl-containing izothiazolines and isothiazolidines from *N*-alkyl(aryl)-C-perfluoroalkyl-C-chlorosulfinimides

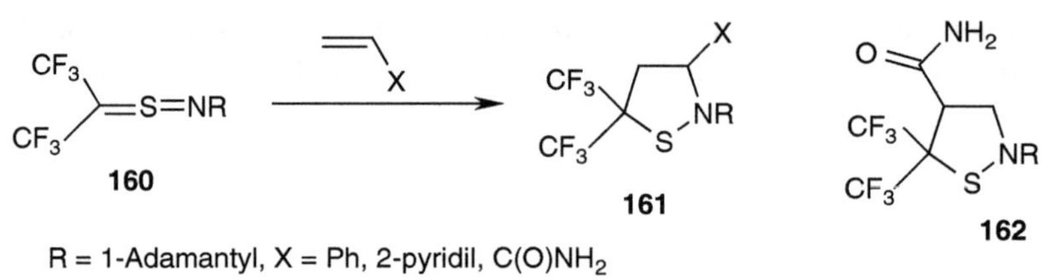

Scheme 50 Synthesis of 5,5-bis(trifluoromethyl)isothiazolidines

5.2 Chlorinated and Brominated Isothiazole Derivatives

The chlorination of 3,5-substituted isothiazoles **163** with *N*-chlorosuccinimide gave corresponding 4-chloroderivatives **164** [95] (Scheme 51).

New convenient methods for the synthesis of 3-trichloromethyl-4,5-dichloro-isothiazole **165**, 3-dichloromethyl-4,5-dichloro-isothiazole **166** [96], and 3-carboxy-4,5-dichloro-isothiazole **167** [97–99] comprising the reaction of 2-nitro-pentachloro-1,3-butadiene with sulfur were proposed.

A number of esters and amides of the acid **167** are potentially biologically active [100].

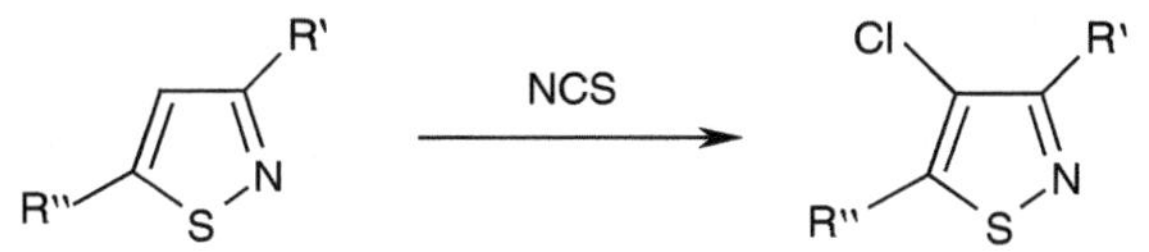

163 **164**

R' = p-ClPh, R'' = Me; R' = Me, R'' = NH_2; R' = Me, R'' = t-$BuMe_2SiNH$

Scheme 51 The chlorination of 3,5-substituted isothiazoles

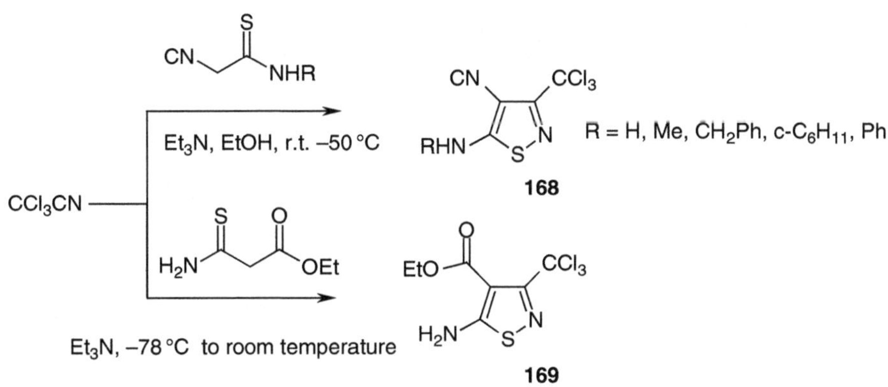

Scheme 52 Reaction of trichloroacetonitrile with 2-cyanothioacetamides or 2-etoxycarbonylthioacetamides

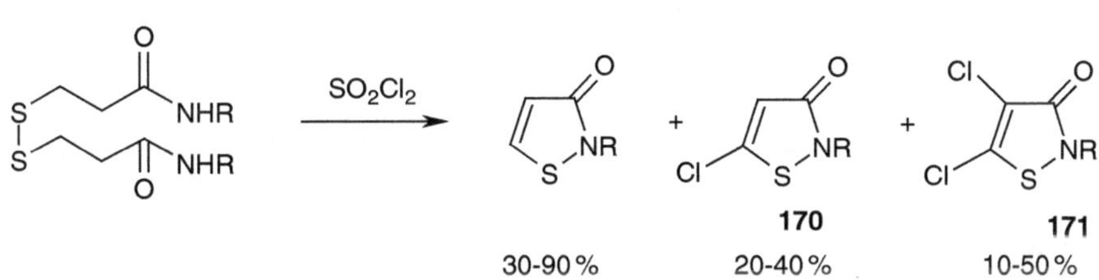

R = Me, Et, C_6H_{11}, CH_2Ph, Ph, p-Cl-$C_6H_4CH_2$, CH_2COOEt, $CH(COOMe)CH_2Ph$

Scheme 53 Synthesis of 5-chloro and 4,5-dichloro-3-alkyl(aryl)amino isothiazolones

4,5-Disubstituted-3-trichloromethylisothiazoles **168**, **169** were prepared from trichloroacetonitrile and 2-cyanothioacetamides or 2-etoxycarbonylthioacetamides [91] (Scheme 52).

5-Chloro and 4,5-dichloro-3-alkyl(aryl)amino isothiazolones **170**, **171** were synthesized starting from dithiopropionic acid amides. These compounds were used for the investigation of antiproliferative effects through inhibition of protein prenylation [101] (Scheme 53).

Iisothiazolones **170, 171** being reacted with $POCl_3$ followed by quenching with NH_3 afforded the 3-aminoderivatives [101, 102], which could be easily oxidized to the corresponding 1,1-dioxides **173** [101] (Scheme 54).

5-Chloro-3-amino-isothiazoles **172** are effective inhibitors of PCAF histone acetyl transferase and cell proliferation and so far interesting for possible therapeutic applications [102, 103].

The bromination of isothiazole 1,1-dioxide **173** proceeds via electrophilic addition–elimination reaction, resulting in the formation of 4-bromo-3-amino-isothiazole dioxide **175** [101] (Scheme 55).

Despite of a great progress in the field of C–C-coupling methodology, only very little was known about isothiazole C–C coupling chemistry [104]. A more detailed study of Suzuki, Stille, Negishi, Ulmann, and Sonogashira C–C coupling at C-3 position of 3-chloro(bromo)-5-chloro(bromo)isothiazole-4-carbonitrile **176** was published in [105]. The reactions are regiospecific and lead to the formation of 3-halo-5-(hetero/aryl, alkenyl ethethyl, alkynyl)isothiazoles.

2-*t*-Butyl-5-chloro-isothiazolone **177** was involved in Suzuki reaction, and isothiazolone **178** prepared was used for the synthesis of new nonpeptidic inhibitors of protein tyrosine phosphatase **179** [106, 107] (Scheme 56).

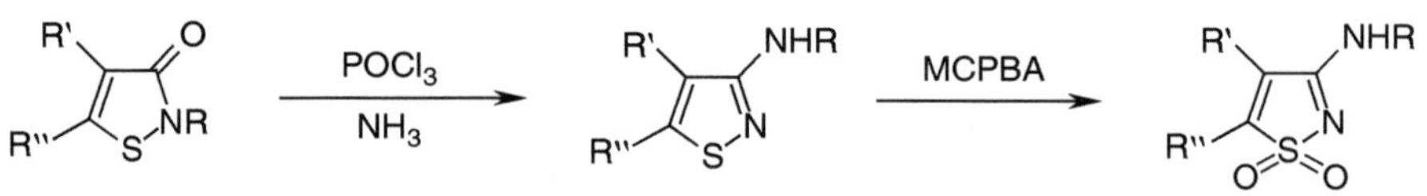

170, 171 **172** **173**

R' = R'' =H; R' = H, R'' = Cl; R' = R'' = Cl

Scheme 54 Synthesis of isothiazolone 1,1-dioxides

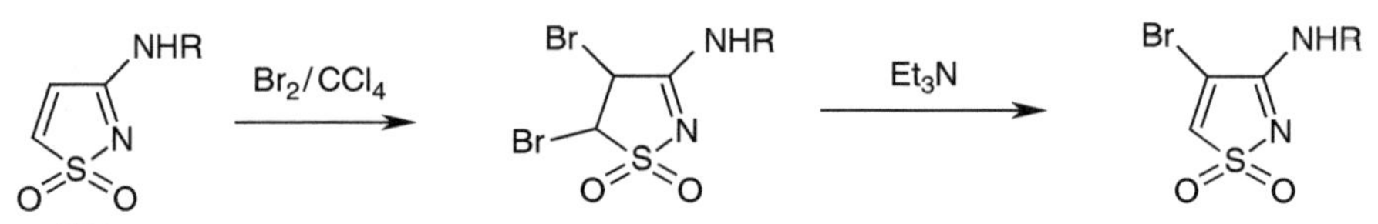

173 **174** **175**

R = CH_2Ph

Scheme 55 Formation of 4-bromo-3-amino-isothiazole dioxide

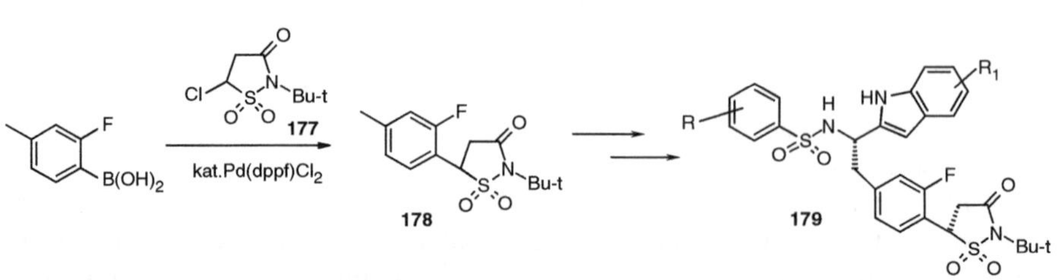

Scheme 56 Suzuki reaction of 2-*t*-butyl-5-chloro-isothiazolone

6 Halogenated Thiopyran Derivatives

6.1 Fluorinated Thiopyran Derivatives

Monofluorinated thiopyrane derivatives 2-fluoro-2H-thiopyrane **180** and 4-fluoro-4H-thiopyrane **181** were obtained as a mixture in 93:7 ratio by the reaction of thiopyrylium tetrafluoroborate or thiopyrylium iodide with strong nucleophilic fluoride sources such as $Me_4N^+F^-$ or AgF [108] (Scheme 57).

2,4,6-Tris(*tert*-butyl)derivate **182** reacts similar in spite of strong steric effect of bulky *t*-butyl groups and give a mixture of both isomers **183**, **184** with a little change in the ratio of isomers (Scheme 58).

The synthesis of new fluoro-containing heterocycle like 3,5-difluoro-thiopyran-2-thione **185** by the reduction of 4-fluoro-5-tetrafluoroethyl-3H-1,2-dithiole-3-thione **186** with sodium sulfide was described in [109] (Scheme 59).

Synthesis and chemical properties of fluoro-containing thiocarbohydrate-based fluorides attracted significant attention during the last decade.

2,3,4,6-Tetra-*O*-acetyl-5-thio-α-D-glucopyranosyl fluoride **187** was prepared by the reaction of penta-*O*-acetyl-5-thio-α-D-glucopyranose **188** with 70% HF-pyridine [110] (Scheme 60).

The fluoro-containing derivatives of 6-thiosialic acid **189** was synthesized by the DAST fluorination of OH-derivative **190** [111]. This compound was used for the glycosylation of alkanols.

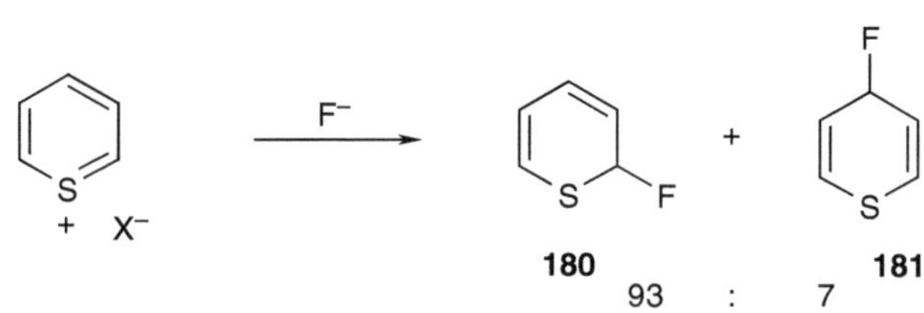

Scheme 57 Synthesis of 2-fluoro-2H-thiopyrane and 4-fluoro-4H-thiopyrane

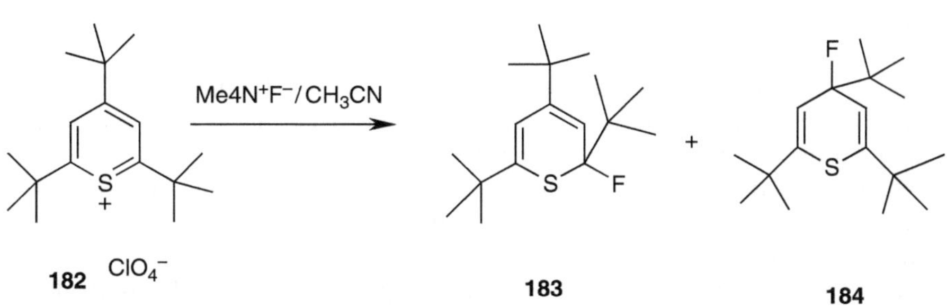

Scheme 58 Fluorination reaction of sterically hindered thiopyrylium salt

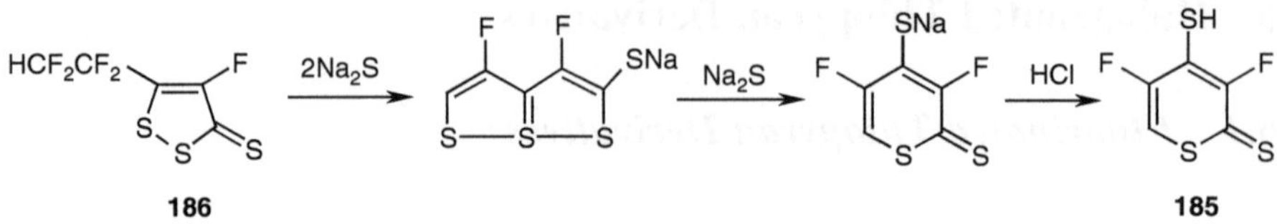

Scheme 59 Synthesis of 3,5-difluoro-thiopyran-2-thione

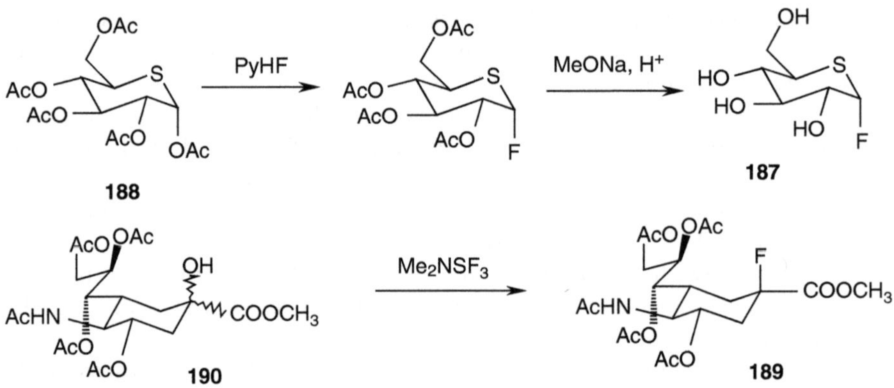

Scheme 60 Synthesis of thiocarbohydrate fluorides

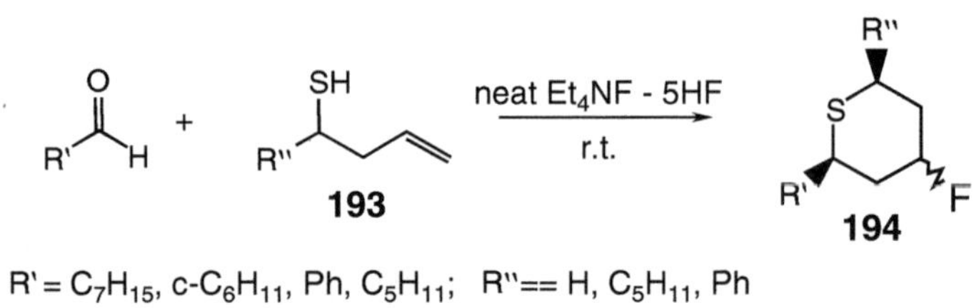

Scheme 61 Synthesis of monofluorinated tetrahydrothiopyrans via the Prins reaction

Fluorination of 2,3-*O*-isopropylidene-1,5-dithio-β-D-xylo-pyranoside **191** with DAST gave after removal of acetone the desired 4-deoxy-4-fluoro derivative of 1,5-dithio-β-D-xylopyranoside **192**, potential orally active antithrombotic agent [112].

Prins cyclization of various aldehydes and homoallylic thiols **193** was successfully performed in Et_4NF 5HF [113] (Scheme 61).

Perfluorinated derivatives of tetrahydrothiopyrane with fragment CF_2SCF_2 **195** and $CF_2SF_4CF_2$ **196** were prepared by electrochemical fluorination of tetrahydrothiopyran [114].

Some recent publications devoted to the preparation of 2-polyfluoroalkyl-2H-thiopyrans **197** via the hetero-Diels–Alder reaction of thioacyl chlorides with 1,3-butadienes [115, 116] (Scheme 62).

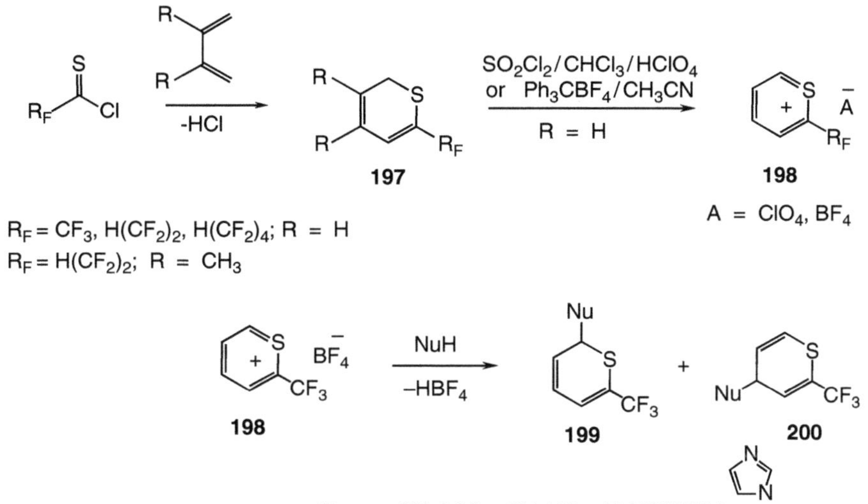

Nu	MeO	i-PrO	N_3	CH_3COS	CH_2NO_2	$H_2NC(O)NH$	
Ratio of	1:0	1:0	1:0	3:1	1:2	1:0	5:1

Scheme 62 Synthesis of 2-polyfluoroalkyl-2H-thiopyrans and 2-polyfluoroalkylthiopyrylium salts

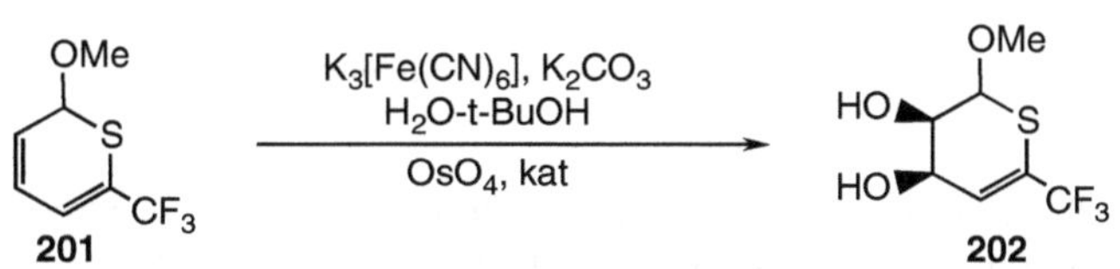

Scheme 63 Dihydroxylation reaction of trifluoromethyl-containing thiopyran derivative

2-Polyfluoroalkylthiopyrylium salts **198** have been synthesized by oxidative aromatization of 2-polyfluoroalkyl-2H-thiopyrans **197** with perchloric acid or triphenylmethane tetrafluoroborate. Addition of nucleophile to 2-trifluoromethylthiothiopyrylium tetrafluoroborate **198** in a basic medium proceeds mainly at the α-position to give the corresponding 2-substituted 6-trifluoromethyl-2H-thiopyrans **199** [116].

Cis-dihydroxylation of **201** affords the new fluorine-containing thiohexenopyranoside derivative **202** [116] (Scheme 63).

Novel class of thiopyrans: 2H-3-acyl-5-phenyl-6-trifluoromethylthiopyrans **203** were obtained from trifluormethylbenzylketone and acrolein or enones [117].

Derivative of 4-trifluoromethyl-2H-thiopyran **204** was prepared by the condensation of α,β-unsaturated ketoesters with alkoxycarbonylthioacetamide [118] (Scheme 64).

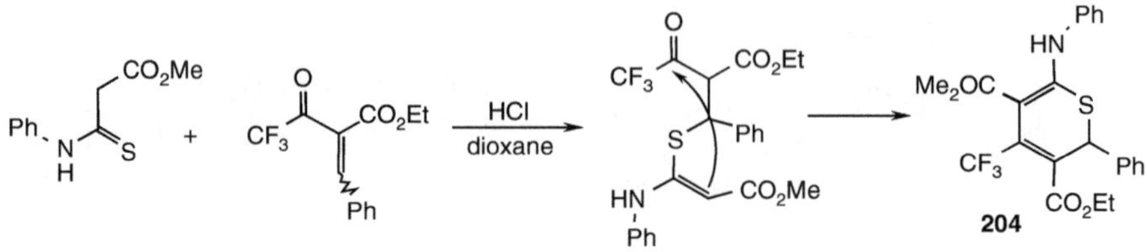

Scheme 64 Condensation reaction of α,β-unsaturated ketoesters with alkoxycarbonylthioacetamide

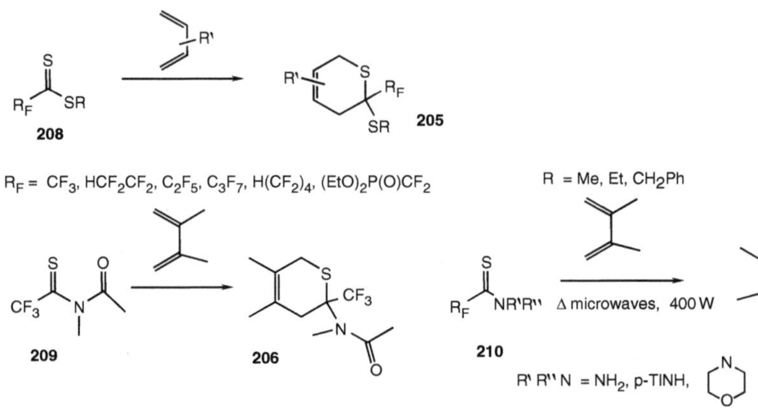

Scheme 65 Synthesis of fluorinated 3,6-dihydro-2H-thiopyrans

Fluorinated 3,6-dihydro-2H-thiopyrans **205**, **206** were synthesized by the reactions of perfluoroalkanedithiocarboxylates **208** [119] or thioimide **209** [120] with 1,3-butadienes (Scheme 65).

Recently [121], the first examples of hetero-Diels–Alder reactions of polyfluoroalkanthiocarboxylic acid amides **210** and 2,3-dimethylbutadiene under microwave heating were demonstrated.

A series of chiral *S*- or *O*-alkylthionoesters **211** have been synthesized from chlorides and corresponding thiols or alcohols. The thia-Diels–Alder reactions of these esters with symmetrical 1,3-dienes proceeds with diastereoselectivity up to 60% [122] (Scheme 66).

Reactivity of fluorinated 3,6-dihydro-2H-thiopyrans toward nucleophilic and oxidizing reagents was investigated on the example of 2,2-bis(trifluoromethyl) thiopyran **213** [123].

Compound **213** has the low reactivity toward *i*-PrMgCl. Reaction led to acyclic compound **214**. The oxidation of **213** with MCPBA results in the formation of epoxy sulfone **215** (Scheme 67).

Reaction of trifluoromethyl-containing enone **216** with ammonium hydrosulfide gave rise to tetrahydrothiopyran **217** [124] (Scheme 68).

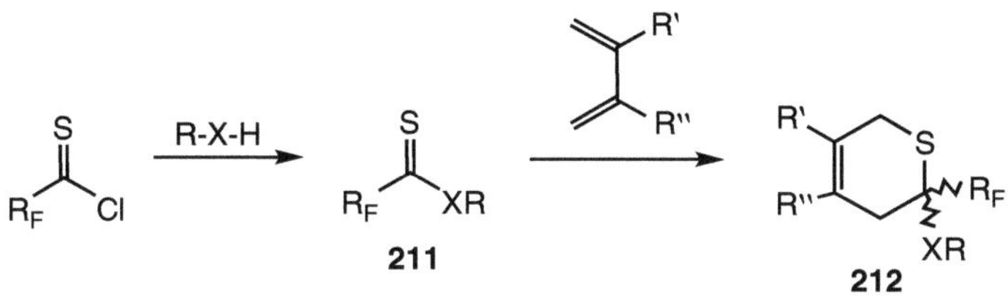

Scheme 66 The thia-Diels–Alder reactions of perfluoro alkanethionocarboxylates with symmetrical 1,3-dienes

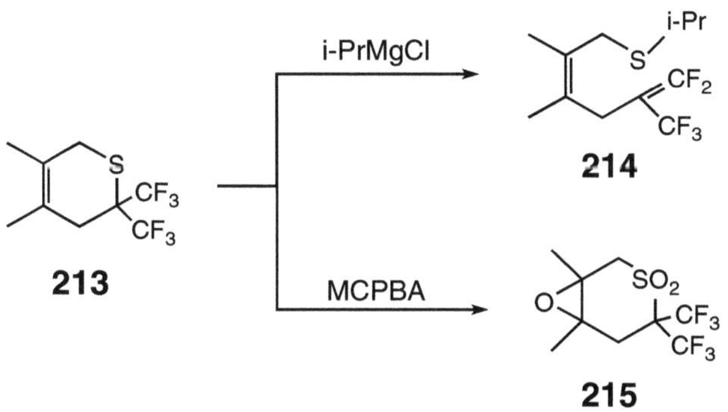

Scheme 67 Reactions of fluorinated 3,6-dihydro-2H-thiopyrans with nucleophilic and oxidizing reagents

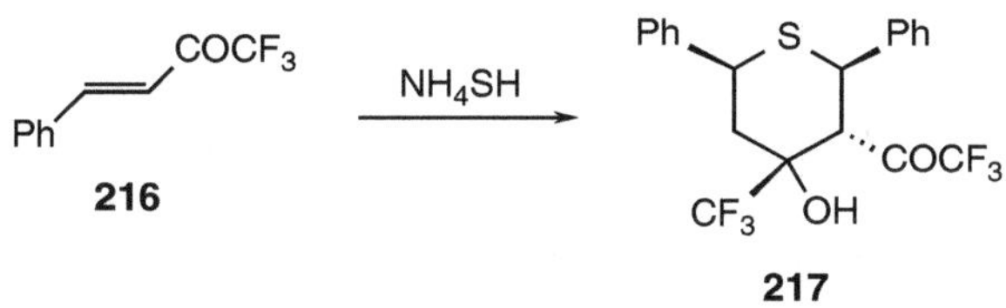

Scheme 68 Synthesis of fluorinated tetrahydrothiopyran from trifluoromethyl-containing enone

6.2 Chlorinated Derivatives of Thiopyran

Chemistry of chlorinated thiopyran derivatives was studied intensively 20–30 years ago [125–130] and will not be discussed in this review.

Among these publications, it is noteworthy to mention only the investigations of α-halogenated sulfones **218** and isolation for the first time of episulfones **219** under the conditions of the Ramberg–Bachlund reaction [131, 132] (Scheme 69).

Later a novel approach to enantiopure cyclopentenes **220** was elaborated based on the ability of α-halogenotetrahydrothiopyranes 1,1-dioxide **218** subjected to the Ramberg–Bachlund rearrangement [133].

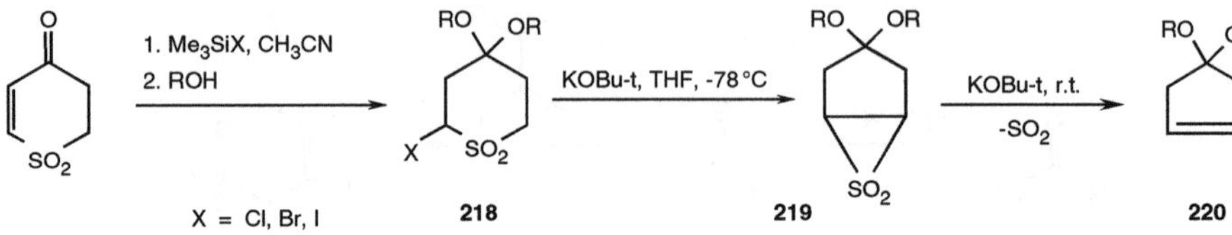

Scheme 69 Ramberg–Bachlund reaction of α-halogenated tetrahydrothiopyran 1,1-dioxides

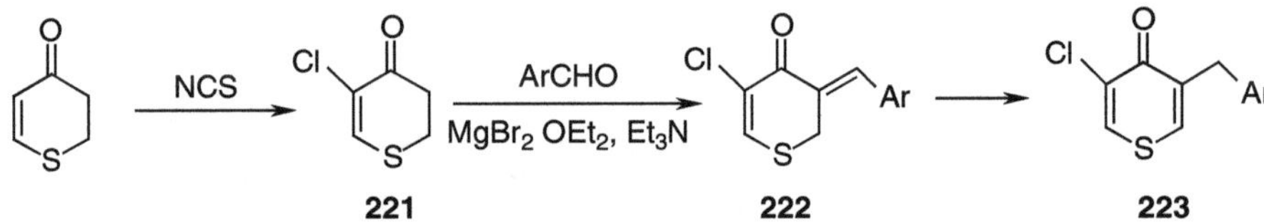

Scheme 70 Synthesis of 5-chloro-2,3-dihydrothiopyran-4-ones

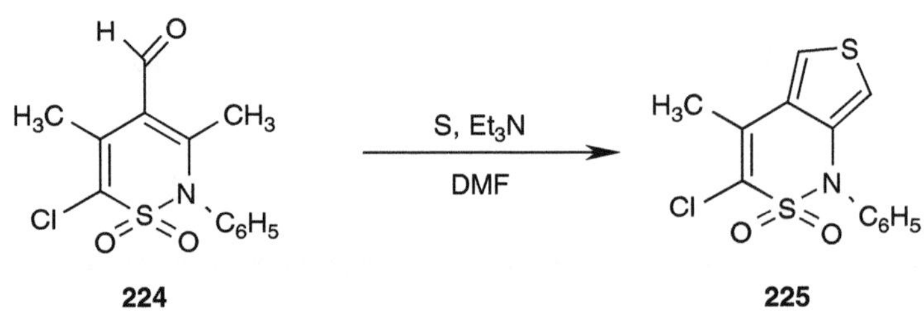

Scheme 71 Synthesis of 2,2-dioxo-1H-thieno[3,4-c][1,2]thiazine

The synthesis of 5-chloro-2,3-dihydrothiopyran-4-ones **221** was described in [134]. This compound is a convenient starting material for the synthesis of novel 3-substituted-4H-thiopyran-4-ones **222**, **223** [134] (Scheme 70).

7 Halogenated Thiazines Derivatives

Halogen-containing 1,2-, 1,3-, and 1,4-thiazines remain the less-investigated class of compounds discussed in this review. Until now this type of compound only sparely documented in the literature. Some literature data are shown below.

6-Chloro-3,5-dimethyl-1,1-dioxo-1,2-thiazine-4-carbaldehyde **224** was used for the first synthesis of 2,2-dioxo-1H-thieno[3,4-c][1,2]thiazine **225** [135] (Scheme 71).

Haloderivatives of 1,3-tiazines are more investigated. The new chiral trifluoromethyl-substituted azines, (*S*)-aryl-4-trifluoromethyl-4H-1,3-thiazines **226** were synthesized by the reaction of chiral aminoketones **227** with phosphorus pentasulfide [136] (Scheme 72).

Reaction of 2,3-dihalogenopropyl amines with CS_2 gave a mixture of isomeric 5-halogeno-3,4,5,6-tetrahydro-1,3-thiazine-2-thiones **228** and 5-halogenomethylthiazolidine-2-thiones **229** [137]. Tetrahydrothiazines **228** could be easily isomerized to **229** upon heating in MeOH or EtOH (Scheme 73).

The cycloaddition of aroyliminochloromethane sulfenyl chlorides **230** with 3,3-dimethyl-2-butene leads to 4,4,5-trimethyl-5,6-dihydro-2-chloro-4H-1,3-thiazine **231** [138]. This multistep reaction includes an addition of sulfenyl chloride to C=C bond, 1,2-schift of Me-group, cycle formation, and final elimination of aroyl group (Scheme 74).

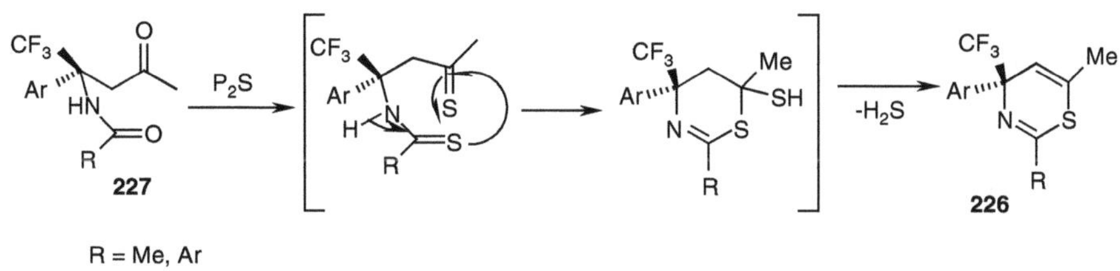

Scheme 72 Synthesis of chiral (*S*)-aryl-4-trifluoromethyl-4H-1,3-thiazines

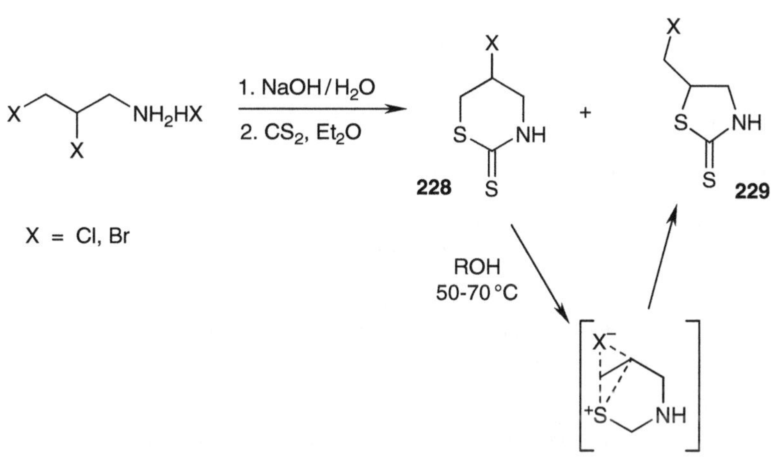

Scheme 73 Formation of a mixture of isomeric 5-halogeno-3,4,5,6-tetrahydro-1,3-thiazine-2-thiones and 5-halogenomethylthiazolidine-2-thiones

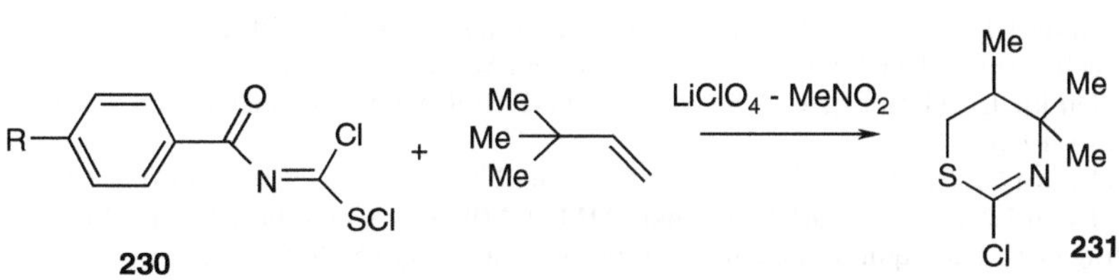

Scheme 74 Synthesis of 4,4,5-trimethyl-5,6-dihydro-2-chloro-4H-1,3-thiazine

The present review updates the synthesis and chemistry of five and six-membered halogeno-containing S and S,N-heterocycles. This class of compounds is a topic of intensive evaluation. Alongside, it should be mentioned that many of the considered heterocycle types are still very rare and very sparely described in the literature. That is why investigation of halogeno-containing heterocycles is still the issue of the day for the modern organic chemistry.

References

1. Kuthan J, Sebek P, Boehm S (1994) Adv Heterocycl Chem 59:179–244
2. Elguero J, Fruchier A, Jagerovich N et al (1995) Org Prep Proc Int 27:33–74
3. Taylor RJK, Casy G (2003) Org React 62:357–475
4. Vorbruggen H, Ruh-Pohlenz C (1995) Org React 55:1–654
5. Hepworth JD, Heron MB (2004) Prog Heterocyc Chem 16:405–430
6. Burger K, Wucherpfenning U, Brunner E (1995) Adv Heterocyc Chem 60:2–64
7. Shermolovich YuG (2009) Fluorinated derivatives of five-membered heterocycles containing oxygen, sulfur, selenium and phosphorus atom. In: Petrov S (ed) Fluorinated heterocyclic compounds: synthesis, chemistry and applications. Wiley, Hoboken, NJ, pp 159–225
8. Dvornikova E, Bechcicka M, Kamieńska-Trela K et al (2003) J Fluor Chem 124:159–168
9. Higashiguchi K, Matsuda K, Asano Y et al (2005) Eur J Org Chem 91–97
10. Cerichelli G, Crestoni ME, Fornarini S (1990) Gazz Chim Ital 120:749–755
11. Chonan T, Takahashi K (2004) Bull Chem Soc Jpn 77:1487–1497
12. Ichikawa J, Wada Y, Fujiwara M et al (2002) Synthesis 13:1917–1936
13. Singh RP, Shreeve JM (2003) J Org Chem 68:6063–6065
14. Singh RP, Shreeve JM (2001) Org Lett 3:2713–2715
15. Singh RP, Twamley B, Shreeve JM (2002) J Org Chem 67:1918–1924
16. Kovregin AN, Sizov AY, Ermolov AF (2002) Izv Akad Nauk Ser Khim (Russ) 6:945–947
17. Murakami S, Ishii H, Fuchigami T (2004) J Fluor Chem 125:609–614
18. Murakami S, Kim S, Ishii H et al (2004) Synlett 5:815–818
19. Facchetti A, Yoon MH, Stern CL et al (2004) J Am Chem Soc 126:13480–13501
20. Facchetti A, Deng Y, Wang A et al (2000) Angew Chem Int Ed 39:4547–4551
21. Bardia B, Rao JT (2004) Asian J Chem 16:511–512
22. Tanemura K, Suzuki T, Nishida Y et al (2003) Chem Lett 32:932–933
23. Voronkov MG, Deryagina N, Perevalova VI (1980) Chem Heterocycl Comp (Engl.Trans) 16:225–227
24. Hull JW Jr, Romer DR, Podhorez DE et al (2007) Beilstein J Org Chem. doi:10.1186/1860-5397-3-23
25. Kruse CG, Poels EK, Jonkers EL et al (1978) J Org Chem 43:3548–3553
26. Delaney PA, Johnston AW (1985) Tetrahedron 41:3845–3851
27. Drux R (1994) J Prakt Chem Chemiker-Zeitung 336:89–90
28. Schmidt DL, Heeschen JP, Klingler TC et al (1985) J Org Chem 50:2840–2847
29. Tsuchihashi G, Iriuchijima S (1970) Bull Chem Soc Jpn 43:2271
30. Belen'kii LI, Gromova GP, Krayushkin MM (1993) Khimiya Geterotsikl Soedin (Russ) 1940–1945
31. Belen'kii LI, Gromova GP, Kolotaev AV et al (2000). Chem Heterocycl Comp 36:256–263.
32. Belenkii LI, Gromova GP, Krayushkin MM (1990) Gazz Chim Ital 120:365–368
33. Brugier D, Outurquin F, Paulmier C (2000) Tetrahedron 56:2985–2993
34. Marchand P, Gulea M, Masson S et al (2000) Org Lett 2:3757–3759
35. Kafarsky P, Lejczak B (1991) Phosphorus Sulfur Silicon 63:193–215

36. Mikolajczyk M, Balczewsky P (1994) β-Phosphoryl substituted organosulfur compounds. In: Block E (ed) Advances in sulfur chemistry, vol 1. JAI, Stamford, pp 41–96
37. Li L, Collard DM (2005) Macromolecules 38:372–378
38. Sakamoto Y, Komatsu S, Suzuki T (2001) J Am Chem Soc 123:4643–4644
39. Nomura A, Sakamaki S, Hongu M et al (2010) J Med Chem 53:6355–6360
40. Oster A, Klein T, Werth R et al (2010) Bioorg Med Chem 18:3494–3505
41. Burger K, Helmreich B (1994) Heterocycles 39:819–832
42. Burger K, Hennig L, Fuchs A et al (2005) Monatsh Chem 136:1763–1779
43. Burger K, Fuchs A, Hennig L et al (2001) Monatsh Chem 132:929–945
44. Brugier D, Outurquin F, Paulmier C (2001) J Chem Soc Perkin Trans I 37–43
45. Willy B, Frank W, Muller TJJ (2010) Org Biomol Chem 8:90–95
46. Tso HH, Yang NC, Chang YM (1995) J Chem Soc Chem Commun 1349–1350
47. Chou T, Chen MM (1987) Heterocycles 26:2829–2834
48. Lee SJ, Lee JC, Peng ML et al (1989) J Chem Soc Chem Commun 1020–1021
49. Kruze CG, Poels EK, van der Gen A (1979) J Org Chem 44:2911–2915
50. Boge A, Brunck J-S, Schwar G et al (1992) Chem Ber 125:1641–1646
51. Birk C, Voss J (1996) Tetrahedron 52:12745–12760
52. Delancy PA, Johnstone RAW, Leonard PA et al (1991) J Chem Soc Perkin Trans I 285–289
53. Alonso AM, Horcajada R, Motevalli M et al (2005) Org Biomol Chem 3:2842–2847
54. Kumar V, Remers WA (1981) J Org Chem 46:4298–4300
55. Yakovlev VV, Shklyarenko AA (2008) Russ J Org Chem 44:629–631
56. Nowak I, Rogers LM, Rogers RD et al (1999) J Fluor Chem 99:73–81
57. Wiebe HA, Braslavsky S, Heichlen J (1972) Can J Chem 50:2721–2724
58. Kobayashi Y, Kumadaki I, Ohsawa A et al (1974) Tetrahedron Lett 15:2841–2844
59. Smith JR, Lemal DM (2000) J Fluor Chem 102:323–332
60. Kobayashi Y, Kawada K, Ando A et al (1984) Tetrahedron Lett 25:1917–1920
61. Fortuna CG, Barresi V, Musumarra G (2010) Bioorg Med Chem 18:4516–4523
62. Boyd DR, Sharma ND, Gunaratne N et al (2003) Org Biomol Chem 1:984–994
63. Bartoli G, Latrofa A, Naso F et al (1972) J Chem Soc Perkin Trans I 2671
64. Cohen F, Koehler MFT, Bergeron F et al (2010) Bioorg Med Chem 20:2229–2233
65. Baba D, Ishii H, Higashiya S et al (2001) J Org Chem 66:7020–7024
66. Burger K, Geith K, Hoss E (1989) Synthesis 352–360
67. Burger K, Geith K, Sewald N (1990) J Fluor Chem 46:105–122
68. Burger K, Geith K, Hubl D (1988) Synthesis 189–194
69. Rudnichenko AV, Timoshenko VM, Chernega AN (2004) J Fluor Chem 125:1351–1356
70. Kvicala J, Hovorkova P, Paleta O (2001) J Fluor Chem 108:1–5
71. Saloutina LV, Zapevalov AY, Kodess MI et al (1999) Mendeleev Commun 231–232
72. Saloutina LV, Zapevalov AY, Kodess MI et al (2000) J Fluor Chem 104:155–165
73. Saloutina LV, Zapevalov AJ, Kodess MI et al (2003) J Fluor Chem 120:41–47
74. Sundermeyer J, Roesky HW, Lautner J et al (1990) Chem Ber 123:433–438
75. Vovk MV, Lebed PS, Polovinko VV et al (2008) Russ J Org Chem 44:1836–1839
76. Pae AN, Lee JE, Kim BH et al (2000) Tetrahedron 56:5657–5665
77. Zhu X, Shi D (2009) Chin J Chem 27:565–568
78. Yu H, Qin Z, Dai H et al (2008) ARKIVOC 16:99–109
79. Long DD, Aggen JB, Christensen BG et al (2008) J Antibiot 61:595–602
80. Yamamoto H, Eikyu Y, Okuda S et al (2002) Bioorg Med Chem 10:1535–1545
81. Ghodsi S, Amini M, Miri R et al (2006) Phosphorus Sulfur Silicon 181:2435–2444
82. Sun W, Zheng YR, Xu CH (2007) J Phys Chem 111:11706–11711
83. Dai H, Li YQ, Du D et al (2008) J Agric Food Chem 56:10805–10810
84. Babii SB, Zyabrev VS, Drach BS (2002) Russ J Gen Chem 72:1730–1735
85. Rajput AP (2002) Asian J Chem 14:807–810
86. Raghavan B, Balasubramanian R, Steele JC (2008) J Med Chem 51:1530–1533
87. Simeon FG, Brown AK, Zoghbi SS (2007) J Med Chem 50:3256–3266

88. Balzani V, Venturi M, Credi (2003) A molecular devices and machines. A journey into the nano world. Wiley-VCH, Weinheim
89. Hodgetts KJ, Kershaw MT (2002) Org Lett 4:1363–1365
90. Kovregin AN, Sizov AY, Ermolov AF (2002) Izv Arad Nauk Ser khim (Russ) 6:1000–1002
91. Zawistoski MP, Decker SM, Griffith DA (2009) Tetrahedron Lett 50:7286–7287
92. Ahlemann JT, Roesky HW, Markovsky LN et al (1995) Heteroatom Chem 6:9–13
93. Solov'ev AV, Chernega AN, van Almsick A et al (2005) Z Anorg Allg Chem 631:2842–2847
94. Roesky HW, May A, Noltemeyer M (1993) J Fluor Chem 62:77–99
95. Ohkata K, Ohyama Y, Akiba K (1994) Heterocycles 37:859–867
96. Kaberdin RV, Potkin VI, Ol'dekop YA (1990) Zhurn Org Khimii (Russ) 26:1560–1566
97. Dikusar EA, Potkin VI, Gadzhily RA et al (2010) Russ J Org Chem 46:59–63
98. Potkin VI, Nechai NI, Kaberdin RV (2007) Russ J Org Chem 43:587–590
99. Shukovskaya NA, Dikusar EA, Potkin VI et al (2009) Russ J Gen Chem 79:1660–1662
100. Nechai NI, Dikusar EA, Potkin VI et al (2004) Russ J Org Chem 40:1009–1014
101. Clerici F, Contini A, Gelmi ML et al (2003) Tetrahedron 59:9399–9408
102. Dekker FJ, Ghizzoni M, van der Meer N et al (2009) Bioorg Med Chem 17:460–466
103. Clerici F, Contini A, Corsini A et al (2006) Eur J Med Chem 41:675–682
104. Schroter S, Stock C, Bach T (2005) Tetrahedron 61:2245–2267
105. Christoforou IC, Koutentis PA (2006) Org Biomol Chem 4:3681–3693
106. Combs AP, Zhu W, Crawley ML et al (2006) J Med Chem 49:3774–3789
107. Combs AP, Glass B, Galya LG et al (2007) Org Lett 9:1279–1282
108. Polescher H, Seppelt K (2003) J Fluor Chem 120:49–58
109. Fesun IM, Rozhenko AB, Timoshenko VM (2006) J Fluor Chem 127:774–779
110. Johnston BD, Indurugalla D, Pinto BM et al (2001) J Am Chem Soc 123:12698–12699
111. Yuasa H, Tanetani Y, Miyazaki T (2008) Lett Org Chem 5:262–264
112. Horton D, Li Y, Barberousse V et al (1993) Carbohydr Res 249:39–48
113. Kishi Y, Inagi S, Fuchigami T (2009) Eur J Org Chem 103–109
114. Abe T, Nagase S, Baba H (1973) Bull Chem Soc Jpn 46:3845–3848
115. Sizov AY, Kovregin AN, Serdyuk RN et al (2006) Russ Chem Bull 55:1200–1208
116. Siry SA, Timoshenko VM (2010) Tetrahedron Lett 51:6406–6409
117. Alvernhe GM, Greif D, Laurent AJ (1995) J Fluor Chem 70:121–125
118. Finkelstein BL, Sethuraman MR (2000) J Heterocycl Chem 37:1583–1586
119. Shermolovich YG, Portella C, Tschenn O (1997) Bull Soc Chim Fr 134:697–702
120. Laduron F, Nyns C, Janousek Z et al (1997) J Prakt Chem 339:697–707
121. Mikhailichenko SS, Bouillon JP, Besson T et al (2010) Tetrahedron Lett 51:990–993
122. Timoshenko VM, Siry SS, Rozhenko AB et al (2010) J Fluor Chem 131:172–183
123. Petrov VA, Lustig S, Marshall W (2007) J Fluor Chem 128:1227–1234
124. Sanin AV, Nenajdenko VG, Kuz'min VS et al (1996) J Org Chem 61:1986–1989
125. Sashida H, Tsuchia T (1985) Chem Pharm Bull 34:3644–3652
126. Sasaki T, Hayakawa K, Nishida S (1982) Tetrahedron 38:75–83
127. Gaoni Y (1981) J Org Chem 46:4502–4510
128. Iriuchijima S, Ishibashi M, Tsuchihashi G (1973) Bull Chem Soc Jpn 46:921–928
129. Boge A, Schwar G, Voss J (1993) Phosphorus Sulfur Silicon 83:175–181
130. Gonzalez FS, Mendoza PG, Aparicio PJL (1988) Carbohyd Res 183:227–240
131. Jeffery SM, Sutherland AG, Pyke SM et al (1993) J Chem Soc Perkin Trans I 2317–2327
132. Ewin RA, Loughlin WA, Pyke SM et al (1993) Synlett 660–662
133. McAllister GD, Taylor JK (2001) Tetrahedron Lett 42:1197–1200
134. Abae MS, Mojtahedi MM, Zahedi MM et al (2007) Synthesis 21:3339–3344
135. Fanghanel E, Bartossek H, Baumeister U (1998)J Heterocycl Chem 35:1449–1454
136. Vovk MV, Golovach NM, Sukach VA (2010) J Fluor Chem 131:229–233
137. Trofimova TP, Fedoseev VM, Tkachenko SE (2002) Chem Heterocycl Comp 38:1533–1534
138. Borisov GN, Osmanov VK, Sokolov IG (2002) Chem Heterocycl Comp 38:1150–1151

Top Heterocycl Chem (2012) 27: 139–184
DOI: 10.1007/7081_2011_62

Published online: 23 September 2011

Heterocyclic Reagents Containing Nitrogen–Halogen Bond: Recent Applications

Satoshi Minakata, Youhei Takeda, and Junpei Hayakawa

Abstract Progress in the chemistry of unique and important reactions utilizing halogens on nitrogens embedded in heterocyclic compounds is discussed. N–X heterocyclic reagents can be useful in halogenation, oxidation, substitution, addition reaction, cyclization, and asymmetric reactions. This review focuses on the recent representative reaction patterns in which N–X heterocyclic reagents are used, with emphasis on developments in this area since 2005.

Keywords Halogenating reagents · Heterocyclic compounds · Nitrogen–halogen bonds · Organic reactions · Synthetic methodologies

Contents

S. Minakata (✉), Y. Takeda and J. Hayakawa
Department of Applied Chemistry, Graduate School of Engineering, Osaka University, Yamadaoka 2-1, Suita, Osaka 565-0871 Japan
e-mail: minakata@chem.eng.osaka-u.ac.jp

Abbreviations

BCIH	Bis(collidine)iodonium hexafluorophosphate
BNP	1,1′-Binaphthyl-2,2′-diyl hydrogenphosphate
BPIT	Bis(pyridine)iodonium tetrafluoroborate
CAN	Cerium(IV)ammonium nitrate
DBDMH	1,3-Dibromo-5,5-dimethylhydantoin
DBU	1,8-Diazabicyclo[5.4.0]undec-7-ene
DHP	3,4-Dihydro-2*H*-pyran
DIDMH	1,3-Diiodo-5,5-dimethylhydantoin
DIPEA	Diisopropylethylamine
DMAP	4-(*N*,*N*-dimethylamino)pyridine
dppm	1,1-Bis(diphenylphosphino)methane
HDMS	1,1,1,3,3,3-Hexamethyldisilazane
IBX	2-Iodoxybenzoic acid
NBS	*N*-Bromosuccinimide
NBSac	*N*-Bromosaccharin
NCS	*N*-Chlorosuccinimide
NCSac	*N*-Chlorosaccharin
NFOBS	*N*-Fluoro-*O*-benzenedisulfonimide
NFPY-BF_4	*N*-Fluoropyridinium tetrafluoroborate
NFPY-OTf	*N*-Fluoropyridinium trifluoromethanesulfonate
NFSI	*N*-Fluorobenzenesulfonimide
NIS	*N*-Iodosuccinimide
NISac	*N*-Iodosaccharin
PMP	*p*-Methoxyphenyl
PTSA	*p*-Toluenesulfonic acid
Py	Pyridine
SDS	Sodium dodecyl sulfate
Selectfluor®	1-Chloromethyl-4-fluoro-1,4-diazoniabicyclo[2.2.2]octane bis(tetrafluoroborate)
TADDOL	α,α,α′,α′-Tetraaryl-2,2-dimethyl-1,3-dioxolan-4,5-dimethanol
TBCA	Tribromoisocyanuric acid
TBDMS	*tert*-Butyldimethylsilyl
TCCA	Trichloroisocyanuric acid

TEMPO	2,2,6,6-Tetramethyl-piperidyl-1-oxy
TFA	Trifluoroacetic acid
TICA	Triiodoisocyanuric acid

1 Introduction

The exploitation of potential reagents for developing new, efficient, and convenient synthetic methods is an art and constituents a challenging process in the field of organic chemistry. Considerable efforts have been made over the years to discover newer, more efficient reagents that can be used more safely. One class of reagents in this area is N-halogenated reagents. A large number of compounds referred to as N-halo reagents are extensively used in organic synthesis. N-Halo reagents are easy to handle, since all of the halogen is consumed, as opposed to only half, the case when elemental halogens are employed. Among these, a heterocycle in which a halogen is attached to the embedded nitrogen is a versatile ability for organic synthesis, because the conformation of the substituents on the nitrogen is fixed by the ring framework. N-Halogenated heterocycles have some unique and specific features such as high activity of the N–X bond and the various modes by which this bond can be split, which result in their widespread application in organic synthesis. Highly reactive intermediates, halogen radicals, or cations can be generated, depending on the conditions used. Therefore, N–X heterocyclic reagents have the potential to participate in important reactions such as halogenation, oxidation, substitution, addition reactions, cyclization as well as asymmetric halogenation. When the halogen is attached to a nitrogen that also contain electron-withdrawing groups, it can act as a Lewis acid for organic transformations.

The scope of the application of N-halogenated heterocycles is so wide that all reactions with these reagents cannot be introduced within the framework of a single review. This review will focus on recent (during the past 5 years) advances in the use of N–X heterocyclic reagents for unique organic reactions based on new concepts and strategies. Representative N–X heterocyclic reagents, which are shown in Fig. 1, are discussed in this review.

2 N–F Heterocyclic Reagents

N-Fluorinated heterocycles act as electrophilic fluorinating agents. Some representative reagents are 1-chloromethyl-4-fluoro-1,4-diazoniabicyclo[2.2.2] octane bis(tetrafluoroborate) (Selectfluor®), *N*-fluoropyridinium salts and *N*-fluorocamphorsultam. In this section, some unique and interesting reactions that take advantage of electrophilic nature of these types of reagents are discussed.

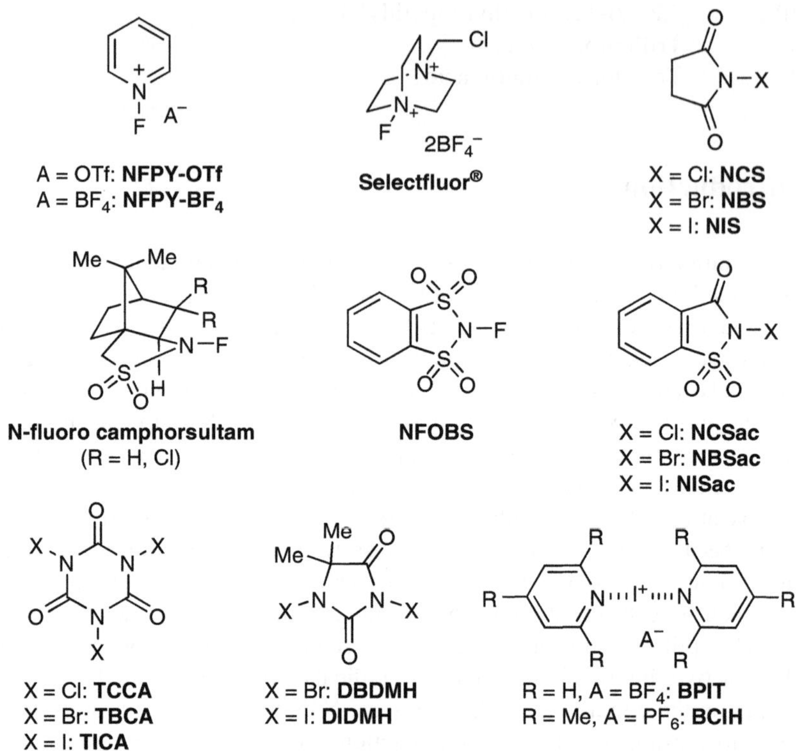

Fig. 1

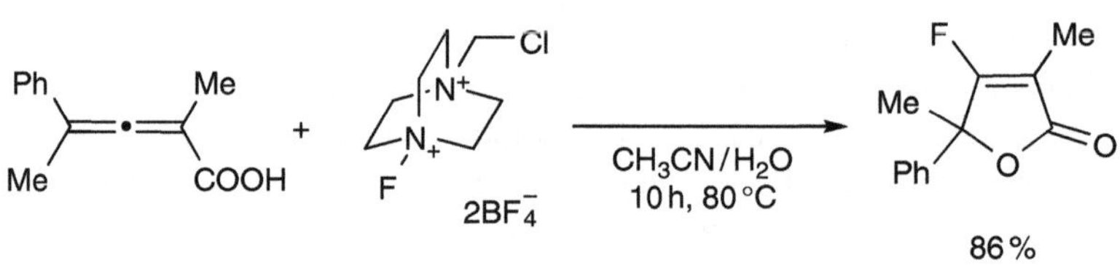

Scheme 1

2.1 *Electrophilic Fluorocyclization*

Monofluorinaton via the electrophilic fluorocyclization reaction of 2,3-allenoic acids with Selectfluor® was developed [1]. The reaction proceeded in MeCN in the presence of 10 equiv of H_2O or even in pure water (Scheme 1).

In contrast to the bromo- and iodocyclization of alkenyl alcohols, it is difficult to realize fluorocyclization because of the weak electrophilicity of the F atom.

Fluorocyclization of allylsilanes using Selectfluor® was successful, giving either *cis* or *trans* fluorocyclized products [2]. The activation of the alkene moiety with a silyl group is critical for the reaction to proceed (Scheme 2).

2.2 Fluorination of Carbonyl Compounds

Numerous enolizable substrates with 1,3-dicarbonyl and monocarbonyl funvtionalities have been efficiently monofluorinated with fluorinating reagents such as Selectfluor® [3], *N*-fluoro-*O*-benzenedisulfonimide (NFOBS) [4, 5], and *N*-fluorosultam derivative (*N*-fluoro-3-ethyl-3-methyl-1,1-dioxo-2,3-dihydro-1*H*-1-λ^6-benzo[e]1,2-thiazin-4-one) [6].

In 2004, Stavber's group reported on the selective and effective fluorination of various types of organic compounds in water using Selectfluor® [7]. More recently, an efficient and environmentally benign synthetic protocol for the direct fluorination of ketones in water was achieved using Selectfluor® and the aqueous SDS micellar system [8]. Although the reaction proceeded in pure water, the addition of SDS improved the efficiency (Scheme 3).

Until 2005, the difluorination of monocarbonyl compounds was difficult to achieve [9–11], but efficient synthetic methods for the preparation of α,α-difluoroketones starting from enamines [12] or imines [13] has now been developed using Selectfluor® (Scheme 4).

An alternative and formal difluorination of the position α to a carbonyl group was realized by the electrophilic fluorination of β-ketosulfones using Selectfluor® followed by base-catalyzed cleavage of the resulting *gem*-difluorinated sulfones (Scheme 5) [14, 15].

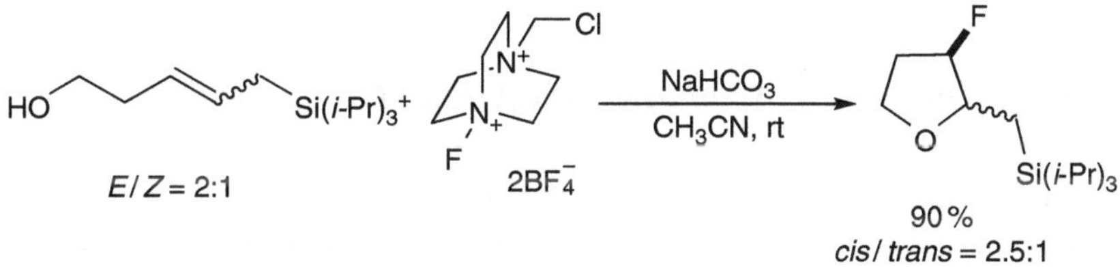

Scheme 2

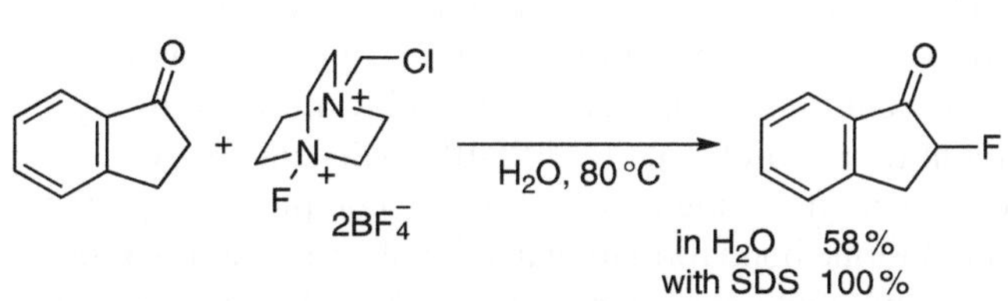

Scheme 3

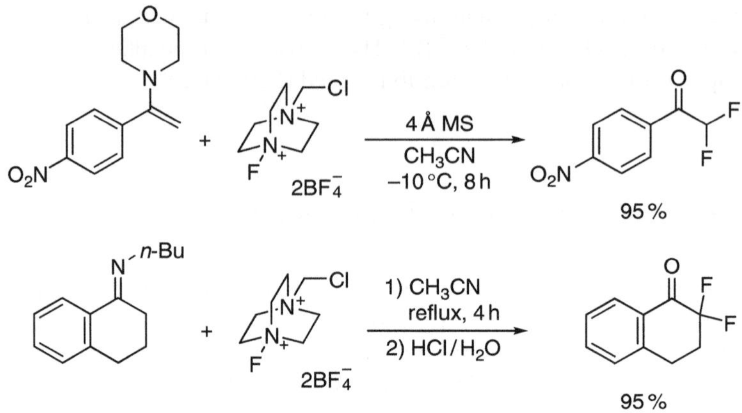

Scheme 4

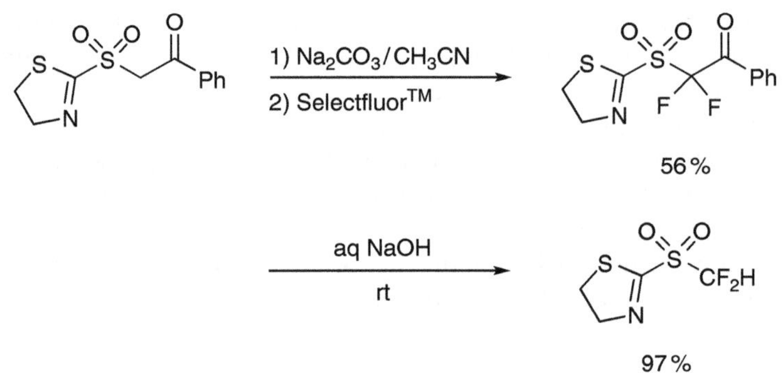

Scheme 5

2.3 *Fluorination of Aromatic and Allene Compounds*

Since the regioselective synthesis of fluorinated aromatics is very difficult, Snieckus and Davis developed a directed *ortho* metalation-based method for the regiospecific introduction of fluoro substituents into aromatic substrates using NFOBS and other fluorinating reagents [16]. In 2006, the first Pd-catalyzed C–H activation/C–F bond-forming reaction leading to the preparation of fluorinated aromatic compounds was achieved by Sanford's group. This reaction proceeded under oxidizing conditions using an electrophilic fluorinating agent such as *N*-fluoropyridinium tetrafluoroborate (NFPY-BF_4) (Scheme 6) [17].

Ritter and co-workers, using carbon–fluorine complexes, reported on new strategy for carbon–fluorine bond formation that involves the fluorination of arylboronic acids. Although this method requires two steps and a stoichiometric amount of palladium complex, the functional group tolerance, broad substrate scope, and

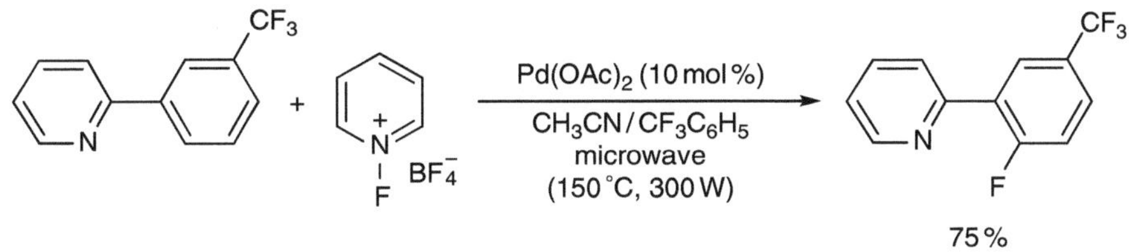

Scheme 6

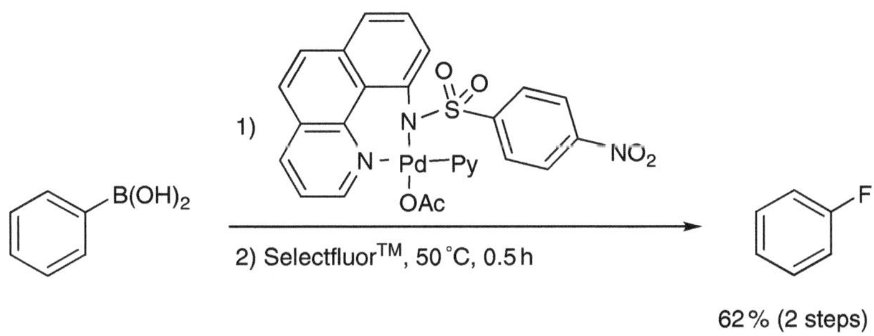

Scheme 7

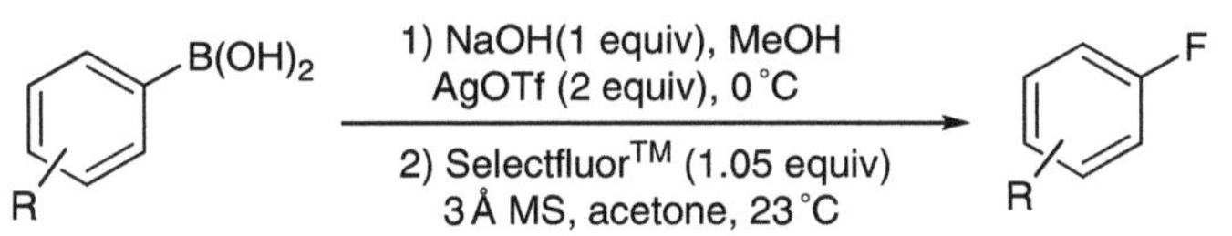

Scheme 8

regiospecificity of the reaction provide a general method for the fluorination of functionalized arenas (Scheme 7) [18, 19].

The regiospecific Ag-mediated fluorination of aryl and alkenylboronic acids and esters was also reported by Ritter's group (Scheme 8) [20].

The group subsequently identified silver as a suitable transition metal for promoting the fluorination of arylstannanes and proposed the intermediacy of bimetallic arylsilver(II) fluorine complexes [21].

On the other hand, the first route to propargylic monofluorides based on the use of electrophilic fluorinating reagent was developed. Allenylsilanes were reacted at room temperature in acetonitrile with Selectfluor® to give secondary propargylic fluorides (Scheme 9) [22].

As another fluorination of allenes, the internal C–C double bond in 3-aryl-1, 2-allenes was highly regioselectively fluorohydroxylated, giving 2-fluoroalken-3-ols by using Selectfluor® (Scheme 10) [23].

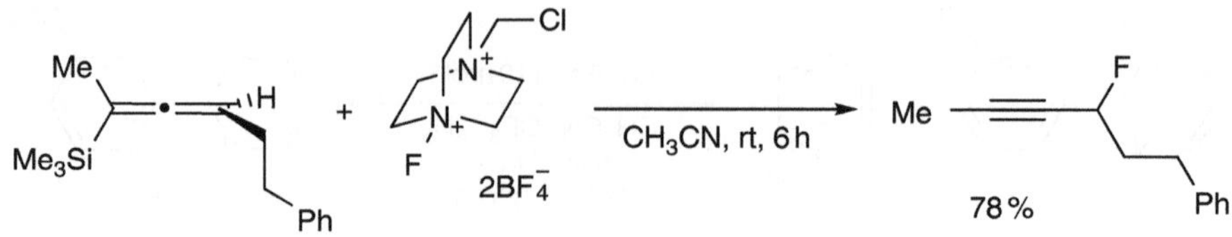

Scheme 9

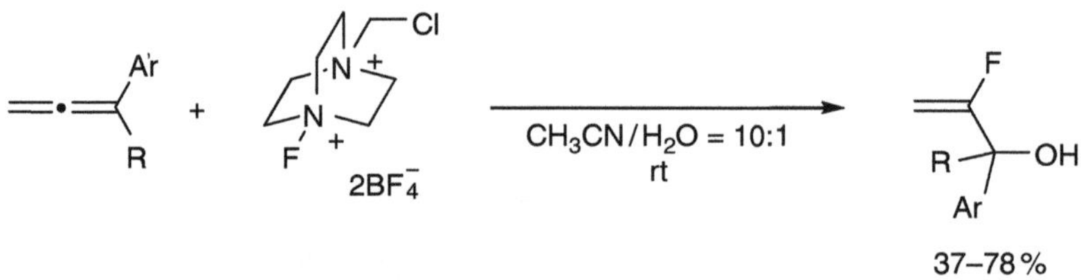

Scheme 10

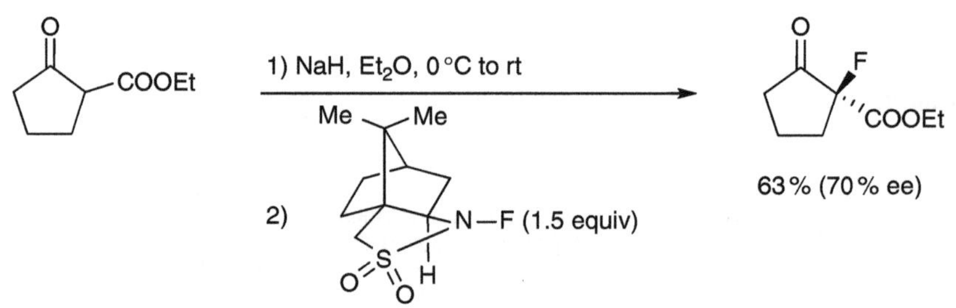

Scheme 11

2.4 Asymmetric Fluorination

In 1998, Lang et al. reported on the first synthesis of an enantioselective fluorinating reagent and realized enantiomeric excesses of up to 70% in reactions of various prochiral metal enolates with the reagents. The optically active reagents were derived from camphor saltam and functioned as enantioselective "F^+"-transfer agents (Scheme 11) [24].

A review on the synthesis of chiral organofluorine compounds was subsequently reported [25], but practical enantioselective fluorinating reagents were not developed at that time. A new asymmetric fluorinating reagent (−)-*N*-fluoro-2,10-(3,3-dichlorocamphorsultam) (**A**) was developed based on the Davis chiral oxaziridine reagent. The reagent was also useful in the asymmetric fluorination of enolates [26, 27]. Takeuchi's group systematically developed a series of efficient asymmetric fluorinating, which resulted in the preparation of *N*-fluoro-3-cyclohexyl-3-methyl-2,3-dihydrobenzo[1,2-*d*]isothiazole 1,1-dioxide (**B**) (Fig. 2) [28].

Fig. 2

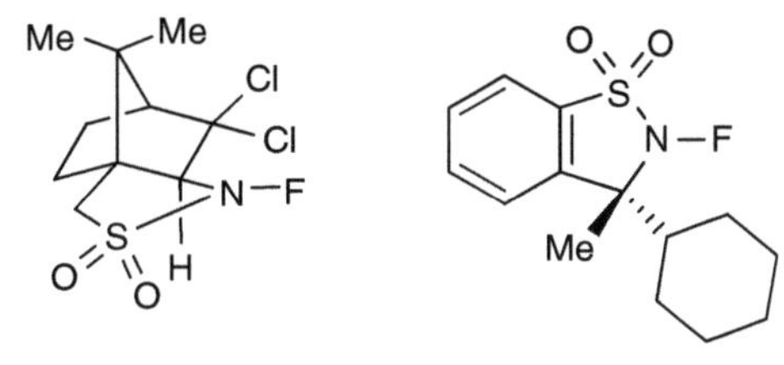

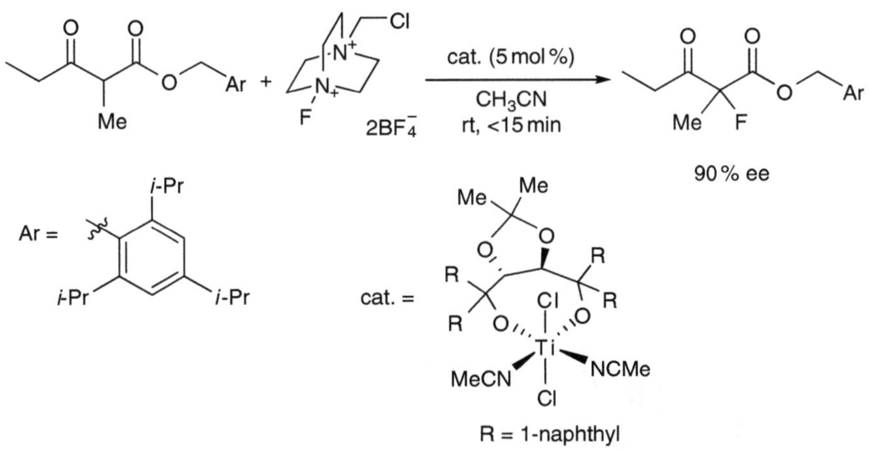

Scheme 12

Although stoichiometric amounts of asymmetric fluorinating reagents were required to achieve agent-controlled enantioselective fluorination, the first successful example of the catalytic asymmetric fluorination of β-ketoesters was reported by Togni's group [29]. They used sterically hindered esters as substrates and Selectfluor® in the presence of a chiral Lewis acid catalyst [$TiCl_2$(TADDOLato)] to prepare α-fluorinated compounds with up to 90% ee (Scheme 12) [30, 31].

Catalytic and asymmetric α-fluorination of either cyclic or acyclic β-ketoesters using heterocyclic fluorinating reagent having N–F bond, NFPY-OTf, was developed by Inanaga's group, in which the catalyst is Sc[(*R*)-F_8BNP]$_3$ (Scheme 13) [32].

New types of enantioselective fluorination reactions were independently revealed by Cahard et al. [33, 34] and Shibata et al. [35, 36], in which cinchona alkaloids/Selectfluor® combinations were used. Cahard's group reported on the synthesis of the first enantiopure *N*-fluoro quaternary ammonium salts of cinchona alkaloids and their application to the asymmetric fluorination of enolates, silyl enol ethers, and α-amino acid derivatives. Shibata's group examined the fluorination of α-substituted silyl enol ethers and acyclic or cyclic α-cyano- or α-ketoesters with quinine/Selectfluor® combinations prepared in situ from quinine and Selectfluor® in MeCN at ambient temperature. The catalytic approach of this enantioselective fluorination mediated by a cinchona alkaloid and Selectfluor® was developed by

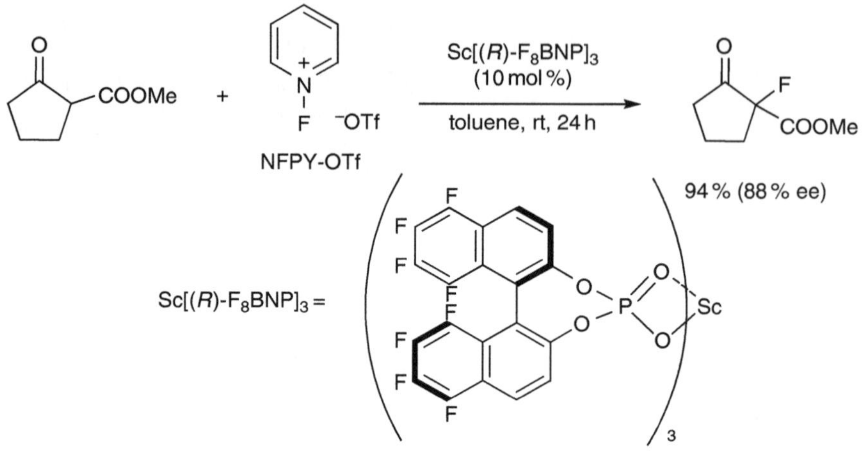

Scheme 13

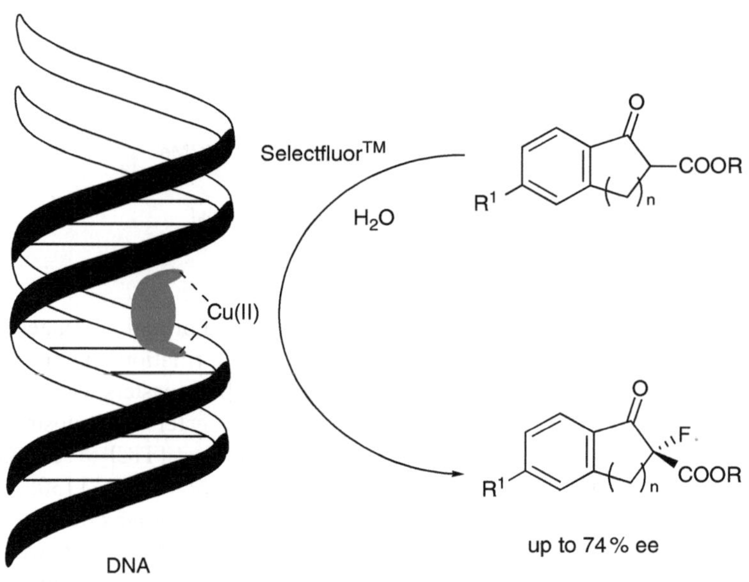

Scheme 14

elegantly employing acyl enol ethers as substrates in the presence of sodium acetate to afford α-fluoroketones in good yields with up to 54% ee [37].

A unique enantioselective fluorination of β-ketoesters with Selectfluor® catalyzed by DNA and a nonchiral ligand-Cu(II) complex was reported by Shibata and Toru. The fluorination of indanone carboxylate was achieved using a combination of a DNA catalyst (st-DNA, salmon testes) and an achiral copper(II) complex, which was prepared by reacting $Cu(NO_3)_2{\cdot}3H_2O$ with 4,4′-dimethyl-2,2′-bipyridine before use (Scheme 14) [38].

Useful fluorination sources such as *N*-fluorobenzenesulfonimide (NFSI), in addition to N–F heterocyclic reagents can be used in catalytic and asymmetric fluorination reactions, but they are not heterocycles and are in this review [39, 40].

2.5 Oxidative Reactions Utilizing Fluoronium Ion

When the fluoronium ion is used, it acts not only as an electrophile but also as a highly reactive oxidizing reagent because of its electronegative characteristics. Therefore, there are many examples of its use as an oxidant. A few examples of oxidative reactions will be treated in this section and unique catalytic processes involving cationic metal (gold) species will be discussed based on some quite recent results.

The combination of Selectfluor® and bromide (KBr) generated bromonium ions can be used in reactions with a wide range of alkenes via addition reaction, brominating ring closure, oxidative bromination, and the Hunsdieker–Borodin reaction (Scheme 15) [41].

The iodination of substituted anisoles and the oxidation of benzyl alcohols proceeds in the presence of iodine and Selectfluor®, wherein water is a crucial component of such reactions. The reactions can also be induced by a fluoronium ion as an oxidant [42]. Selectfluor® represents a useful oxidizing reagent for the sulfenylation of indoles at the 3-position under mild conditions in short reaction times (Scheme 16) [43].

An *N*-fluoroheterocyclic salt was found to be a versatile organocatalyst that mediates reactions between ethyl diazoacetate and *N*-arylimines or *N*-trimethylsilylimines to afford a wide range of aziridines. The authors presume that fluoronium ions catalytically activate the imines (Scheme 17) [44].

Scheme 15

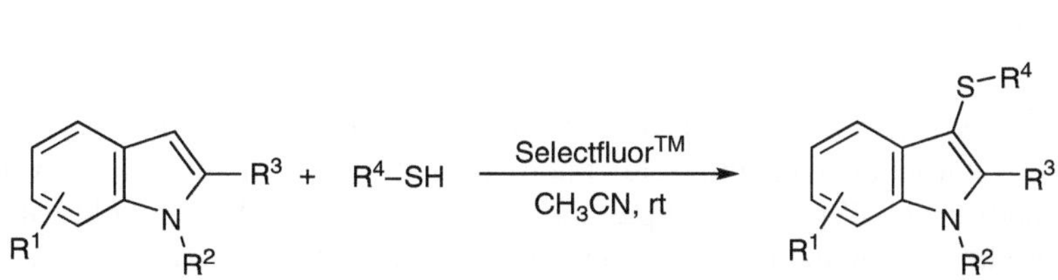

Scheme 16

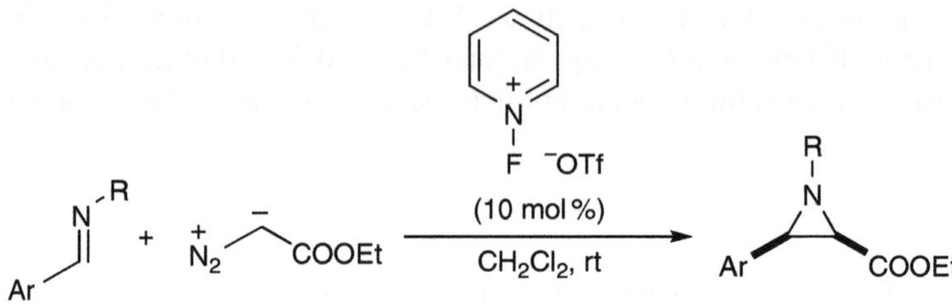

Scheme 17

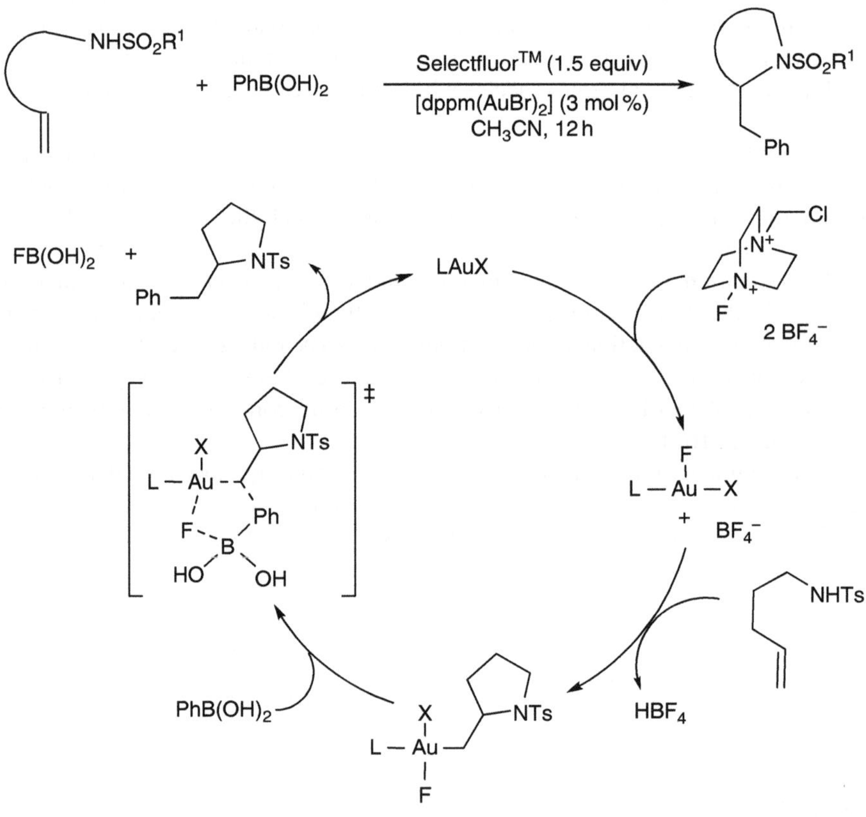

Scheme 18

Toste and Zang independently reported on gold-catalyzed oxidative reactions wherein fluoronium ions and Selectfluor® played important roles in activating a gold catalyst [45–48]. For example, the Toste group curried out a gold-catalyzed aminoarylation reaction of alkenes and arylboronic acids. The reaction is proposed to proceed through a redox cycle involving the initial oxidation of Au(I) into Au (III) by Selectfluor® (Scheme 18).

3 N–X (X = Cl, Br, I) Heterocyclic Reagents

3.1 α-Halogenation of Carbonyl Compounds

Classically, the NBS α-bromination of ketones proceeds via a radical process, but recently, Lewis acids were also employed to promote the α-halogenation of carbonyl compounds. For example, Yang and co-workers reported that, using NBS combined with a Lewis acid, $Mg(ClO_4)_2$, the mild and fast α-bromination of a wide range of functionalized 1,3-dicarbonyl compounds was achieved. This protocol was particularly applicable to the α-monobromination of α-unsubstituted β-ketoesters (Scheme 19) [49].

The reaction was extended to the α-chlorination and iodination of 1,3-dicarbonyl compounds with NCS and NIS, respectively. As alternative Lewis acids for halogenations, Amberlysr-15® [50], $AlCl_3$ [51], and *p*-toluenesulfonic acid (PTSA) [52], and so on were found to permit the halogenation of the α-position of carbonyl compounds. The UV-vis irradiation induced α-bromination of ketones using NBS without any catalyst or radical initiator has been also reported [53].

3.2 Halogenation of Aromatic Compounds

The unique bromination of activated aromatic rings was achieved using tribromoisocyanuric acid (TBCA) or a combination of trichloroisocyanuric acid (TCCA) and sodium bromide. In general reactions with TCCA/Br^- are faster than with TBCA, due to the in situ generation of Br^+, which is more electrophilic than the bromine atom in TBCA (Scheme 20) [54].

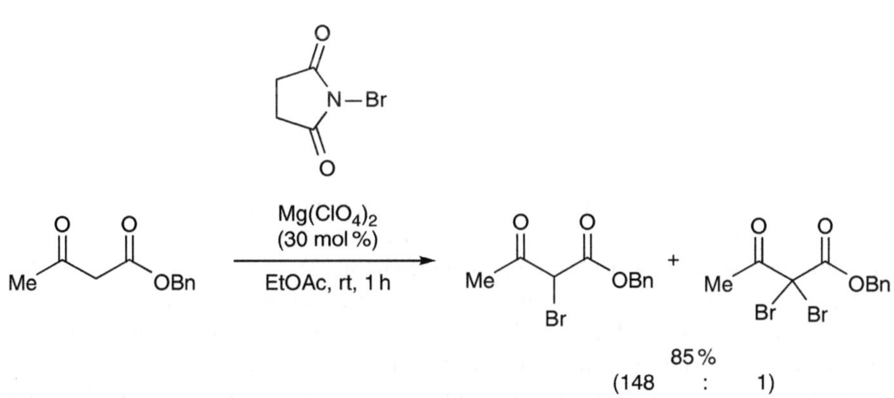

Scheme 19

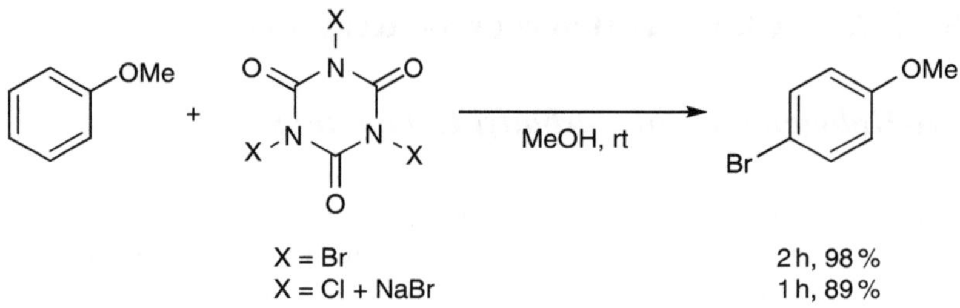

Scheme 20

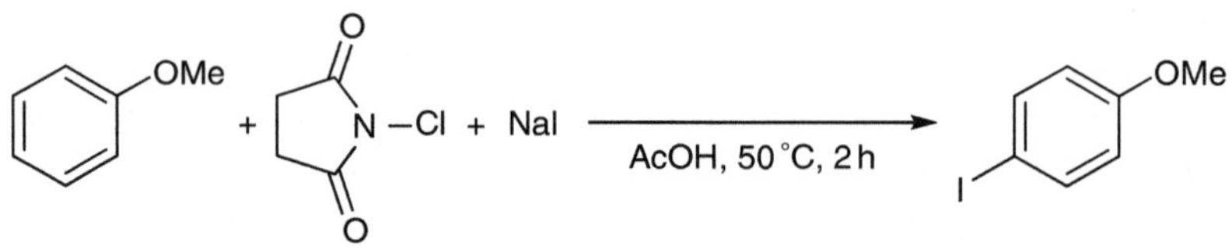

Scheme 21

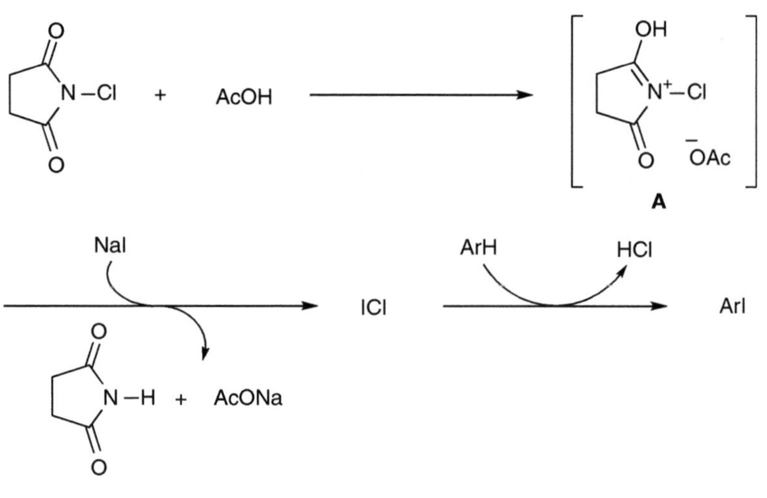

Scheme 22

The iodination of electron-rich aromatic compounds using NCS and NaI was developed by process similar to that described above. For example, anisole was smoothly iodinated to afford 4-iodoanisole (Scheme 21).

A possible pathway for the iodination of aromatic compounds is as follows. NCS is activated by AcOH and the iodide ion attacks the electrophilic chlorine atom of intermediate **A**, resulting in the generation of ICl (Scheme 22) [55].

The halogenation of electron-rich aromatics can be accomplished using UV [56] or ultrasonic [57] irradiation, without the need to an additive, initiator, or catalyst.

Scheme 23

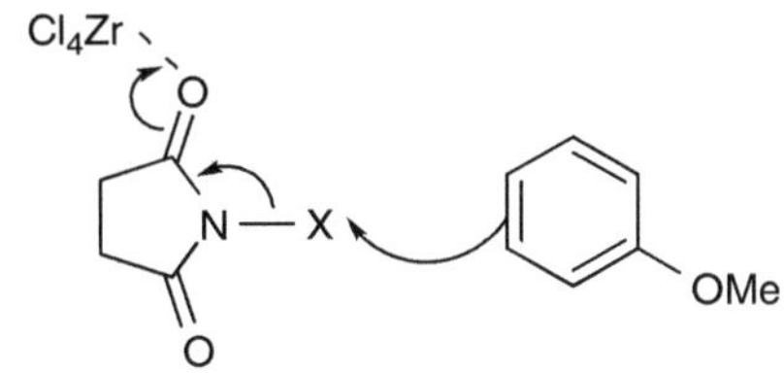

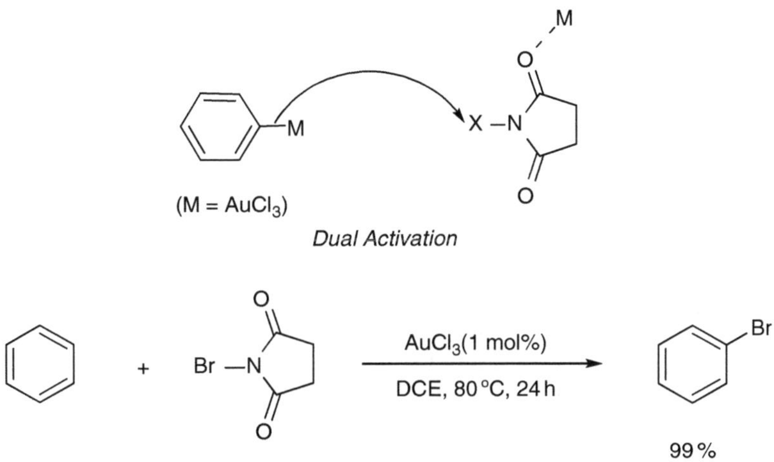

Scheme 24

Although severe reaction conditions were usually required to activate or enhance the halogenating ability of NXS (X = Cl, Br, I), the halogenation of aromatic compounds with NXS catalyzed by $ZrCl_4$ under mild conditions was developed by Yamamoto and co-workers (Scheme 23) [58].

The halogenation of unactivated aromatics proceeds only in the presence of strong Lewis acids or Brønsted acids [59–67]. In these reactions, the conditions are generally very harsh and involve high-catalyst loading [64, 67], high-reaction temperatures [63], or strongly acidic solutions [61]. Wang et al. overcome these issues by utilizing $AuCl_3$ as a catalyst to activate both NXS and CAr–H bond [68]. When benzene was treated with NBS in the presence of a catalytic amount of AuCl (1 mol%) in dichloromethane, bromobenzene was produced in quantitative yield (Scheme 24).

A mild palladium-catalyzed, ligand-directed C–H bond halogenation in arenas using *N*-halosuccinimides was reported by the Sanford group [69]. Although a few groups have demonstrated that similar Pd(II)-catalyzed reactions can be terminated with an oxidative-halogen bond-forming step using X_2 and CuX_2, so far [70, 71], the method offers the possibility of a practical transformation (Scheme 25).

Moreover, Barluenga and co-workers developed a unique *ortho* iodination of β-and γ-aryl alkylamine derivatives [72]. In this reaction, the length of the tether between the arene ring and the trifluoroacetamide moiety drastically affects the selectivity. Excellent results were observed for the iodination when the spacer contained two or three methylene groups (Scheme 26).

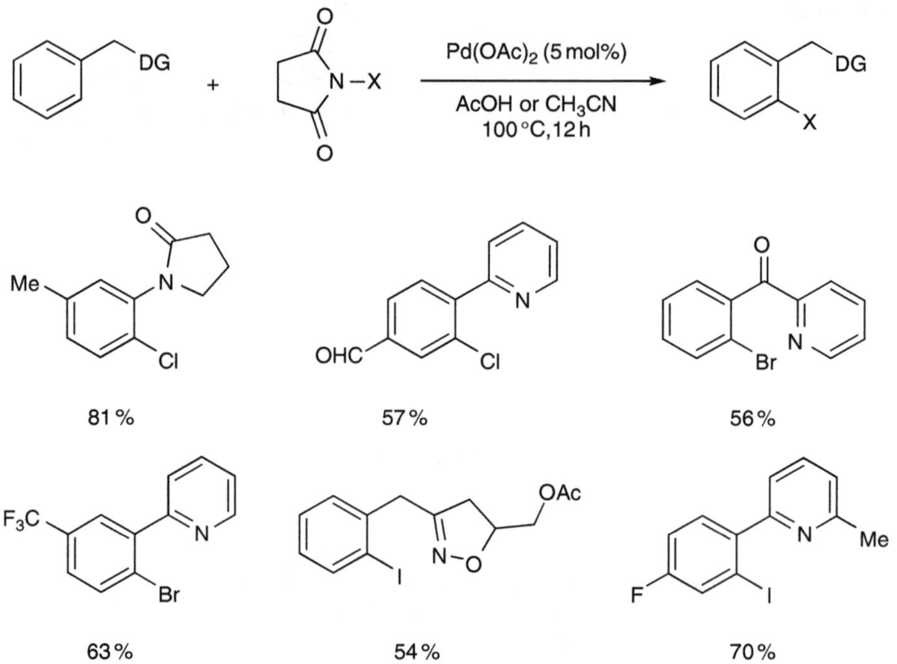

Scheme 25

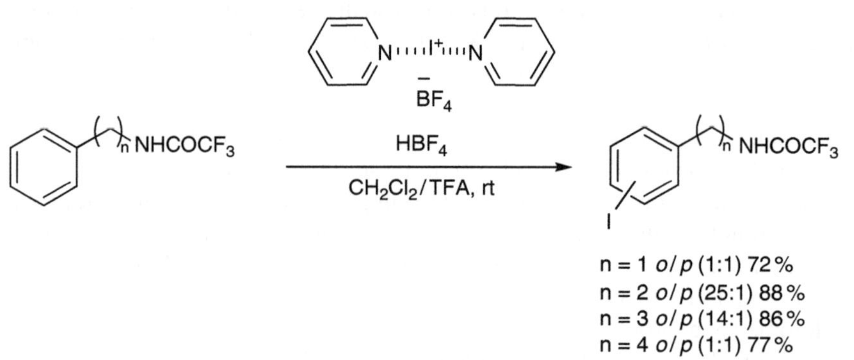

Scheme 26

3.3 *Addition Reactions to C–C Double Bonds*

Vicinal functionalized compounds are valuable synthetic intermediates for use in constructing a variety of functional materials and biologically active compounds. The direct functionaliztion of carbon–carbon double bonds with halogen and

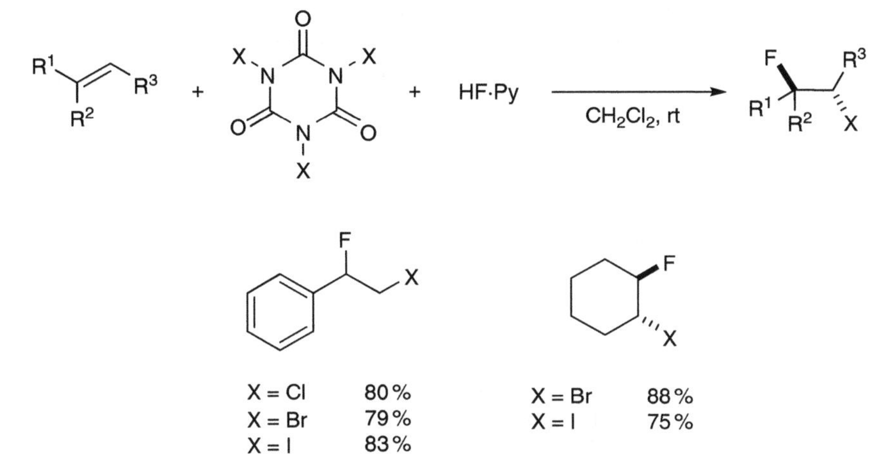

Scheme 27

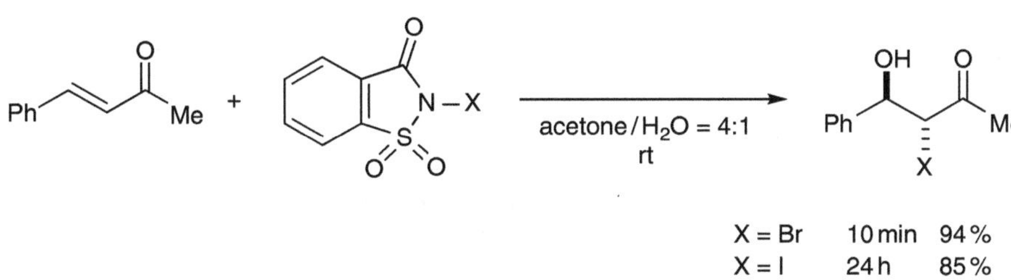

Scheme 28

heteroatoms is a significant approach to the development of useful building blocks for organic synthesis. In this section, recent developments in addition reactions of halogen atoms and heteroatoms (F, O, N) to C–C double bonds are discussed

The halofluorination of alkenes using *N*-chlorosaccharin (NCSac) or trihaloisocyanuric acids and HF·pyridine was reported. The chlorofluorination of alkenes using an NCSac-HF·pyridine system resulted in the formation of vicinal chlorofluoroalkanes. The reaction proceeded with high stereo- and regioselecrivity [73]. The same reaction, using trihaloisocyanuric acids, has also been reported. This addition was also regioselective leading to Markovnikov-oriented products and halofluorinated adducts followed by antiaddition (Scheme 27) [74].

Utilizing *N*-bromo and *N*-iodosaccharin (NISac), the introduction of halogen and oxygen atoms into relatively unreactive, electron-poor alkenes was reported by Dolenc et al. in 2005 [75]. Although the reaction rate for iodination is very slow, bromination reactions are very rapid. Since the bromination of related compounds with NBS was reported to be substantially slower [76], *N*-bromosaccharin (NBSac) should be a potent halogenation reagent that would be suitable for use in addition reactions (Scheme 28).

In related research, alkoxy- or hydroxyiodination of simple alkenes and enol ethers using triiodoisocyanuric acid (TICA) was developed by Mattos and Esteves. The iodinating reagent was prepared by an improved process as shown in Scheme 29 [77].

Even though it is a relatively weak chlorinating reagent, NCS was employed in the alkoxychlorination of alkenes, the addition of the organocatalyst, thiourea, was found to improve the reaction efficiency (Scheme 30).

A key step in the reaction mechanism may involve hydrogen bonds formation between the amide oxygen and thiourea protons as shown in Scheme 31 [78].

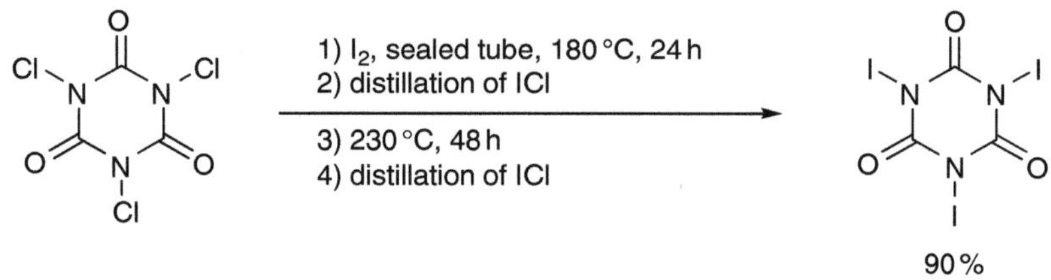

Scheme 29

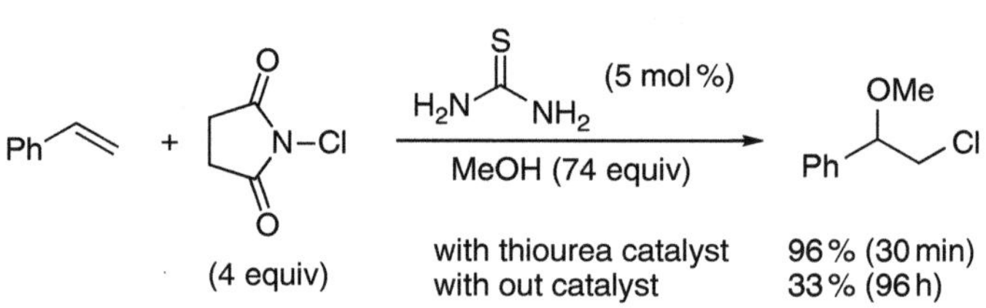

Scheme 30

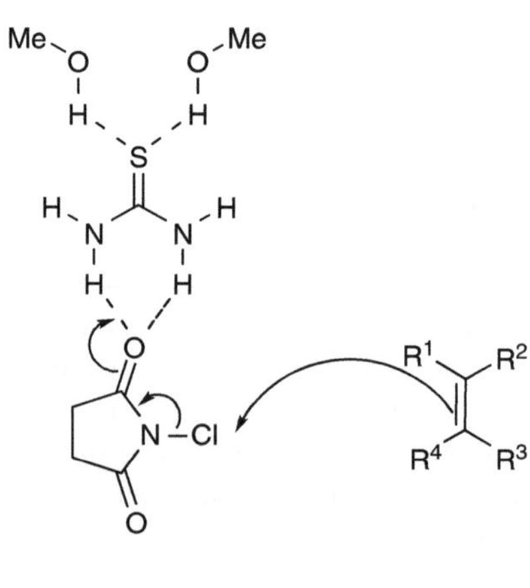

Scheme 31

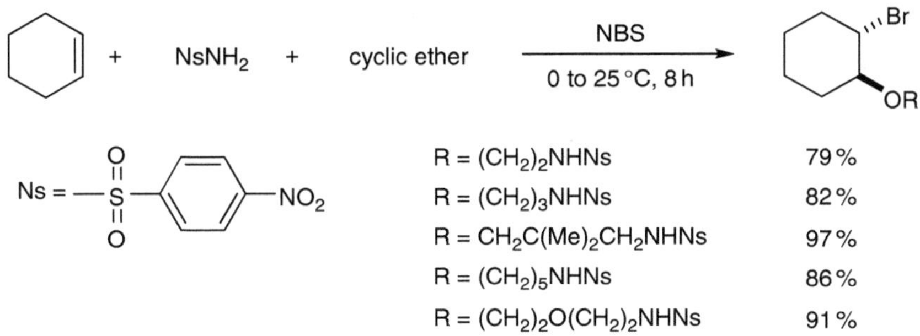

Scheme 32

As a unique, the general and efficient alkoxybromination, regio- and stereoselective functionalization of alkenes using cyclic ethers, amines, and NBS was achieved by Yeung's group in 2010. The alkenes and cyclic ether partners were flexible and could be used in the synthesis of different types of amino ether derivatives. Although the reaction could be applied to various alkenes, examples of the synthesis of amino ethers using cyclic ethers from cyclohexene are shown in Scheme 32 [79].

The haloamination of alkenes has been extensively investigated by many research groups, thus an excellent review by Li et al. appeared in 2006 [80]. Some unique reports thereafter will be dealt with in this section. In a series of reactions in this area, NBS was successfully utilized to introduce bromine atoms and nitrogen units into C–C double bonds. Hajra et al. developed the zinc triflate catalyzed 1,2-bromoazidation of alkenes using NBS and $TMSN_3$. This catalytic process was a highly regioselective, stereoselective, and a high-yield method for the synthesis of anti-1,2-bromoazides from a wide range of alkenes, including α,β-unsaturated carbonyl compounds [81, 82].

In 2008, inexpensive $FeCl_2$ was found to be good catalyst for the amidobromination of alkenes using simple amides (carboxamides and sulfonamides) and NBS. A proposed mechanism for the reaction is as follows. The reaction of alkenes with NBS or species **A** produces the cyclic bromonium ion **B** and anionic iron complexes **C** or **D**. A nucleophilic attack of complex **E** or **D** on **B** then affords the adduct [83] (Scheme 33).

The metal-free aminochlorination of β-nitrostyrene was achieved by Li and co-workers, wherein an N–Cl heterocyclic species is generated in situ via the reaction of *N,N*-dichlorosulfonamide and 4-dimethylaminopyridine (DMAP) (Scheme 34) [84].

As an alternative nonmetal process for amidohalogenation reactions, the hypervalent iodine-catalyzed aminobromination of alkenes was reported by Wang's group under solvent-free conditions [85] or in organic solvents [86]. Although 50–75 mol% of $PhI(OAc)_2$ was required in these procedures, the amount of catalyst could be successfully reduced to 20 mol% when water was used as a solvent (Scheme 35) [87].

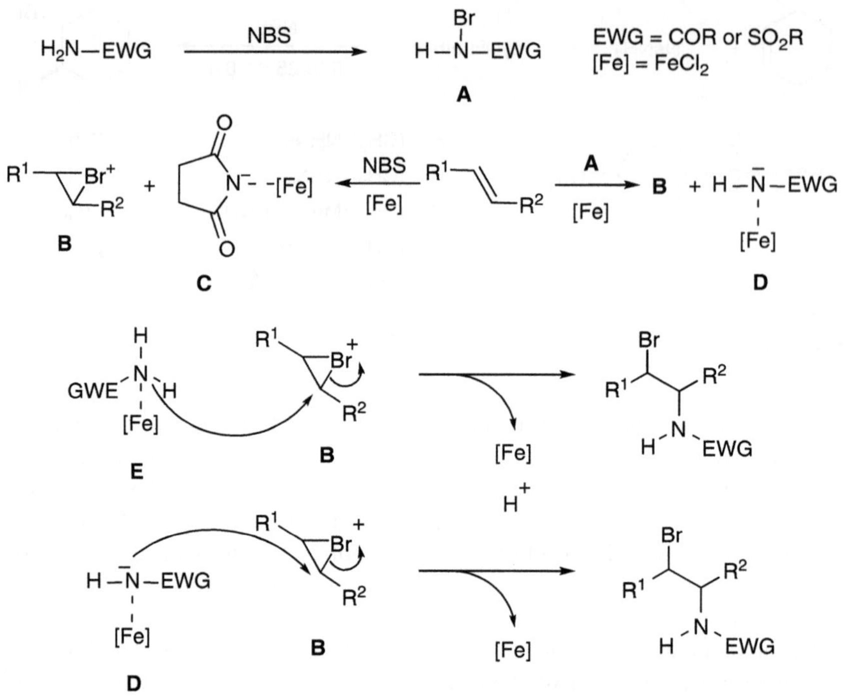

Scheme 33

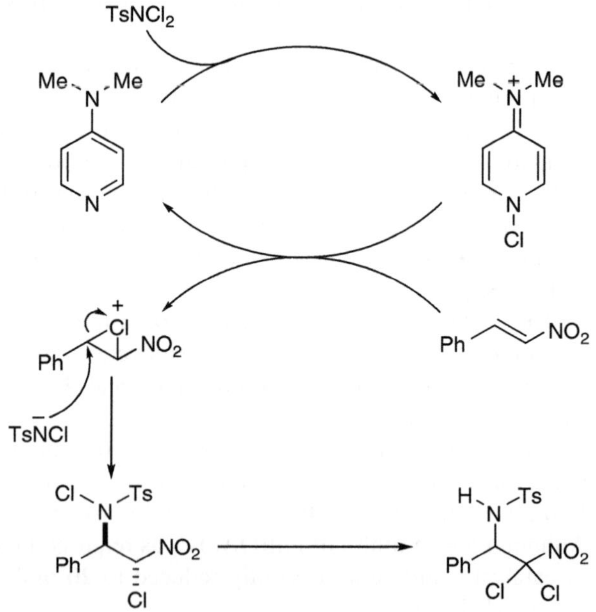

Scheme 34

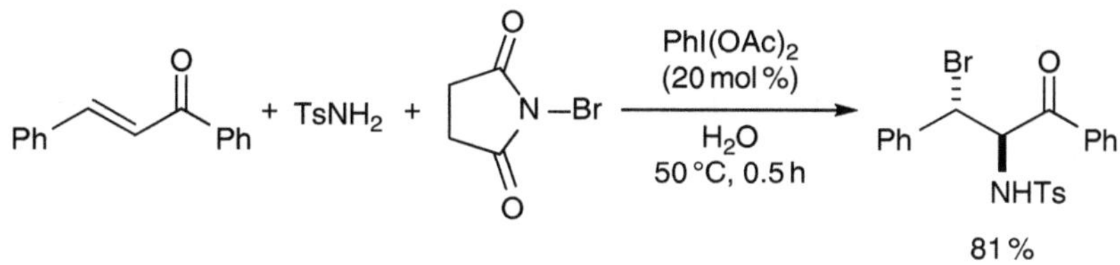

Scheme 35

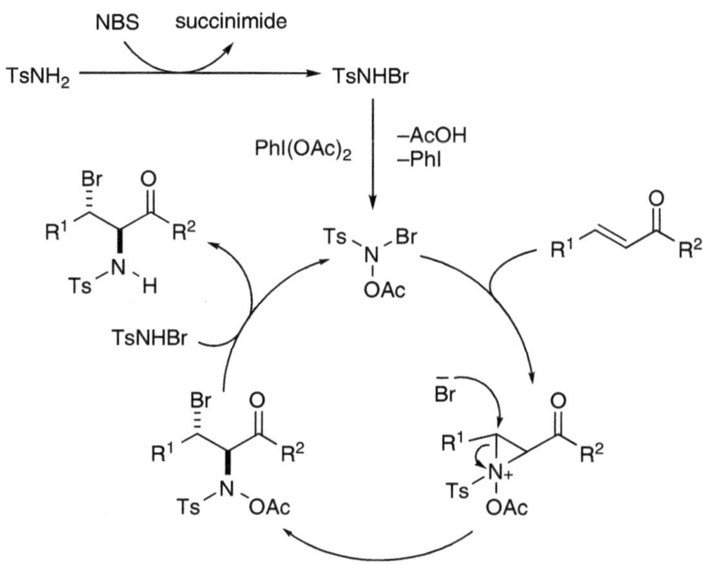

Scheme 36

Although the role of $PhI(OAc)_2$ and the exact mechanism of the reaction is not quite clear at this time, a tentative pathway is proposed as follows. NBS reacts with $TsNH_2$ to generate *N*-bromo-*p*-toluenesulfonamide. The TsNHBr is then oxidized by $PhI(OAc)_2$ to give *N*-acetoxy-*N*-bromo-*p*-toluenesulfonamide, which then is attacked by the alkene to generate an aziridinium intermediate. The aziridinium cation is immediately attacked by a bromide ion via an SN2 reaction to afford the adduct in a regio- and stereoselective manner (Scheme 36).

In 2009, Wei and co-workers found KI to be an efficient alternation catalyst for this transformation. Compared to heavy metal catalysts, KI is easily removed from the reaction mixture by simple manipulation. A series of alkenes, including α,β-unsaturated carbonyl compounds (chalcone derivatives and cinnamates) and simple alkenes (styrene, indene, cyclohexene, and 1-hexene) could be employed in the reaction (Scheme 37) [88].

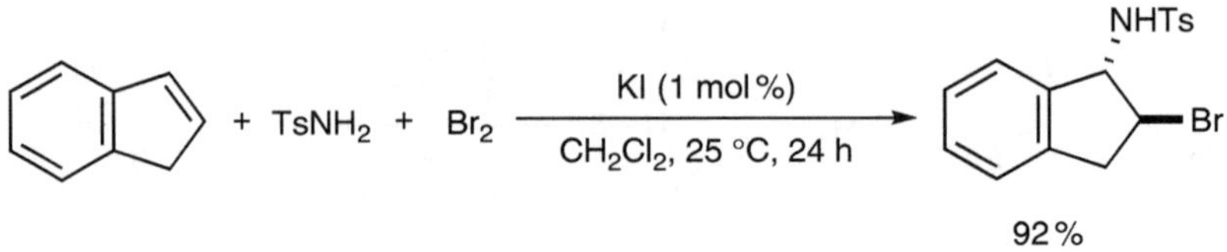

Scheme 37

Scheme 38

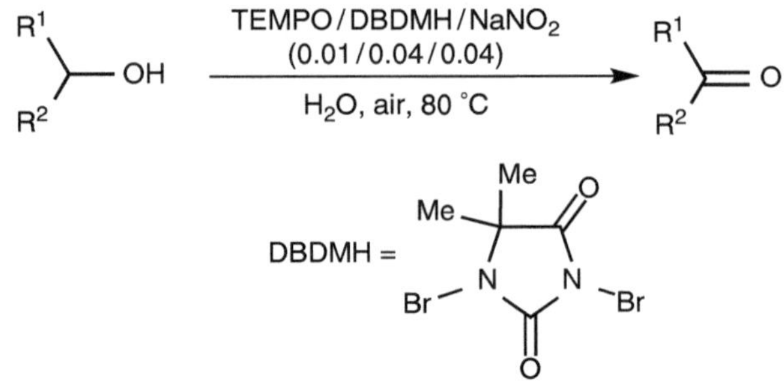

Wei's group also developed the metal [89, 90] and nonmetal [91] powder-catalyzed aminobromination of alkenes with $TsNH_2$ and NBS. Only 1 mol% catalyst is sufficient to initiate the reaction, among these catalysts, silicon powder is very unique and no report involving the use of elemental silicon as a catalyst in organic synthesis has been reported to date.

3.4 *Oxidation Reactions*

N–X Heterocyclic reagents function as efficient oxidants of organic molecules. A highly efficient catalytic system that does not involve the use of transition metals in water was developed for the aerobic oxidation of benzylic alcohols. The system consists of 1 mol% of 2,2,6,6-tetramethyl-piperidyl-1-oxy (TEMPO) as a catalyst, with a catalytic amount of 1,3-dibromo-5,5-dimethylhydantoin (DBDMH) and $NaNO_2$ as cocatalysts (Scheme 38) [92].

The chemoselective oxidation of benzylic and secondary alcohols proceeded in the presence of a catalytic amount of KBr, trichloroisocyanuric acid, and wet SiO_2. The oxidation of bromide ions by HOCl derived from TCCA with water generates HOBr, which is a key species for the oxidation of alcohols [93].

The oxidation of primary alcohols to carboxylic acids was achieved by aerobic photooxidation using NBS as the catalyst (Scheme 39) [94].

Various primary alcohols and primary, secondary, and tertiary amines were oxidatively converted into the corresponding nitriles using 1,3-diiodo-5, 5-dimethylhydantoin (DIDMH) in aqueous ammonia (NH_3) [95–97]. Togo and

Scheme 39

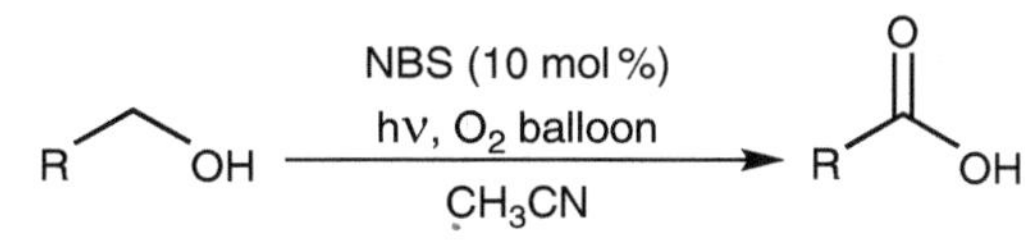

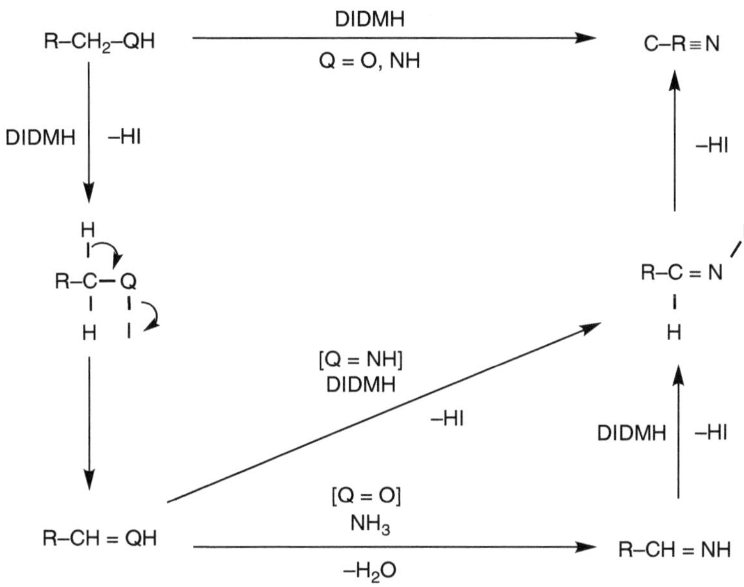

Scheme 40

co-workers greatly expanded the utility of the reagent for oxidation reactions. A plausible pathway for the synthesis of nitriles is as follows. The initial O-iodination and N-iodination of alcohols and amines with DIDMH occurs to generate the *O*-iodo and *N*-iodo compounds, followed by the β-elimination of HI by ammonia to form an aldehyde and an aldimine, respectively. The aldehyde reacts with ammonia to give the aldimine. The aldimine then reacts with DIDMH to give the *N*-iodo aldimine, followed by the β-elimination of HI by ammonia to afford the corresponding nitrile (Scheme 40).

An analogous reaction was reported by Veisi using TCCA as a halogenating reagent for the formation of nitriles from alcohols, amines, aldehydes, and benzyl halides [98].

In general, the direct conversion of alkenes to α-haloketones is synthetically more advantageous than reactions involving ketones. A variety of alkenes were shown to be converted into the corresponding α-bromo/iodoketones when reacted with NBS/NIS and hypervalent iodine, IBX (Scheme 41) [99].

The halonium species derived from the reaction of alkenes with NBS would be attacked by IBX, leading to alkoxyperiodinanes, which subsequently collapse to α-halocarbonyl compounds (Scheme 42).

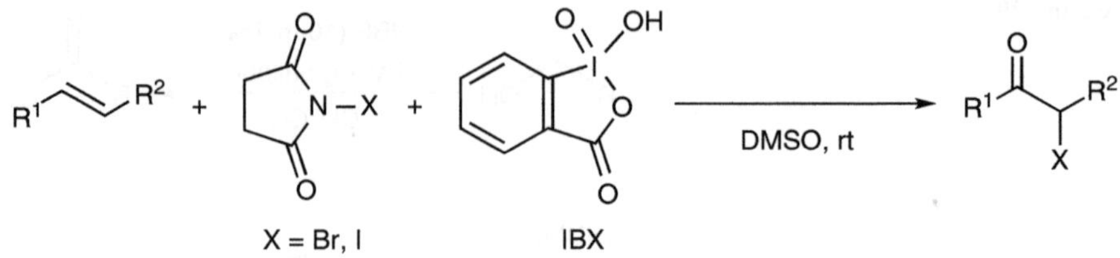

Scheme 41

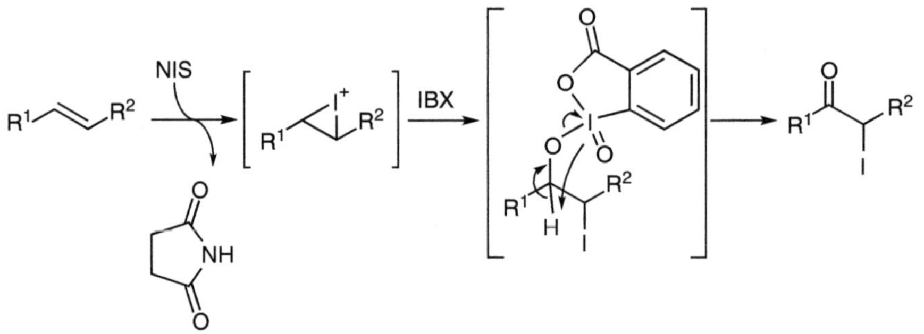

Scheme 42

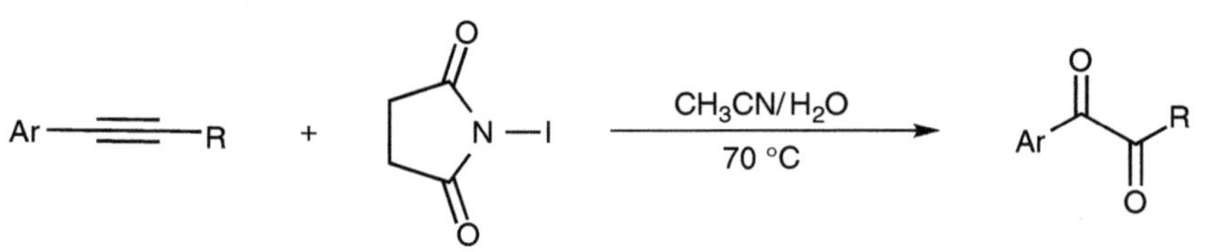

Scheme 43

As an example of the oxidation of alkynes with N–X heterocyclic reagents, an approach to preparing α-diketones via the reaction of internal aryl alkynes with NIS/water was developed by Fu and co-workers (Scheme 43) [100].

When terminal alkynes were used as substrates and were treated with CuI, NBS, and diiopropylethylamine (DIPEA), a Glaser-type coupling reaction proceeded under mild conditions. Alkynes with sensitive groups such as acetal and ketal, TBDMS, ester, and amide functional groups were found to react smoothly to give the functionalized 1,3-diynes (Scheme 44) [101].

A particular type of oxidation with N–X heterocyclic reagents will be discussed below. A combination of cerium(IV)ammonium nitrate (CAN) and NBS was used for the synthesis of a variety of α-hydroxy ketones and α-amino ketones from epoxides and aziridines, respectively (Scheme 45) [102].

The oxidation of thioacetates or thiocarbamates using a combination of NCS and dilute hydrochloric acid proceeded smoothly, leading to the production of the

Scheme 44

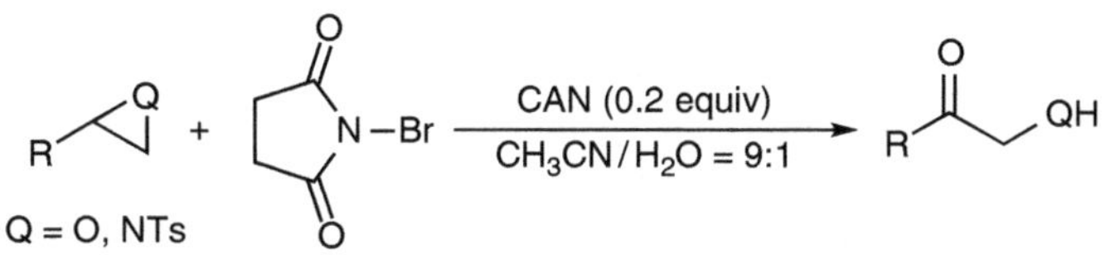

Scheme 45

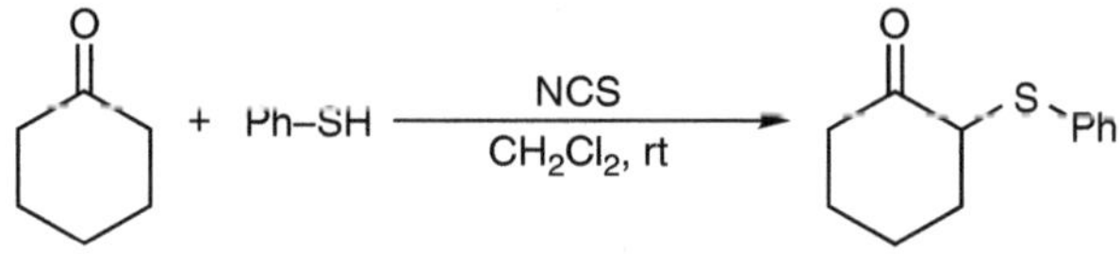

Scheme 46

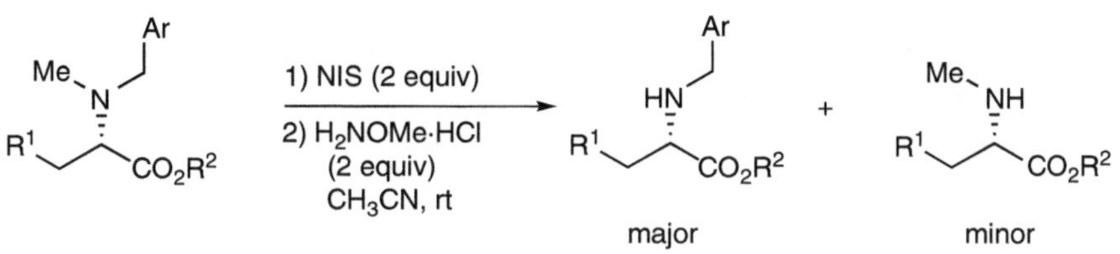

Scheme 47

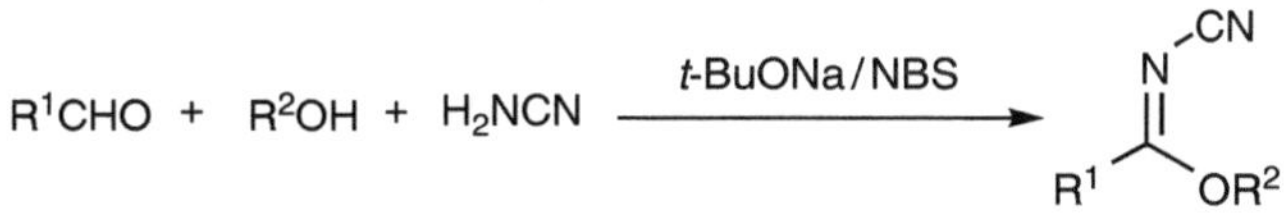

Scheme 48

corresponding sulfonyl chlorides [103]. On the other hand, a combination of aromatic thiols and NCS induced the sulfonylation of ketones that contained α-hydrogens (Scheme 46) [104].

The selective demethylation of *N*-methyl group in *N*-methyl-*N*-arylmethyl-α-amino esters was accomplished by the oxidation of the amino group in a unique use of NIS in an acetonitrile system followed by treatment with *O*-methylhydroxyamine hydrochloride (Scheme 47) [105].

In 2009, He and co-workers reported on one-pot procedure for the cyanoimidation of aldehydes to form *N*-cyanoimidates using NBS as an oxidant. The resulting *N*-cyanobenzimidate was a good precursor for the synthesis of 1,2,4-triazoles via a reaction with hydrazine derivatives (Scheme 48) [106].

3.5 Substitution Reactions

Halogenation through substitution reactions can also be accomplished using N–X heterocyclic reagents. Readily available alcohols were converted into the corresponding bromides and iodides by NBSac and NISac in combination with triphenylphosphine under neutral conditions (Scheme 49) [107].

Although the reaction proceeded smoothly and efficiently, no details of the stereochemistry of the substitution reaction were given. The stereochemical outcome of halogenation using PPh_3-NCS could be controlled to give either an inversion of configuration (through SN2 reaction) or a retention of configuration (through a double SN2 reaction) by having an appropriate substituent at the β-position from the hydroxy group (Scheme 50) [108].

The authors assumed that stereoselective chlorination takes place, as shown in Scheme 51.

A silyl group was also displaced with halogen atoms. Synthetically useful (*Z*)-2-aryl-1-silylethenes were successfully synthesized via the one-pot monoiodo-desilylation/isomerization of 2-aryl-1,1-bis(silyl)ethenes using NCS or BPIT (Scheme 52) [109].

The trimethylsilyl group on alkynes can be substituted with a N–Cl heterocyclic reagent. 1-Chloroalkyne derivatives were readily prepared from trimethylsilyl-protected alkynes using TCCA and $AgNO_3$. The method is selective for trimethylsilyl-protected alkynes relative to triisopropylsilyl-protected derivatives (Scheme 53) [110].

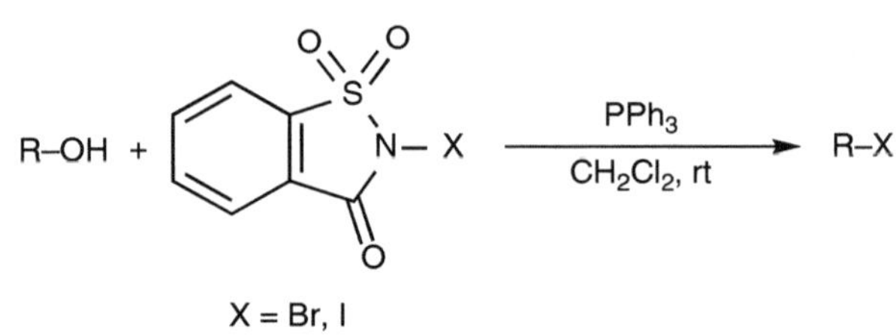

Scheme 49

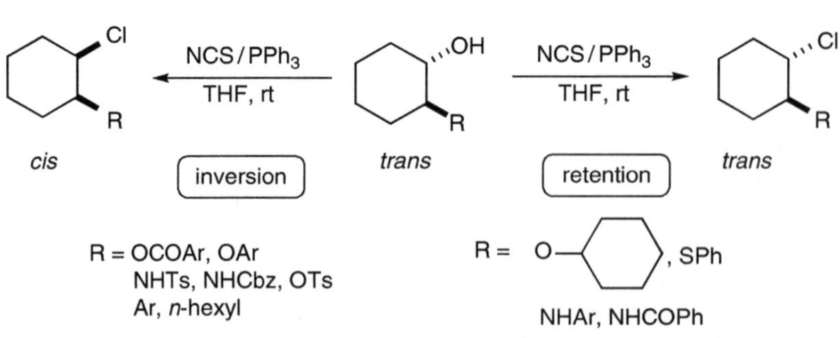

Scheme 50

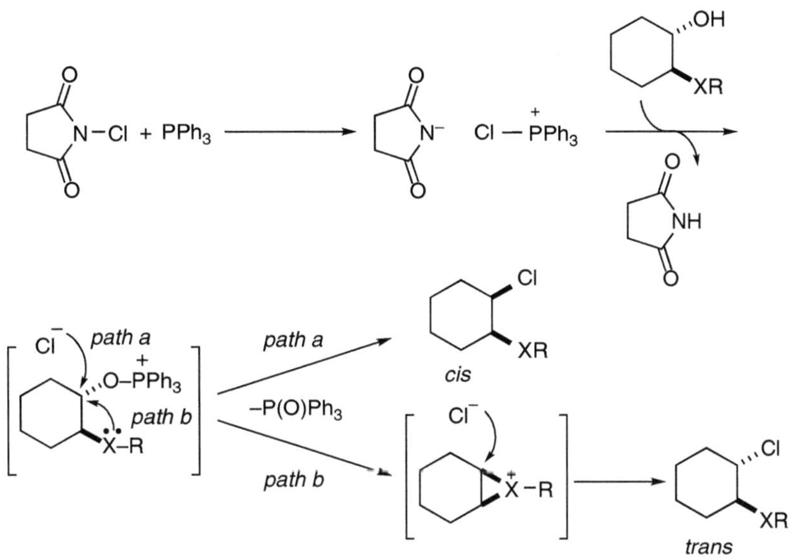

Scheme 51

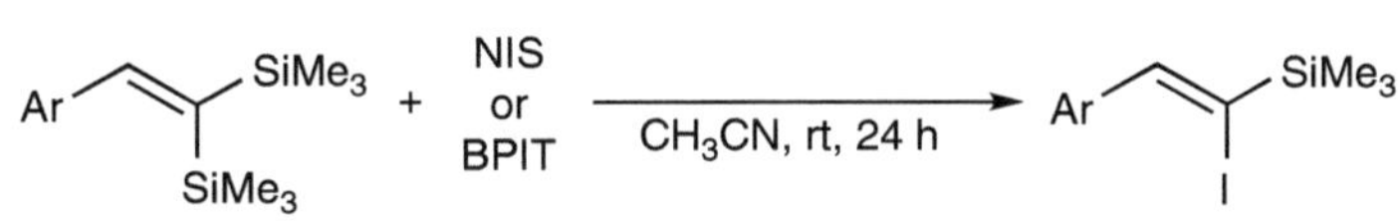

Scheme 52

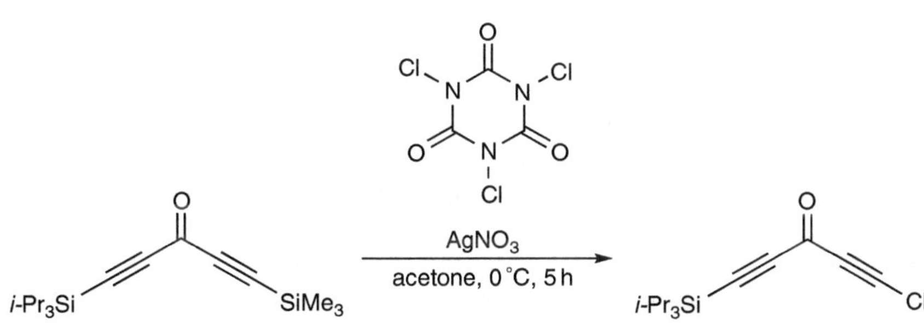

Scheme 53

3.6 Cyclization Reactions

Intra- and intermolecular cyclization reactions using N–X heterocyclic reagents are discussed below in this section. The mode of the reactions is electrophilic or oxidative cyclization and involves halonium or radical species.

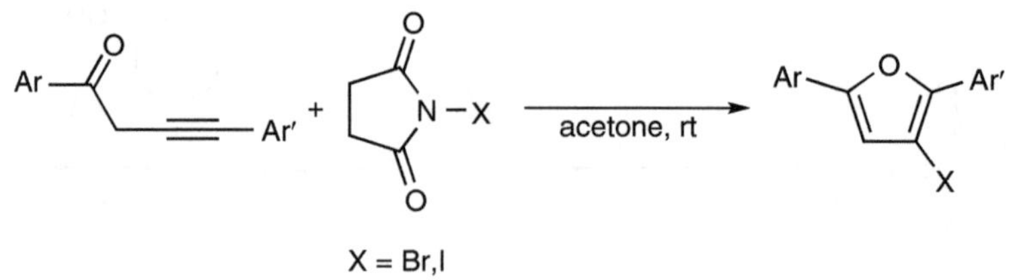

Scheme 54

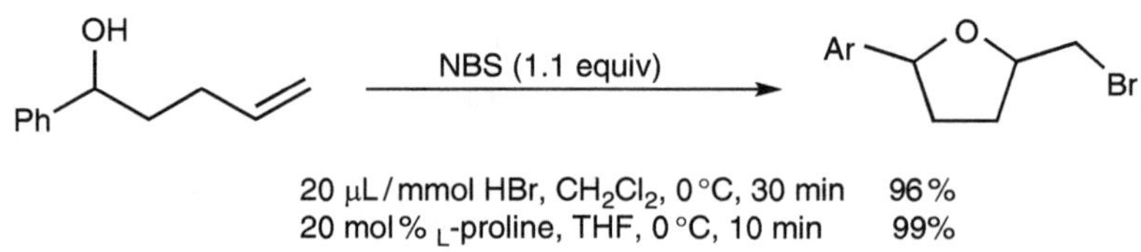

Scheme 55

The 5-*endo-dig* electrophilic cyclization of 1,4-diaryl but-3-yn-1-ones with NBS or NIS, in the absence of base, gave 3-halo-2,5-diarylfuranes (Scheme 54) [111].

The 5-*exo-dig* cyclization of 5-hydroxypentenes was reported to be catalyzed by L-proline using NBS, giving a β-brominated tetrahydrofuran. Although the reaction is also catalyzed by HBr, a catalytic amount of L-proline accelerated the bromocyclization (Scheme 55) [112].

As methods involving radical species in the cyclization step, the following two procedures were developed by Togo's group. 3,4-Dihydro-2,1-benzothiazine 2, 2-dioxides were synthesized by the reaction of *N*-methyl-2-arylethanesulfonamides with DIDMH under irradiation conditions. The reaction proceeds through the formation of an N–I bond in sulfonamides with DIDMH and the generation of sulfonamidyl radicals via the hemolytic cleavage of the N–I bond induces cyclization, followed by oxidation to aromatic rings with DIDMH (Scheme 56) [113].

The reaction system was applied to the preparation of chroman derivatives from 3-aryl-1-propanols using DIDMH under irradiation conditions [114]. A potential intermolecular cyclization of alkenes and NCSac was developed. Namely, NCSac was shown to undergo electrophilic Ritter-type reactions with alkenes in acetonitrile, and the resulting β-chloro sulfonylamidines could be ring opened and cyclized to give imidazolines (Scheme 57) [115].

A highly efficient procedure for the conversion of alkenes and CO_2 into cyclic carbonates in water using NBS and DBU was developed by Li's group. In the case of styrene, the reaction proceeded excellently, affording styrene carbonate in 89% yield after a 3 h reaction. Based on the above results, as a more environmentally benign process, a catalytic amount of bromide ion together with aqueous H_2O_2 instead of NBS could be used to produce cyclic carbonates (Scheme 58) [116].

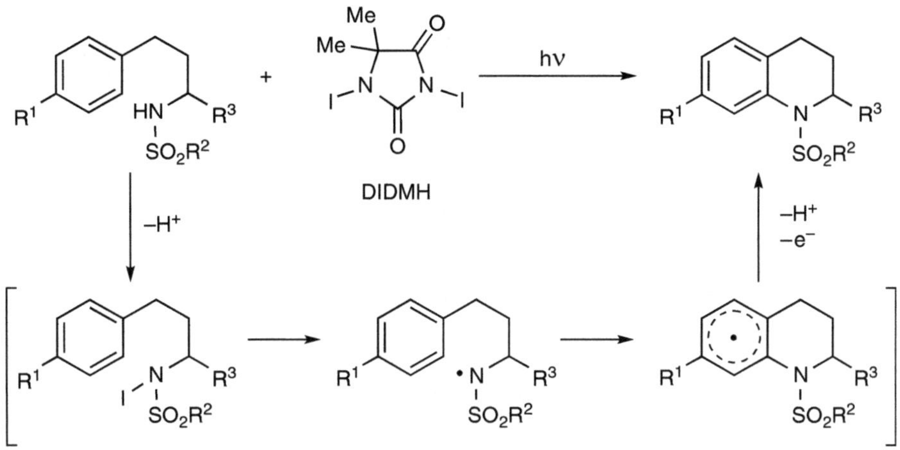

Scheme 56

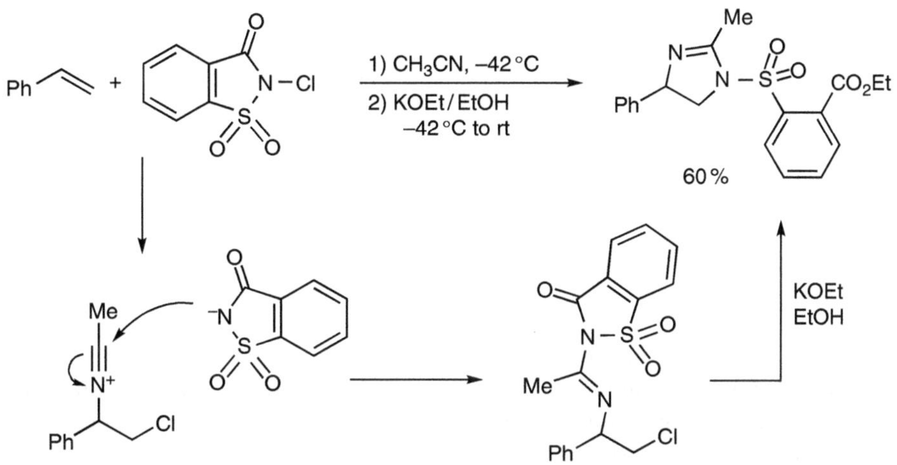

Scheme 57

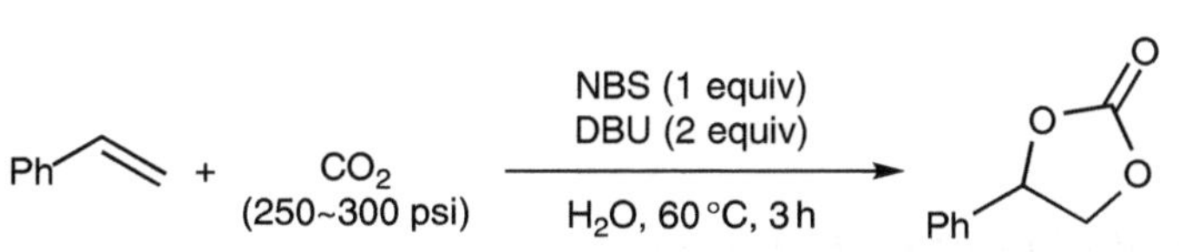

Scheme 58

The one-pot synthesis of oxazolines based on the haloamidation of alkenes with NBS and a stoichiometric amount of water was revealed. The reaction is catalyzed by the Lewis acid, $Cu(OTf)_2$ (Scheme 59) [117].

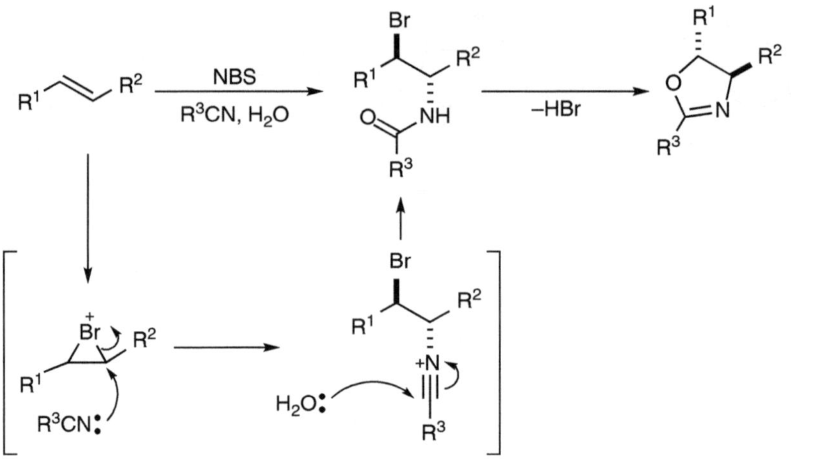

Scheme 59

Scheme 60

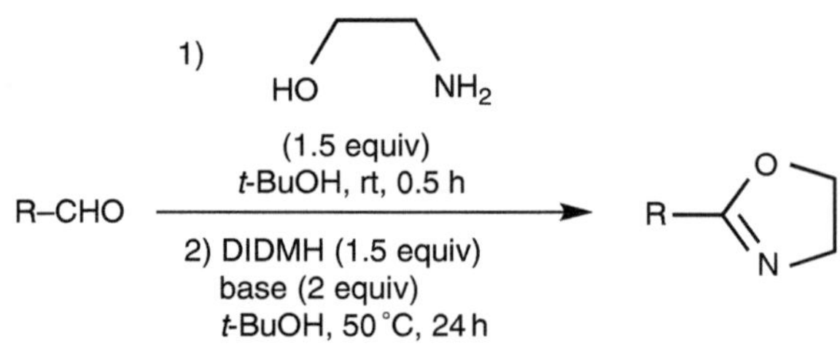

As an alternative method for the synthesis of 2-oxazolines, the reaction of aldehydes with 2-aminoethanol is well known, and DIDMH was recently found to be good reagent for the reaction. 2-Aryl-2-oxazoliness could be prepared in good yields by the reaction of aromatic aldehydes with 2-aminoethanol using DIDMH. Aliphatic aldehydes were also applicable for the reaction. The method has been used for the synthesis of chiral 2-substituted 2-oxazolines (Scheme 60) [118].

3.7 *Amidation of C–H Bonds*

Although many methods are available for the amidation of benzylic sp^3 C–H bonds, the substituent on the nitrogen is restricted to a tosyl group because the nitrogen sources are PhI=NTs or chloramines-T. In order to overcome this drawback, an impressive and air-stable $FeCl_2$/NBS catalyst/oxidant system for catalyzing the

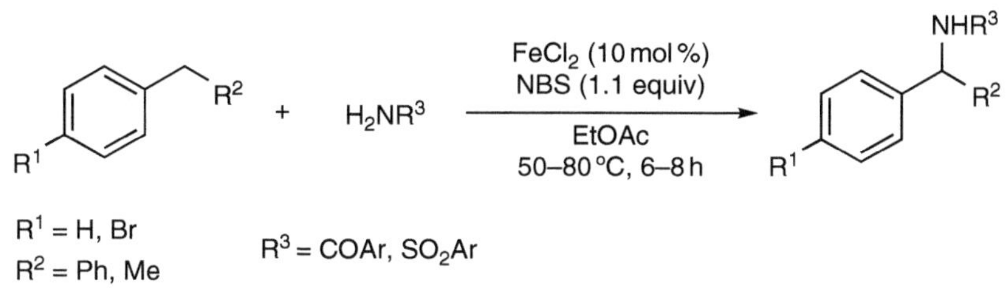

Scheme 61

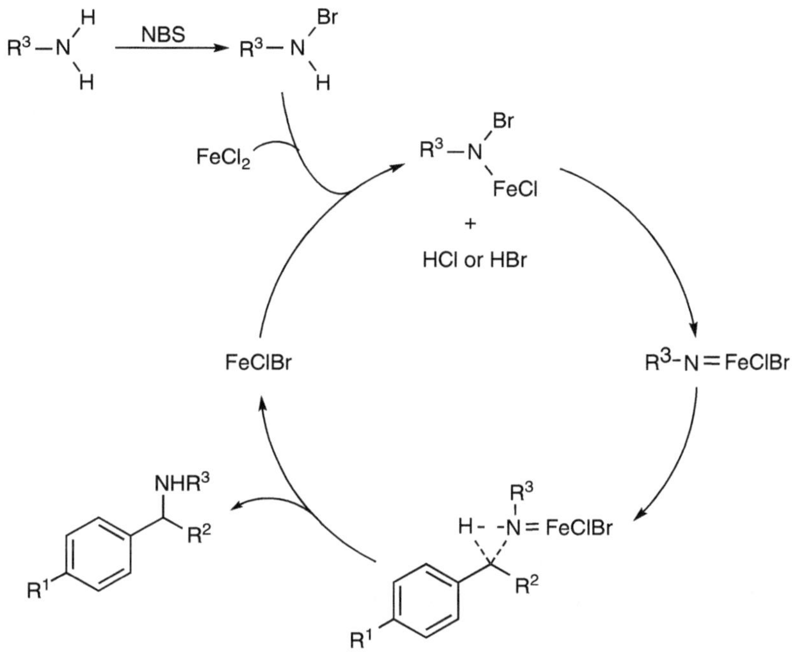

Scheme 62

amidation of the benzilic position was developed by Fu's group, in which benzamide and arylsulfonamides were applicable for the reaction (Scheme 61).

The reaction of sulfonamide with NBS in EtOAc gave *N*-bromosulfonamide. Diphenylmethane and $FeCl_2$ were added to the resulting solution, leading to the production of the amidation product in 60% yield. The following reaction pathway was proposed. The reaction of NBS with the amide produces N-brominated amide **A**, treatment of **A** with $FeCl_2$ give **B**, which is then transferred into the iron–nitrene complex **C**. The reaction of **C** with benzylic C–H bonds forms intermediate **D**, and removal of the iron salt in **D** affords the product (Scheme 62) [119].

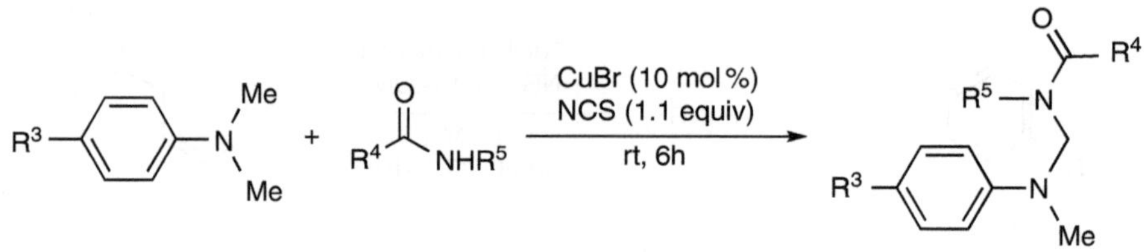

Scheme 63

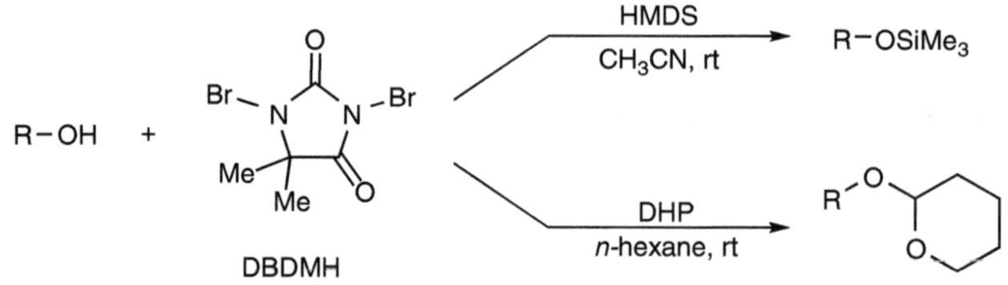

Scheme 64

The use of a copper catalyst (CuBr) instead of $FeCl_2$ could expand the scope of the reaction, i.e., amidation of C–H bonds adjacent to a nitrogen atom. In this reaction, NCS was found to be relatively superior to NBS (Scheme 63) [120].

Metal catalysts are required for the above reaction system, but the combination of sulfonamide and DIDMH could realize the sulfonylamidation of alkylbenzenes at the benzylic position [121].

3.8 Protection and Deprotection Reactions

N–X Heterocyclic reagents serve as good agents in protection and deprotection reactions. Trimethylsilylation and tetrahydropyranylation are popular methods that are in widespread use for the protection of hydroxyl groups of alcohols.

DBDMH was used in the efficient and chemoselective trimethylsilylation and tetrahydropyranylation of benzylic and primary and secondary aliphatic alcohols under mild conditions (Scheme 64).

Most alcohols that are protected with HMDS are completely consumed within 1 min, giving the desired silyl ethers in excellent yields. The reaction of alcohols with dihydropyran also proceeded within 5–60 min [122].

For deprotection reactions of amines protected with a *p*-methoxyphenyl (PMP) group, TCCA was found to be effective in liberating the amine. The authors found that periodic acid (H_5IO_6) also functions as an oxidant in the deprotection of PMP groups (Scheme 65) [123].

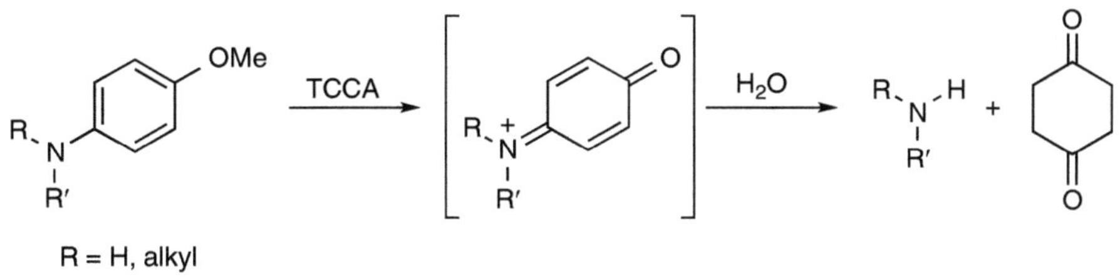

Scheme 65

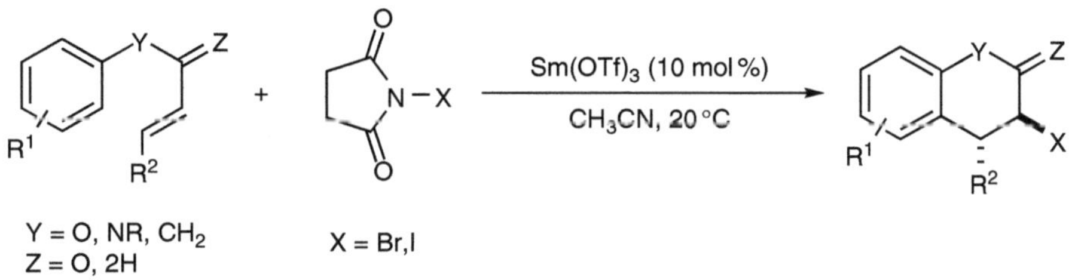

Scheme 66

3.9 Iodocyclization Reactions

Intramolecular halocyclization, especially iodocyclization, is a powerful tool for the construction of ring systems. NIS and BPIT are potential iodonium ion sources for the cyclization.

The Lewis acid, $Sm(OTf)_3$, which catalyze the intramolecular halo-arylation of tethered alkenes was developed using *N*-halosuccinimide (NBS and NIS). The catalytic method permitted the regio- and stereoselective synthesis of annulated arene heterocycles and carbocycles such as 2-chromanones, chromans, tetrahydroquinolines, and tetralins containing a halofunctionality (Scheme 66) [124].

Intra- and intermolecular iodoarylation reactions of allenes using BPIT and HBF_4 were discovered by Barluenga and co-workers. This approach permitted some new, basic frameworks to be prepared (Scheme 67) [125].

As an alternate iodocarbocyclization, an electrophilic 6-*exo* and 5-*endo* cyclization starting from 1,5-enynes was independently reported. Kirsch et al. reported that 1,5-enynes are suitable substrates in iodonium-induced carbocyclization reactions, resulting in the formation of six-membered carbocycles. Treatment of a variety of 1,5-enynes with NIS (3 equiv) under aerobic conditions gave iodobenzenes (Scheme 68) [126].

At almost the same time, Toullec and Michelet reported on the electrophilic 5-*endo*-selective iodocarbocyclization of 1,5-enynes. In this reaction, 1,1-dimethyl substituents are important for realizing this unprecedented 5-*endo* process.

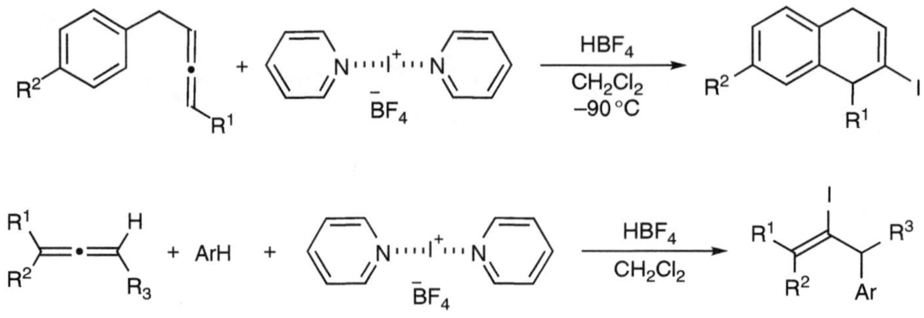

Scheme 67

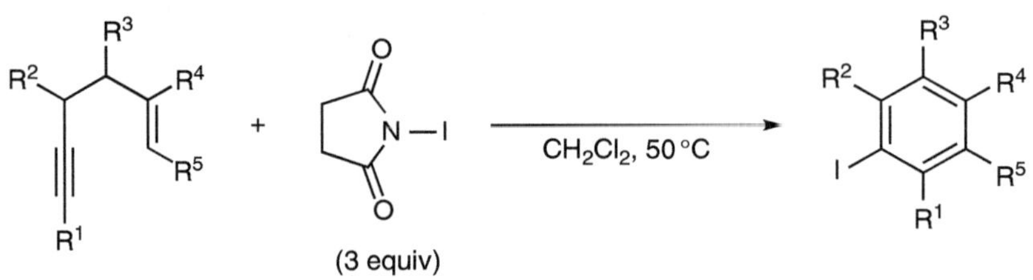

Scheme 68

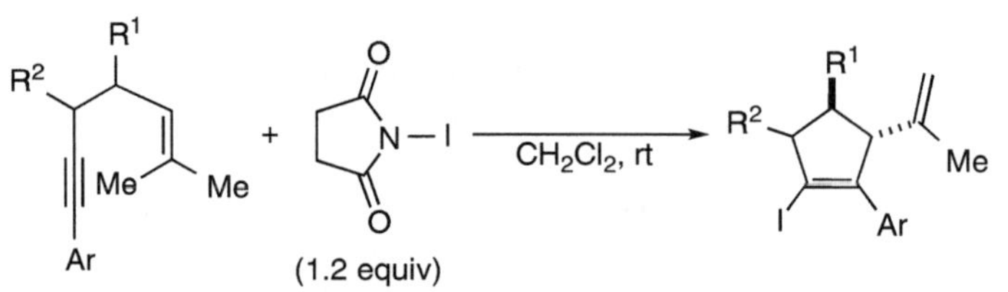

Scheme 69

The cyclization is also promoted by NIS and was found to be completely diastereoselective (Scheme 69) [127].

The iodoaminocyclization of noncyclic N-protected alkenylamines has been studied extensively, but the reaction of N-nonprotected alkenylamines has received only limited attention. The NIS-promoted cyclization of bishomoallylic secondary amines containing a 2-allyl-*N*-benzylcyclohexylamine unit was achieved by Diaba and Bonjoch [128].

As a unique iodoaminocyclization, switchable access to dihydropyrazoles and pyrazoles was developed from hydrazides with a tethered alkynyl moiety by reagent-controlled iodocyclization. When $I(coll)_2PF_6$ (BCIH) was employed in reactions with hydrazides, dihydropyrazoles were selectively produced. In contrast

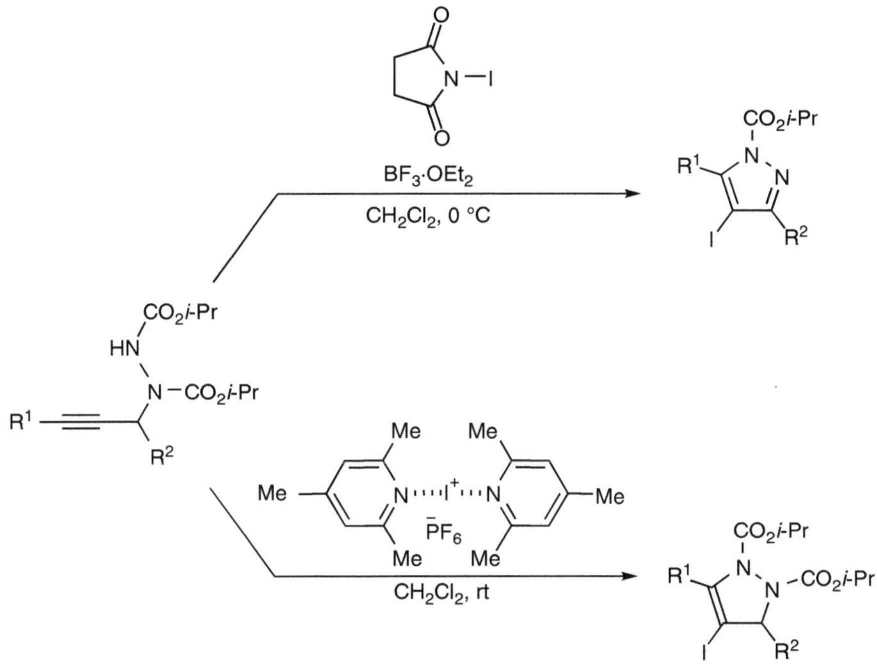

Scheme 70

to this phenomenon, the use of NIS and $BF_3{\cdot}OEt_2$ afforded only pyrazoles (Scheme 70) [129].

3.10 Asymmetric Halogenation

Asymmetric halogenation reactions using an N–X heterocyclic reagent are discussed below. The reaction pattern can be divided into enantioselective iodolactonization, oxidative kinetic resolution and oxidative asymmetric desymmetrization, asymmetric bromoamidation, and asymmetric halogenation α to a carbonyl group.

Reagent-induced asymmetric halolactonization reactions have been investigated since 1992. In most cases, only poor to moderate enantioselectivities were obtained. The salen-Co(II) complex was reported to catalyze the enantioselective iodolactonization of 4-pentenoic acid derivatives by Cao's group. The method afforded the highest enantioselectivity among the currently known reagent-controlled iodolactonization protocols (Scheme 71) [130].

The asymmetric oxidation of 1,2-diols using NBS with copper(II) triflate and (*R*, *R*)-Ph-Box was exploited by Onomura and co-workers. When a meso-1,2-diol

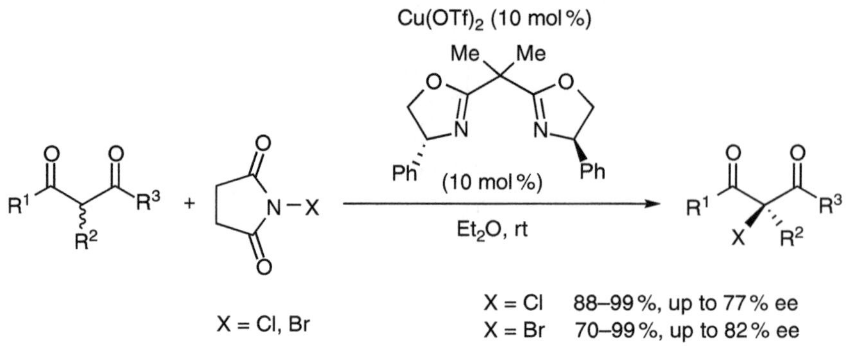

Scheme 71

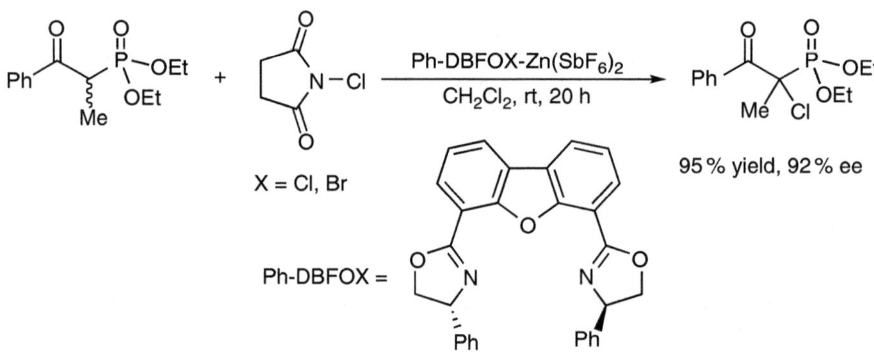

Scheme 72

was employed in the reaction, effective desymmetrization was induced to give the (*R*)-benzoin in good enentioselectivity (Scheme 72).

This method could be applied to the kinetic resolution of dl-hydrobenzoin. In this reaction, the (*S*)-benzoin was obtained along with the optically active hydrobenzoin (Scheme 73) [131].

Onomura's group also developed a method for the unique kinetic resolution of racemic aminoaldehydes by oxidation with NIS. Racemic aminoaldehydes were converted into chiral amino acid methyl esters and aminoaldehyde dimethyl acetals [132].

The first catalytic regio- and enantioselective bromoamination of chalcones by chiral N,N′-dioxide/scandium(II) complexes was achieved by Feng and co-workers. Treatment of chalcone derivatives with sulfonamides and NBS in the presence of a catalytic amount of a chiral ligand and $Sc(OTf)_3$ afforded β-amido-α-bromoketones in excellent yields and enantio- and diastereoselectivities (Scheme 74) [133].

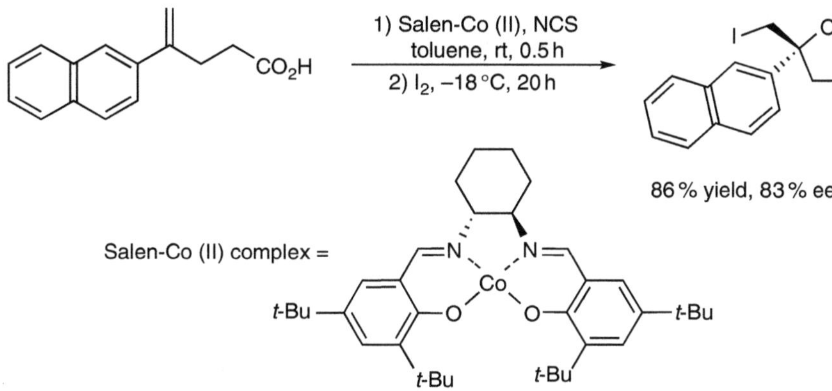

Scheme 73

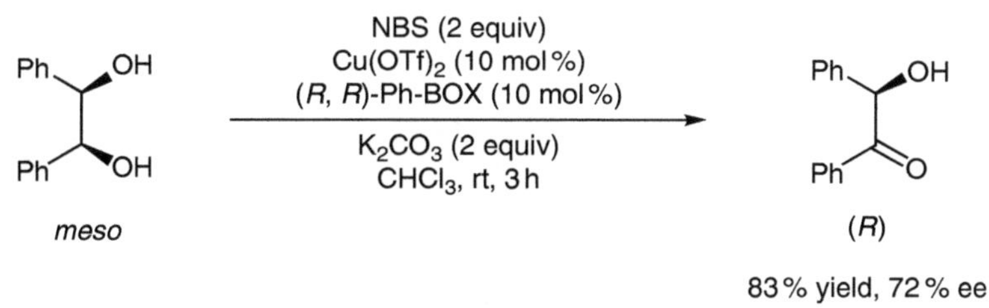

Scheme 74

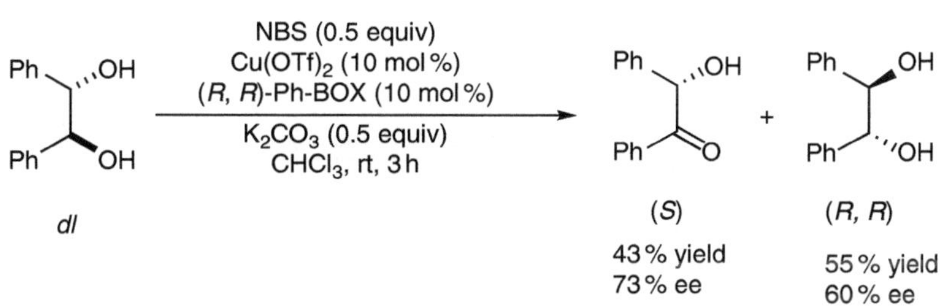

Scheme 75

There are many reports of the asymmetric halogenation at the α-position of carbonyl compounds. Most involve the use of metal catalysts, but organocatalytic reactions have recently been a subject of extensive investigation. In the case of recent metal catalyzed asymmetric halogenations, Jørgensen and co-workers utilized chiral bisoxazoline ligands for the bromination and chlorination of β-ketoesters and β-keto phosphonates. The asymmetric α-halogenation of β-ketoesters proceeded with enantioselectivity, when chiral bisoxazoline copper (II) complexes and NBS or NCS was used (Scheme 75) [134].

Good enantioselective chlorination of β-keto phosphonates was achieved in the case of the $Zn(SbF_6)_2$ catalyst and Ph-DBFOX ligand using NCS (Scheme 76) [135].

The first organocatalytic (C_2-symmetric diphenylpyrrolidine catalyst) enantioselective α-bromination of aldehydes and ketones was reported in 2005. However, in this reaction, 4,4-dibromo-2,6-di-*tert*-butyl-cyclohexa-2,5-dione, which is not N–X heterocyclic reagent, was used as the bromonium source for inducing excellent enantioselectivity [136].

The direct asymmetric iodination of aldehydes using a bifunctional amino alcohol organocatalyst was developed by Maruoka's group. This method represents a rare example of the catalytic and highly enantioselective synthesis of optically active α-iodoaldehydes (Scheme 77) [137].

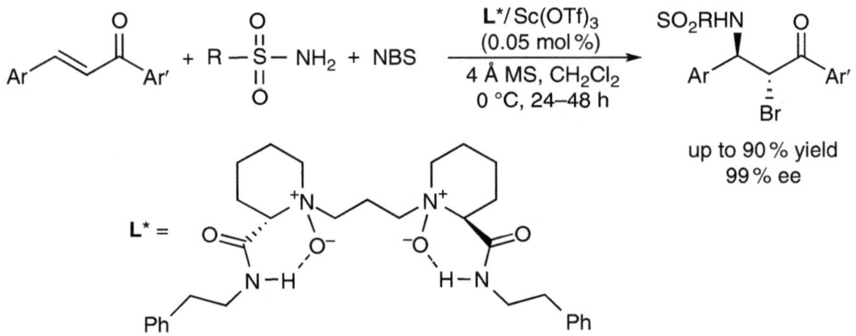

Scheme 76

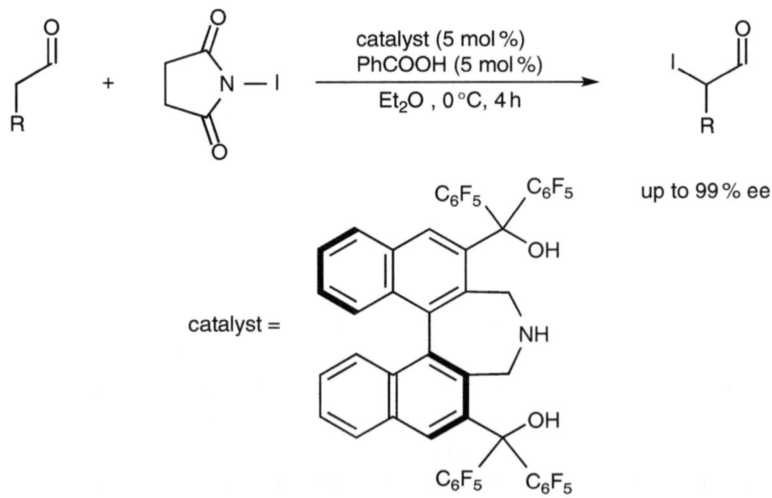

Scheme 77

(*S*)-Pyrrolidine-thiourea was found to act as bifunctional organocatalyst for the enantioselective α-chlorination of aldehydes using NCS as a chloronium source. The catalyst could be easily recovered by fluorous solid-phase extraction in good yield and high purity (Scheme 78) [138].

The N,N′-dioxide ligand described in this section for the asymmetric bromoamination of chalcones was used for the enantioselective α-chlorination of cyclic β-ketoesters. Although a minor tuning of the catalyst was required, highly efficient asymmetric chlorination could be realized (Scheme 79) [139].

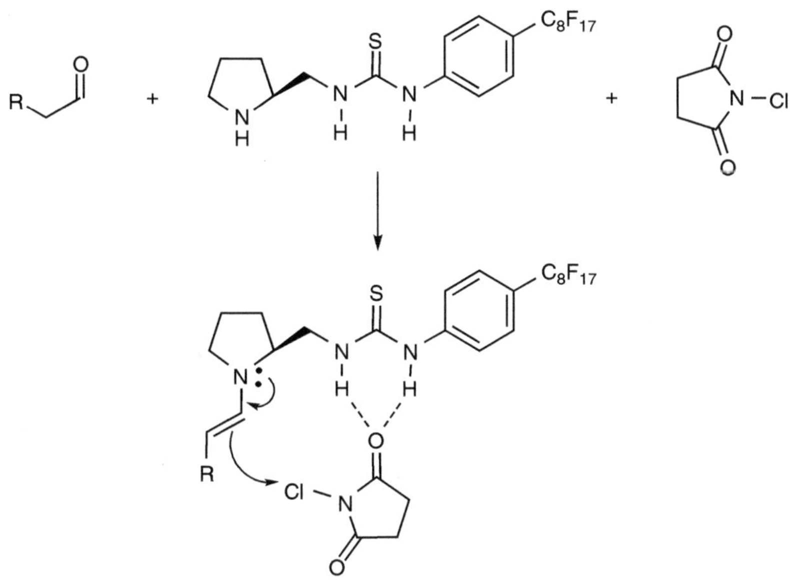

Scheme 78

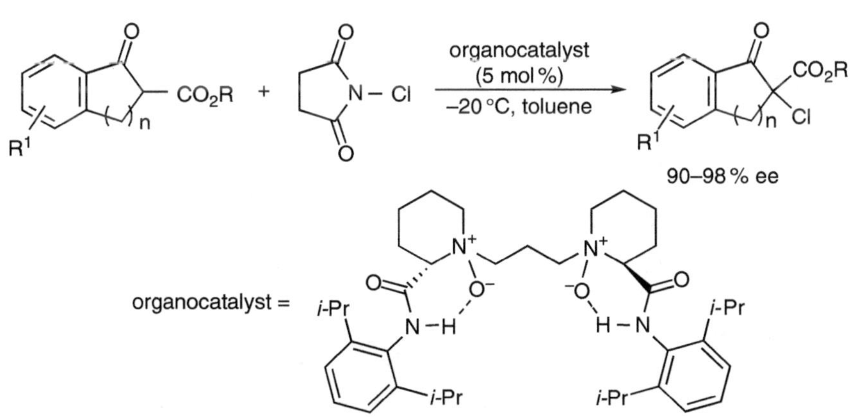

Scheme 79

3.11 NBS-Catalyzed Reactions

In this section, recent development of unique NBS-catalyzed reactions is discussed. In 2005, Sudalai and co-workers reported that NBS efficiently catalyzed the hydroamination and hydroalkoxylation of electron-rich styrene derivatives using tosylamides, carbamates, and alcohols as nucleophiles to give amino and ether derivatives, respectively (Scheme 80).

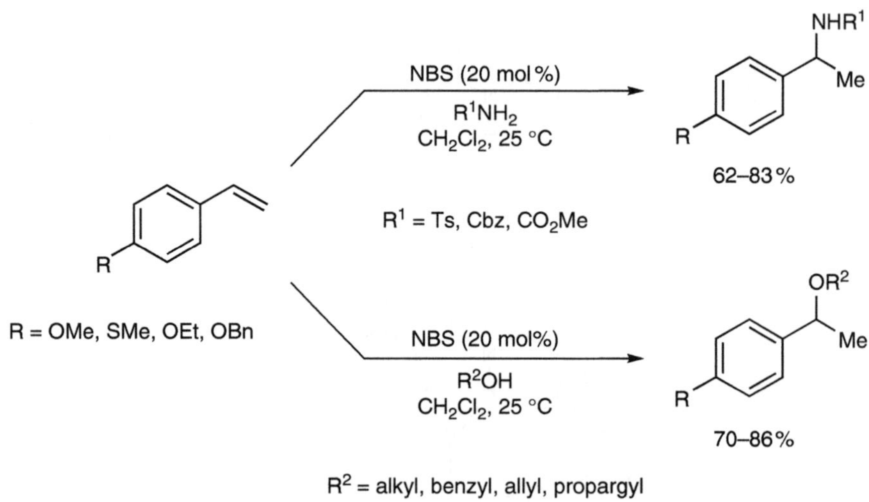

Scheme 80

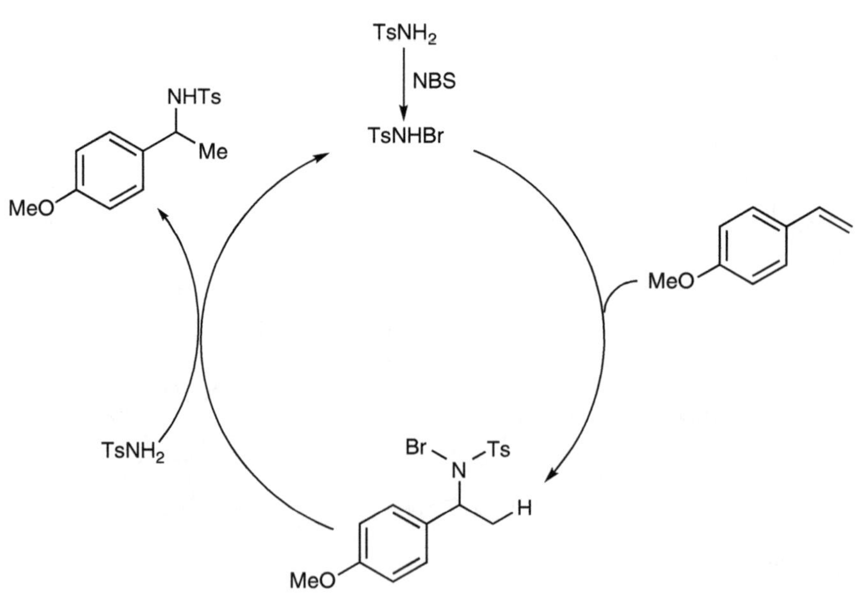

Scheme 81

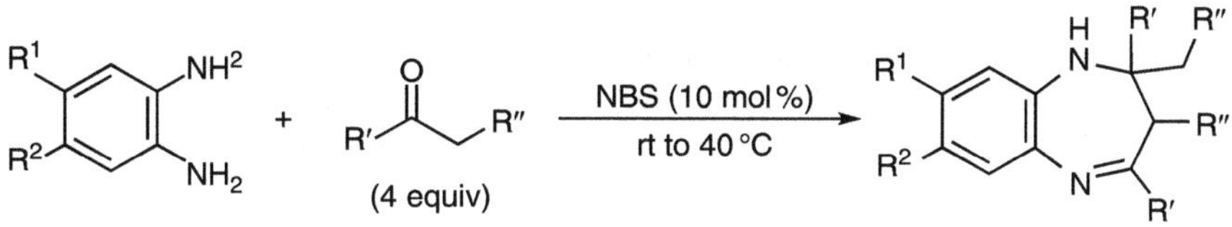

Scheme 82

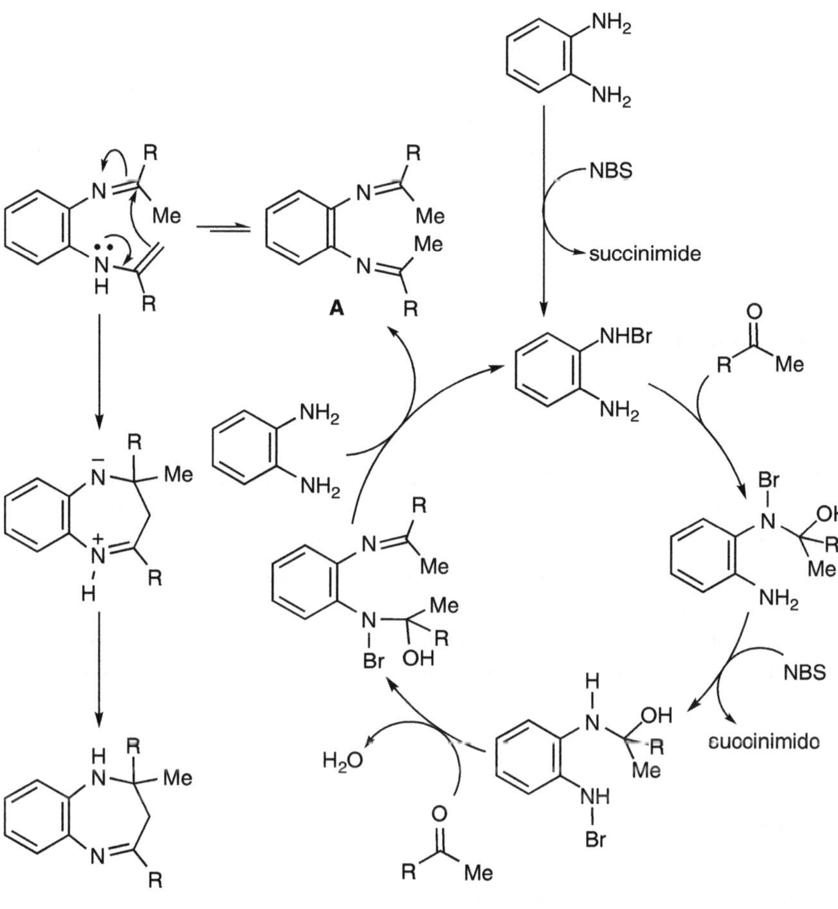

Scheme 83

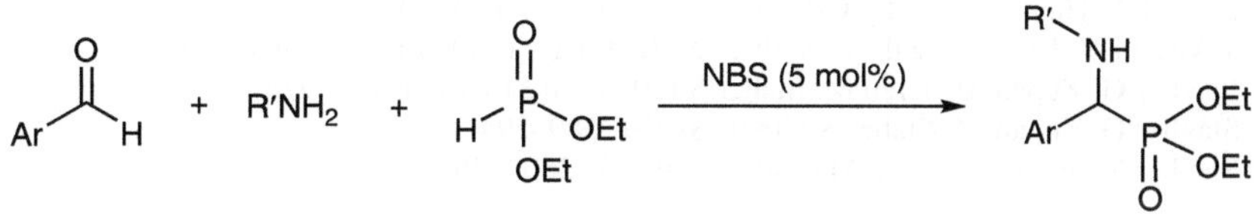

Scheme 84

Tosylamide ($TsNH_2$) reacted with NBS to generate TsNHBr, the subsequent interaction between TsNHBr and 4-methoxystyrene leads to the protonation of a styrene double bond in a Markovnikov fashion. Finally, the regeneration of the species TsNHBr takes place by the reaction of intermediate **A** with $TsNH_2$, giving the desired product (Scheme 81) [140].

The synthesis of 1,5-benzodiazepine was also found to be catalyzed by NBS under mild conditions. NBS catalyzed the condensation of aromatic as well as aliphatic ketones with substituted *o*-phenylenediamines (Scheme 82).

The mechanism for the condensation reaction involves an intermolecular imine–enamine cyclization promoted by NBS. The amino group of *o*-phenylenediamine is first activated by NBS and then attacks the carbonyl group of the ketone, leading to the formation of diimine **A**. An isomeric enamine **B** then undergoes cyclization to afford the seven-membered ring (Scheme 83) [141].

Three-component reactions of aromatic aldehydes, amines, and diethyl phosphite catalyzed by NBS gave a-amino phosphonates under solvent-free conditions (Scheme 84) [142].

4 Conclusions

This review provides a broad overview of synthetic utility of N–X heterocyclic reagents. The versatility and potential of these reagents can be further enhanced and modified by a combination of Lewis acids and bases, oxidants, metal catalysts, and optically active compounds. These reagents have multiple reactivity such as electrophilic and nucleophilic properties and free-radical reactivity depending on the substrates, solvent, catalysts, additives, and reaction conditions that are employed. We strongly hope that the present review with up to date information on this important subject will provide encouragement for active researchers in this field to develop and apply N–X heterocyclic reagents in synthetic organic reactions in the future [143].

References

1. Zhou C, Ma Z, Gu Z, Fu C, Ma S (2008) J Org Chem 73:772–774
2. Wilkinson SC, Lozano O, Schuler M, Pacheco MC, Salmon R, Gouverneur V (2009) Angew Chem Int Ed 48:7083–7086
3. Banks RE, Lawrence NJ, Popplewell AL (1994) J Chem Soc Chem Commun 343–344
4. Davis FA, Han W (1991) Tetrahedron Lett 32:1631–1634
5. Davis FA, Han W, Murphy CK (1995) J Org Chem 60:4730–4737
6. Takeuchi Y, Liu Z, Suzuki E, Shibata N, Kirk KL (1999) J Fluor Chem 97:65–67
7. Stavber G, Zupan M, Jereb M, Staber S (2004) Org Lett 6:4973–4976
8. Stavber G, Zupan M, Staber S (2009) Synlett 589–594
9. Stavber S, Jereb M, Zupan M (2002) Synthesis 2609–2615

10. Tsushima T, Kawada K, Tsuji T (1982) J Org Chem 47:1107–1110
11. Stavber S, Zupan M (1996) Tetrahedron Lett 37:3591–3594
12. Peng W, Shreeve JK (2005) J Org Chem 70:5760–5763
13. Pravst I, Zuoan M, Stavber S (2005) Synthesis 3140–3146
14. Lodhmani-Khouzani H, Poorheravi MR, Sadeghi MMM, Caggiano L, Jackson RFW (2008) Tetrahedron 64:7419–7425
15. Lodhmani-Khouzani H, Hajiheidari D (2010) J Fluor Chem 131:561–569
16. Snieckus V, Beaulieu F, Mohri K, Han W, Murphy CK, Davis FA (1994) Tetrahedron Lett 35:3465–3468
17. Hull KL, Anani WQ, Sanford MS (2006) J Am Chem Soc 128:7134–7135
18. Furuya T, Martin H, Ritter T (2008) Angew Chem Int Ed 47:5993–5996
19. Furuya T, Ritter T (2008) J Am Chem Soc 130:10060–10061
20. Furuya T, Ritter T (2009) Org Lett 11:2860–2863
21. Furuya T, Strom AE, Ritter T (2009) J Am Chem Soc 131:1662–1663
22. Carroll L, Pacheco C, Garcia J, Gouverneur V (2006) Chem Commun 4113–4115
23. Zhou C, Li J, Lu B, Fu C, Ma S (2008) Org Lett 10:581–583
24. Differding E, Lang RW (1988) Tetrahedron Lett 29:6087–6090
25. Bravo P, Resnati G (1990) Tetrahedron Asymmetry 10:661–692
26. Davis FA, Zhou P, Murphy CK (1993) Tetrahedron Lett 34:3971–3974
27. Davis FA, Zhou P, Murphy CK, Sundarababu G, Qi H, Przeslawski RM, Chen B-C, Carroll PJ (1998) J Org Chem 63:2273–2280
28. Takeuchi Y, Suzuki T, Satoh A, Shiragami T, Shibata N (1999) J Org Chem 64:5708–5711
29. Hintermann L, Togni A (2000) Angew Chem Int Ed 39:4359–4362
30. Piana S, Devillers I, Togni A, Rothlisberger U (2002) Angew Chem Int Ed 41:979–982
31. Frantz R, Hintermann L, Perseghini M, Broggini D, Togni A (2003) Org Lett 5:1709–1712
32. Suzuki S, Furuno H, Yokoyama Y, Inanaga J (2006) Tetrahedron Asymmetry 17:504–507
33. Cahard D, Audouard C, Plaquevent J-C, Roques N (2000) Org Lett 2:3699–3701
34. Mohar B, Baudoux J, Plaquevent J-C, Cahard D (2001) Angew Chem Int Ed 40:4214–4216
35. Shibata N, Suzuki E, Takeuchi Y (2000) J Am Chem Soc 122:10728–10729
36. Shibata N, Suzuki E, Asahi T, Shiro M (2001) J Am Chem Soc 123:7001–7009
37. Fukuzumi T, Shibata N, Sugiura M, Nakamura S, Toru T (2006) J Fluor Chem 127:548–551
38. Shibata N, Yasui H, Nakamura S, Toru T (2007) Synlett 1153–1157
39. Ibrahim H, Togni A (2004) Chem Commun 1147
40. Ma J-M, Cahard D (2004) Chem Rev 104:6119–6146
41. Ye C, Shreeve JM (2004) J Org Chem 69:8561–8563
42. Pavlinac J, Zupan M, Stavber S (2006) J Org Chem 71:1027–1032
43. Yadav JS, Reddy BVS, Reddy YJ (2007) Tetrahedron Lett 48:7034–7037
44. Bew SP, Fairhurst SA, Hughes DL, Legentil L, Liddle J, Pesce P, Nigudkar S, Wilson MA (2009) Org Lett 11:4552–4555
45. Melhado AD, Brenzovich WE Jr, Lackner AD, Toste FD (2010) J Am Chem Soc 132:8885–8887
46. Brenzovich WE Jr, Benitez D, Lackner AD, Shunatona HP, Tkatchouk E, Goddard WA III, Toste FD (2010) Angew Chem Int Ed 49:5519–5522
47. Zhang G, Peng Y, Cui L, Zhang L (2009) Angew Chem Int Ed 48:3112–3115
48. Zhang G, Cui L, Wang Y, Zhang L (2010) J Am Chem Soc 132:1474–1475
49. Yang D, Yan Y-L, Lui B (2002) J Org Chem 67:7429–7431
50. Meshram HM, Reddy PN, Sadashiv K, Yadav JS (2005) Tetrahedron Lett 46:623–626
51. Cui H-F, Dong K-Y, Nie J, Zheng Y, Ma J-A (2010) Tetrahedron Lett 51:2374–2377
52. Pravst I, Zupan M, Stavber S (2006) Tetrahedron Lett 47:4707–4710
53. Arbuj SS, Waghmode SB, Ramaswamy AV (2007) Tetrahedron Lett 48:1411–1415
54. de Almeida, LS, Esteves PM, de Mattos MCS (2006) Synthesis 221–223
55. Yamamoto T, Toyota K, Morita N (2010) Tetrahedron Lett 51:1364–1366
56. Chhattise PK, Ramaswamy AV, Waghmode SB (2008) Tetrahedron Lett 49:189–194

57. Arsenyan P, Paegle E, Belyakov S (2010) Tetrahedron Lett 51:205–208
58. Zhang Y, Shibatomi K, Yamamoto H (2005) Synlett 2837–2841
59. Schmid H (1946) Helv Chim Acta 29:1144–1151
60. Roberts JC, Roffey P (1966) J Chem Soc C 160–163
61. Lambert FL, Ellis WD, Parry RJ (1965) J Org Chem 30:304–306
62. Coleman RS, Grant EB (1991) J Org Chem 56:1357–1359
63. Paul V, Sudalai A, Daniel T, Srinjvasan KV (1994) Tetrahedron Lett 35:7055–7056
64. Zanka A, Kubota A (1999) Synlett 1984–1986
65. Tanemura K, Suzuki T, Nishida Y, Satsumabayashi K, Horaguchi T (2003) Chem Lett 32:932–933
66. Olah GA, Wang Q, Sandford G, Prakash GKS (1993) J Org Chem 58:3194–3195
67. Prakash GKS, Mathew T, Hoole D, Esteves PM, Wang Q, Rasul G, Olah GA (2004) J Am Chem Soc 126:15770–15776
68. Mo F, Yan JM, Qiu D, Li F, Zhang Y, Wang J (2010) Angew Chem Int Ed 49:2028–2032
69. Kalyani D, Dick AR, Anani WQ, Sanford MS (2006) Org Lett 8:2523–2526
70. Wan X, Ma Z, Li B, Zhang K, Cao S, Zhang S, Shi Z (2006) J Am Chem Soc 128:7416–7417
71. Chen X, Hao X-S, Goodhue CE, Yu JQ (2006) J Am Chem Soc 128:6790–6791
72. Barluenga J, Álvarez-Gutiérrez JM, Ballesteros A, González JM (2007) Angew Chem Int Ed 46:1281–1283
73. Dolenc D, Sket B (1995) Synlett 327–328
74. Crespo LTC, da S Ribeiro R, de Mattos MCS, Esteves PM (2010) Synthesis 2379–2382
75. Urankar D, Rutar I, Modec B, Dolenc D (2005) Eur J Org Chem 2349–2353
76. Shin C, Yonezawa Y, Unoki K, Toshimura J (1979) Bull Chem Soc Jpn 52:1657–1660
77. Rodrigo da S. Riberio, Esteves PM, de Mattos MCS (2007) Tetrahedron Lett 48:8747–8751
78. Bentley P, Mei Y, Du J (2008) Tetrahedron Lett 49:2653–2655
79. Zhou L, Tan CK, Zhou J, Yeung Y-Y (2010) J Am Chem Soc 132:10245–10247
80. Li G, Kotti SRSS, Timmons C (2007) Eur J Org Chem 2745–2758
81. Hajra S, Sinha D, Bhowmick M (2006) Tetrahedron Lett 47:7017–7019
82. Hajra S, Sinha D, Bhowmick M (2007) J Org Chem 72:1852–1855
83. Wang Z, Zhang Y, Fu H, Jiang Y, Zhao Y (2008) Synlett 2667–2670
84. Zhi S, Han J, Lin C, An G, Pan Y, Li G (2008) Synthesis 1570–1574
85. Wu X-L, Xia J-J, Wang G-W (2008) Org Biomol Chem 6:548–553
86. Wu X-L, Wang G-W (2008) Eur J Org Chem 6239–6246
87. Wu X-L, Wang G-W (2009) Tetrahedron 65:8802–8807
88. Wei J-F, Zhang L-H, Chen Z-G, Shi X-Y, Cao J-J (2009) Org Biomol Chem 7:3280–3284
89. Chen Z-G, Wei J-F, Li R-T, Shi X-Y, Zhao P-F (2009) J Org Chem 74:1371–1373
90. Chen Z-G, Wei J-F, Wang M-Z, Zhou L-Y, Zhang C-J, Shi X-Y (2009) Adv Synth Catal 351:2358–2368
91. Wei J-F, Chen Z-G, Lei W, Zhang L-H, Wang M-Z, Shi X-Y, Ji R-T (2009) Org Lett 11:4216–4219
92. Liu R, Dong C, Liang X, Wang X, Hu X (2005) J Org Chem 70:729–731
93. Zolfigol MA, Shirini F, Choghamarani AG (2006) Synthesis 2043–2046
94. Kuwabara K, Itoh A (2006) Synthesis 1949–1952
95. Iida S, Togo H (2007) Synlett 407–410
96. Iida S, Ohmura R, Togo H (2009) Tetrahedron 65:6257–6262
97. Iida S, Togo H (2007) Tetrahedron 63:8274–8281
98. Veisi H (2010) Synthesis 2631–2635
99. Moorthy JN, Senapati K, Singhal N (2009) Tetrahedron Lett 50:2493–2496
100. Niu M, Fu H, Jiang Y, Zhao Y (2008) Synthesis 2879–2882
101. Li L, Wang J, Zhang G, Liu Q (2009) Tetrahedron Lett 50:4033–4036
102. Surendra K, Krishnaveni NS, Rao KR (2005) Tetrahedron Lett 46:4111–4113
103. Nishiguchi A, Maeda K, Miki S (2006) Synthesis 4131–4134
104. Yadav JS, Reddy BVS, Jain R, Baishya G (2008) Tetrahedron Lett 49:3015–3018

105. Katoh T, Watanabe T, Nishitani M, Ozeki M, Kajimoto T, Node M (2008) Tetrahedron Lett 49:598–600
106. Yin P, Ma W-B, Chen Y, Huang W-C, Deng Y, He L (2009) Org Lett 11:5482–5485
107. Firouzabadi H, Iranpoor N, Ebrahimzadeh F (2006) Tetrahedron Lett 47:1771–1775
108. Jaseer EA, Naidu AB, Kumar SS, Rao RK, Thakur KG, Sekar G (2007) Chem Commun 867–869
109. Pawluc P, Madalska M, Hreczycho G, Marciniec B (2008) Synthesis 3687–3691
110. Vilhelmsen MH, Andersson AS, Nielsen MB (2009) Synthesis 1469–1472
111. Sniady A, Wheeler KA, Dembinski R (2005) Org Lett 7:1769–1772
112. Lee AS-Y, Taso K-W, Chang Y-T, Chu S-F (2007) Tetrahedron Lett 48:6790–6793
113. Moroda A, Furuyama S, Togo H (2009) Synlett 1336–1340
114. Fukuyama S, Togo H (2010) Synlett 2325–2329
115. Booker-Milburn KI, Guly DJ, Cox B, Procopiou PA (2003) Org Lett 5:3313–3315
116. Eghbali N, Li C-J (2007) Green Chem 9:213–215
117. Hajra S, Bar S, Sinha D, Maji B (2008) J Org Chem 73:4320–4322
118. Takahashi S, Togo H (2009) Synthesis 2329–2332
119. Wang Z, Zhang Y, Fu H, Jiang Y, Zhao Y (2008) Org Lett 10:1863–1866
120. Liu X, Zhang Y, Wang L, Fu H, Jiang Y, Zhao Y (2008) J Org Chem 73:6207–6212
121. Baba H, Togo H (2010) Tetrahedron Lett 51:2063–2066
122. Shirini F, Zolfigol MA, Paktinat M (2006) Synthesis 4252–4256
123. Verkade JMM, van Hemert LJC, Quaedflieg PJLM, Alsters PL, van Delft FL, Rutjes FPJT (2006) Tetrahedron Lett 47:8109–8113
124. Hajra S, Maji B, Karmakar A (2005) Tetrahedron Lett 46:8599–8603
125. Barluenga J, Campos-Gómez E, Minatti A, Rodríguez D, González JM (2009) Chem Eur J 15:8946–8950
126. Crone B, Kirsch SF, Umland K-D (2010) Angew Chem Int Ed 49:4661–4664
127. Pradal A, Nasr A, Toullec PY, Michelet V (2010) Org Lett 12:5222–5225
128. Diaba F, Ricou E, Bonjoch J (2007) Org Lett 9:2633–2636
129. Okitsu T, Sato K, Wada A (2010) Org Lett 12:3506–3509
130. Ning Z, Jin R, Ding J, Gao L (2009) Synlett 2291–2294
131. Onomura O, Arimoto H, Matsumura Y, Demizu Y (2007) Tetrahedron Lett 48:8668–8672
132. Minato D, Nagasue Y, Demizu Y, Onomura O (2008) Angew Chem Int Ed 47:9458–9461
133. Cai Y, Liu X, Hui Y, Jiang J, Wang W, Chen W, Lin L, Feng X (2010) Angew Chem Int Ed 49:6160–6164
134. Marigo M, Kumaragurubaran N, Jørgensen KA (2004) Chem Eur J 10:2133–2137
135. Bernardi L, Jorgensen KA (2005) Chem Commun 1324–1326
136. Bertelsen S, Halland N, Bachmann S, Marigo M, Braunton A, Jorgensen KA (2005) Chem Commun 4821–4823
137. Kano T, Ueda M, Maruoka K (2008) J Am Chem Soc 130:3728–3729
138. Wang L, Cai C, Curran DP, Zhang W (2010) Synlett 433–436
139. Cai Y, Wang W, Shen K, Wang J, Hu X, Lin L, Liu X, Feng X (2010) Chem Commun 1250–1252
140. Kumar S, Sudalai A (2005) Org Lett 7:855–857
141. Kuo C-W, More SV, Yao C-F (2006) Tetrahedron Lett 47:8523–8528
142. Wu J, Sun W, Sun X, Xia H-G (2006) Green Chem 8:365–367
143. Orazi OO, Corral RA, Bertorello HE (1965) J Org Chem 30:1101–1104

Top Heterocycl Chem (2012) 27: 185–218
DOI: 10.1007/7081_2011_64

Published online: 3 November 2011

Recent Progress on the Halogen Dance Reaction on Heterocycles

Michael Schnürch

Abstract The decoration of heterocycles with functional groups and substituents is of great importance for many fields of chemistry. In this context, selective functionalizations of certain positions in a given heterocyclic system are highly desirable and sought after. One way to achieve such selectivity is the prefunctionalization of defined positions with halides, which can then be further transformed to other functional groups and substituents. Halogen Dance reactions proved to be a valuable tool and it was demonstrated in many cases that positions in heterocyclic systems can be activated for subsequent reactions by selective migration of a halide in the corresponding position, which are otherwise extremely difficult to access. Within this chapter, the progress in halogen dance reactions of the last decade will be discussed. Most reported examples relied on the well-established halogen dance induced by organic lithium bases such as LDA or LTMP. However, also several unusual halogen migrations under neutral or even acidic conditions have been reported and will be presented.

Keywords Halogen dance · Halogen dance prevention · Halogen migration · Lithium bases

Contents

M. Schnürch
Institute of Applied Synthetic Chemistry, Vienna University of Technology, Getreidemarkt 9-163/OC, Vienna, Austria
e-mail: mschnuerch@ioc.tuwien.ac.at

Abbreviations

BCHD	Base catalyzed halogen dance
Bn	Benzyl
Boc	*t*-Butoxycarbonyl
BuLi	Butyllithium
DMF	Dimethylformamide
E+	Electrophile
HD	Halogen dance
LDA	Lithiumdiisopropylamide
LIDAKOR	Lithium diisopropylamide-potassium *tert*-butoxide
LTMP	Lithium-2,2,6,6-tetramethylpiperidine
LTMP	Lithiumtetramethylpiperidide
NBS	*N*-Bromosuccinimide
NIS	*N*-Iodosuccinimide
PMB	*p*-Methoxybenzyl
pv	Pivaloyl
rt	Room temperature
TBDMS	*t*-Butyl-dimethylsilyl
THF	Tetrahydrofuran
THP	Tetrahydropyran
TIPS	Triisopropylsilyl
TMS	Trimethylsilyl
TMSCl	Trimethylsilylchloride

1 Introduction

Heterocyclic compounds are of immense importance in many fields of chemistry in particular and science in general. Numerous applications of heterocyclic compounds have been reported, for example, as pharmaceuticals, agrochemicals, organic materials, and many more [1–3]. Hence, the synthesis of heterocycles and more importantly substituted heterocycles is a very important and ever growing research field. In the early years of heterocyclic chemistry, the de novo synthesis of the heterocyclic scaffolds was the most important topic and this goal was often achieved using cyclization strategies [2, 3]. However, often enough these cyclization methods suffered from limitations, most importantly, that the synthesis was limited to compounds bearing certain substituents in defined positions due to the nature of the precursors that had to be applied for the cyclization reaction. Additionally, the library synthesis of differently decorated heterocyclic compounds is often a lengthy process using cyclization strategies since the building blocks to be cyclized have to be synthesized as well. In this regard, halogenated heterocycles are extremely important compounds since they offer the possibility to further transform the halide substituent into a wide variety of other functional groups. For example, via metal–halogen exchange and subsequent addition of electrophiles functional groups can be introduced. Additionally, halides serve as leaving groups for transition metal catalyzed cross-coupling [4–6] or C–H activation reactions [7–11] to form new C–C or also C-heteroatom bonds.

To get to a certain substitution pattern on a given heterocyclic scaffold, it is important to introduce halides in the desired position of the specific ring system. Here, halogen dance (HD) reactions can play an important role, since they allow shifting halides into a position that cannot be halogenated directly. By intelligently designing HD processes, unusual substitution patterns are accessible that are extremely difficult to access via other methods. Since the first observation of an HD reaction [12, 13], many more reports of that process were published and the topic repeatedly reviewed [14–18]. Such reactions have been referred to in the literature by different names such as halogen scrambling, halogen migration, halogen isomerization, base-catalyzed halogen dance (BCHD), or HD, which is by now the commonly used term.

Within this chapter, the progress of HD reactions will be critically discussed with a focus on examples from the new millennium. In the first section, the mechanism of the HD reaction will be briefly presented before the following sections will report on specific examples. Besides work on simple heterocyclic systems, a special emphasis will be laid on the application of HD reactions in the synthesis of complex molecules, e.g., natural products.

Before starting the discussion on HD reactions on heterocycles, it has to be emphasized that HD reactions on carbocycles are possible as well. Indeed, some of the most important contributions to elucidate the mechanism were carried out on halogenated benzenes [19–22].

2 Mechanism and Factors Influencing Halogen Dance Reactions

2.1 *Mechanism of Halogen Dance Reactions*

2.1.1 Historical Development

In 1951, Vaitiekunas reacted 2-bromothiophene with sodium acetylide in liquid ammonia. Surprisingly, the expected 2-acetylenethiophene was not obtained, but instead 4-bromothiophene was isolated [12] accompanied by remaining starting material and a third tar-like fraction, which was believed to be a mixture of di- and tribromothiophenes. Looking back at this report with our knowledge on halogen migration reactions, this was the first report of an HD. Although the author tried to elucidate the mechanism in follow up work [13] this task could not be accomplished at that time. Over the next decade, further investigations were undertaken but the mechanism remained unclear [23–26]. In 1961, Gronowitz found that when reacting various bromothiophenes with *n*-BuLi at −70°C, metal–halogen exchange took place exclusively. In contrast, when the reaction solution was warmed to room temperature (rt) it became evident that scrambling of the brominated and lithiated positions occurred and mixtures of products were obtained after carbonylation [27]. Among others, a mechanism was proposed that corresponded to the *homotransmetalation mechanism* later described on pyridines by Queguenier (see Sect. 2.1.2).

Twenty years after the first observation, it was finally Bunnett who was able to elucidate the mechanism. In a series of papers, he described reactions of different halobenzenes with different bases (mainly potassium anilide), which allowed him to propose a mechanism that is still accepted nowadays [19–22]. Due to his findings, an aryne mechanism could be ruled out and alternatively a mechanism was suggested comprising repetitive nucleophilic displacements of phenyl anions on halides by which all observed products could be explained. It was also he who later coined the expression of BCHD [15].

Since Bunnetts investigations were carried out on benzene systems using amide bases, the mechanism shall be explained on a heterocyclic system and under the use of a nowadays commonly applied lithium base to better fit the subject of this review. Since the first HD was reported on thiophene, it is only logic to use this ring system for the explanation of the mechanism. To mention specific reaction conditions is not necessary in this case since conditions can be found in the examples in the following sections.

It has to be mentioned that primarily bromine and iodine undergo HD reactions. Only few examples of HD reactions of chlorine have been observed; however, only low yields or mixtures of products have been reported [28, 29].

2.1.2 Exemplified Mechanism of a HD Reaction

When 2-bromothiophene **1** is reacted with LDA, the initial lithiation takes place in position 3 since LDA is a nonnucleophilic base that cannot undergo metal–halogen exchange and the electronegative bromine directs lithiation in 3-position (Scheme 1). If the so formed 2-bromo-3-lithiothiophene **3** and starting material **1** are present simultaneously in the reaction mixture, these two species can react with each other. In contrast to LDA, **3** can undergo a metal–halogen exchange, which leads to 2,3-dibromothiophene **4** and 2-lithiothiophene **2**. Again these two intermediates can react with each other giving rise to 3-bromo-2-lithiothiophene **5**, the most stable lithiated species and again to the starting material **1**, which can reenter the cycle of migration reactions. Additionally, 3-bromo-2-lithiothiophene **5**

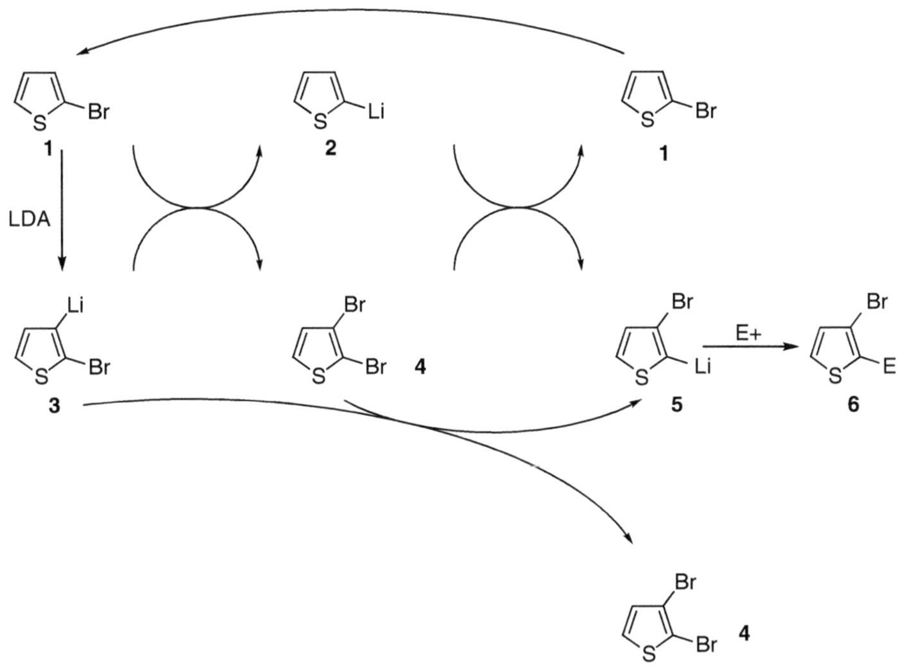

Scheme 1 Mechanism of a HD reaction exemplified on 2-bromothiophene 1

can be produced from the reaction of **3** with **4**. In principle, all lithiated species can react with the unlithiated species present in the reaction mixture. Ultimately, the most stable lithiated intermediate will predominate and can be quenched with electrophiles to give 3-bromo-2-substituted products **6** or can be hydrolyzed simply to 3-bromothiophene.

In most reported cases this mechanism is operable. In recent years an acid catalyzed halide migration has been reported, which is discussed in Sect. 5.

Another way of initiating an HD reaction is via metal–halogen exchange. This method shall be explained on one specific example from the pyridine series (Scheme 2). When 3-bromo-2-chloropyridine **7** was reacted with 0.5 equiv. *n*-BuLi one half of the starting material is converted to 2-chloro-3-lithiopyridine **8** [30]. The substoichiometric amount of base ensures the simultaneous presence of starting material and initially lithiated species. Subsequently, the intermediate **8** and starting material **7** can further react with each other. If they react via a metal–halogen exchange, which is probably the main reaction taking place, the two species simply interconvert into each other. However, also a deprotonation reaction leading to species **9** and **10** can take place. The so obtained 3-bromo-2-chloro-4-lithiopyridine **9** can then undergo a metal–halogen exchange reaction with **7**, which results in the formation of **11** and 3,4-dibromo-2-chloro-pyridine **12**. When then **11** and **12** react with each other, the starting material **7** is regenerated and the most

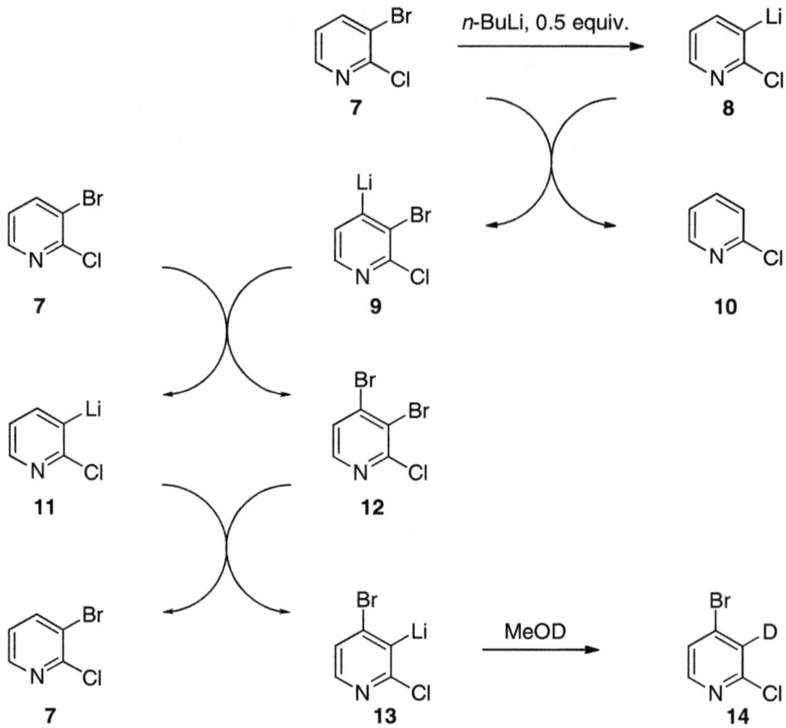

Scheme 2 Homotransmetallation mechanism exemplified on 3-bromo-2-chloropyridine

stable lithiated species, 4-bromo-2-chloro-3-lithiopyridine **13**, which was trapped in this example with MeOD to give **14**.

It has to be emphasized that 0.5 equiv. of the halide-starting material are sacrificed leading to a by-product, in this case 2-chloropyridine. Hence, a maximum yield of 50% HD product based on halide can be obtained. This type of reaction was investigated in detail by Queguiner who termed this pathway *homotransmetallation*.

2.2 *Factors Influencing Halogen Dance Reactions*

There are several reaction parameters that can be altered in order to gain control over an HD reaction. Often, it is also desirable to suppress an HD reaction and also here general guidelines have been proposed that can help to get the desired results. Of course, not always complete control between HD and HD prevention is possible. In the following, the most important parameters and trends for controlling HD reactions are briefly discussed.

2.2.1 Choice of Base

Obviously the choice of base plays a crucial role in the induction of an HD reaction since the reactivity of the base influences the rate of the initial metalation. Hence, switching between bases can lead to different outcomes of the reaction since also the position of lithiation or, more generally, metalation can change when changing the base. The main approach for many years was to induce migrations by K- or Na-amides such as KNH_2, $NaNH_2$, or ArNHK in varying amounts (2–6 equivalents) although principally catalytic amounts (cf. BCHD) are sufficient to cause HD reactions. But such basic reagents have considerable drawbacks. For example, (hetero) arylamines can be formed via exchange reactions or due to the reaction equilibria involved. Additionally, due to the relatively low basicity of amides no complete conversion of starting materials can be achieved resulting in complex product mixtures.

Nowadays, lithiating reagents such as LDA, LTMP, simply *n*-BuLi, and in some cases LIDAKOR (lithium diisopropylamide-potassium tert-butoxide) are the reagents of choice for selective control and prevention of HD. Since these (with the exception of *n*-BuLi) are non-nucleophilic bases, no amino substituents, as in the case of amide bases, are introduced. LDA, LTMP, and LIDAKOR initiate an HD cascade reaction by deprotonation of the most acidic (hetero) aryl-position, the latter (mainly) by a metal–halogen exchange reaction. Independent of the choice of base, the reaction conditions need to be chosen in a way that the simultaneous presence of metalated (or deprotonated) and un-metalated compounds is favored.

2.2.2 Influence of Temperature

Not only the choice of base but also the temperature influences the rate of the initial metalation and hence the outcome of the HD process. At low temperatures, the initial metalation step is slow and such conditions allow the simultaneous presence of metalated and un-metalated starting material. Higher temperatures lead to faster initial lithiation – and considering the mechanism of LDA induced HD reactions – should favor HD prevention since the simultaneous presence of metalated and un-metalated starting material is disfavored. It has to be taken into account that at elevated temperatures the stability of lithiating reagents is of course limited and HD prevention can still be difficult to achieve.

2.2.3 Influence of the Solvent

The reactivity of bases can be tuned by the solvent applied. Hence, also the solvent can play an important role in determining if an HD reaction can proceed or is prevented. LDA, for example, is less reactive in Et_2O than in THF, thus giving a choice for the metalation of aryl protons depending on their acidity. The control and prevention of HD is also influenced by the character of the solvent. Eventually exclusive HD is observed in THF, whereas THP has tendencies to favor HD prevention.

2.2.4 Influence of the Electrophile

When a lithiated species shall be trapped with an electrophile on a system that can potentially give an HD reaction, there are two possibilities. Either the HD cascade is given enough time to form the most stable lithiated intermediate, or this process is prevented by adding an electrophile before a rearrangement can take place. Especially in the latter case, attempts to obtain HD-prevention products, the electrophile plays an important role. As a simplification for the purpose of discussing this topic, we can divide electrophiles in two classes, "fast" reacting and "slow" reacting electrophiles. TMSCl, MeOH, MeOD, ketones, and aldehydes may be considered as rapid electrophiles, whereas DMF, CH_3I, and CH_3SSCH_3 may behave as slow electrophiles.

Highly reactive, ("fast") electrophiles, which remove the lithium intermediate so fast, that the target product formed cannot perform side reactions with other reactive species, favor the formation of HD-prevention products since in such a case, the simultaneous presence of metalated and un-metalated species is prevented. On the other hand, a slow reaction of the metalated species with the electrophile favors exactly the opposite. Of course, complete HD or HD-prevention are only extreme cases. Also mixtures of prevention and migration products may be obtained, although prior to their addition only initially lithiated intermediates had been formed.

2.2.5 Amounts and Order of Reagent Addition

The last parameters that can affect product distribution in HD reactions are the amounts and order of reagent addition. To get to HD products, it is necessary that metalated and un-metalated species are present simultaneously. This can be guaranteed by slowly adding the base to the solution of the starting material. Alternatively, rapid addition of starting compound to the lithium base favors halogen migration, provided that the initial lithiation is not too rapid. Otherwise substantial amounts of prevention product can be formed. In some cases also mixing of starting compound with 80% of the required amount of LDA, followed by the residual 20% after ca. 30 min can be successful in obtaining HD products.

In contrary, reversing the order of events by adding the halide substrate to a solution of the base, it is more likely that complete deprotonation of the halide occurs immediately. Hence, the two species required for migration are not present simultaneously and an HD reaction can be suppressed.

Finally, the stoichiometry of reagents plays an equally crucial role: If a substoichiometric amount of base is used, the requirements to start an HD cascade

Table 1 Influence of reactions conditions on HD reactions

	HD	HD prevention
Temperature	Low temperatures	"High" temperatures
Amount of base	No excess of base	Excess of base
Order of reagent addition	Addition of base to the halide	Addition of halide to the base
Solvent	THF	THP
Electrophile	Slow reacting electrophile	Fast reacting electrophile

can be fulfilled, even if the way of mixing base and halide would typically prevent an HD. On the other hand, when a great excess of base is added quickly to a solution of the halide the HD process might still be inhibited.

2.3 Halogen Dance Versus Halogen Dance Prevention

In the following Table 1, the conditions usually leading to HD products or HD-prevention products are briefly summarized. It is to mention that also exceptions to these general guidelines have been reported [31, 32]. One important contribution to this topic was reported on halobenzenes, which will not be discussed in detail in this chapter but shall be mentioned at this point [33].

3 Recent Progress on the Halogen Dance Reaction on Five-Membered Heterocycles

3.1 Thiophene

Thiophene was the first system on which an HD reaction was observed [12, 13]. Since then, many examples have been reported and the method was successfully used, for example, in the synthesis of compounds with interesting properties in material science. In recent years, mainly bromine migrations have been reported and only one example described an iodine dance. In the absence of any directing group, the most stable position for lithium in thiophene is position 2, and, hence, an HD reaction induced on suitable 2-bromothiophene derivatives lead to the corresponding 3-bromothiophene compounds. The examples reported in the literature deal almost exclusively with this type of migration.

A simple example that demonstrates this behavior was reported in the HD reaction of 2,5-dibromothiophene **15** [34] (Scheme 3). Using 1 equiv. of LDA and subsequent quenching with TMSCl led to 2,4-dibromo-5-trimethylsilylthiophene **16**. When 2 equivalents of LDA and TMSCl were applied 3,4-dibromo-2,5-di(trimethylsily)thiophene **17** was obtained.

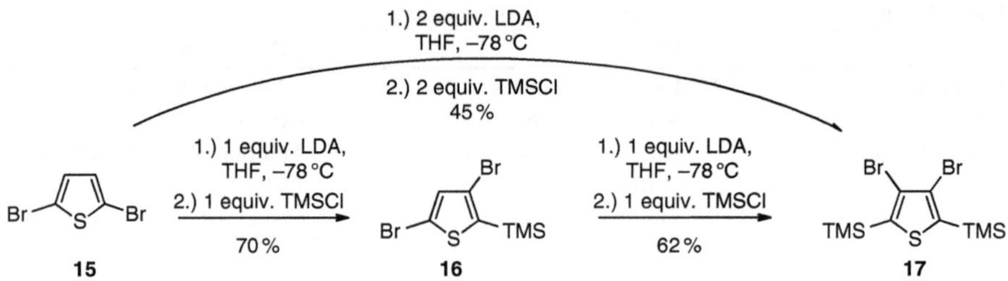

Scheme 3 Single and double HD on 2,5-dibromothiophene

Benhida and co-workers used the HD reaction of 2-bromothiophene **1** to access 2-substituted 3,5-dibromothiophenes **19** [31, 35]. In order to get to an HD product, 2 equiv. of **1** have to be applied since 1 equiv. has to be sacrificed in the course of the reaction (Scheme 4). Since **1** is cheap and readily available, this is not a major drawback in this case. Temperature played a crucial role in getting high yields of HD products. At −78°C, no HD reaction took place and 2-bromo-5-lithiothiophene was formed exclusively and could be trapped with a number of electrophiles in high yields (compounds **18a–f**). However, when the reaction was carried out at room temperature, the rearrangement process took place and compounds such as **19a–f** were obtained demonstrating that depending on the reaction conditions, either HD or HD-prevention products can be obtained (Table 2). Also in the case where 1 equiv. of halide is sacrificed, the reaction proceeds via a typical homotransmetallation HD mechanism with 2,3,5-tribromothiophene **22** as key intermediate (Scheme 4). An alternative mechanism was initially considered but was ruled out by experimental evidence. This method was also exploited in a stereo- and regio-controlled synthesis of bromothiophenyl C-nucleosides [34].

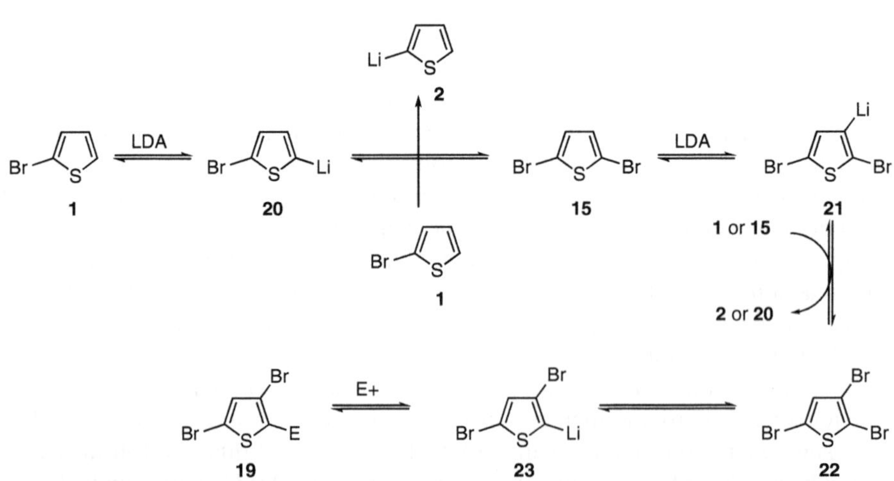

Scheme 4 Mechanism of the homotransmetalation of 2-bromothiophene

Table 2 Homotransmetalation of 2-bromothiophene

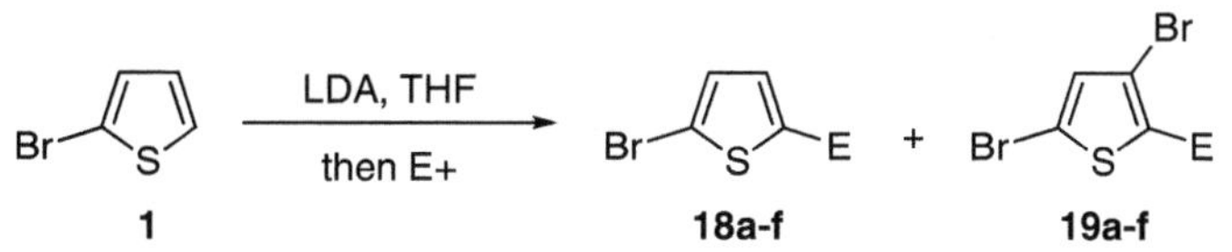

Entry[a]	E+	T (°C)	Product	Yield (%)[b]
1	4-OMe-C_6H_4CHO	−78	18a	95
2	4-OMe-C_6H_4CHO	rt	19a	92
3	C_6H_5-CH_2CH_2CHO	−78	18b	91
4	C_6H_5-CH_2CH_2CHO	0 or rt	19b	90
5	i-Pr-CH_2CHO	−78	18c	88
6	i-Pr-CH_2CHO	rt	19c	90
7	Cyclopentanone	−78	18d	93
8	Cyclopentanone	0 or rt	19d	91
9	Cyclohexanone	−78	18e	90
10	Cyclohexanone	0 or rt	19e	87
11	Bu_3SnCl	rt	18f	87
12	ICH_2CH_2I (I+)	rt	19f	90

[a]Conditions: **1** (2.2 equiv.), LDA (2.2 equiv.), electrophile (1 equiv.)
[b]Isolated yields

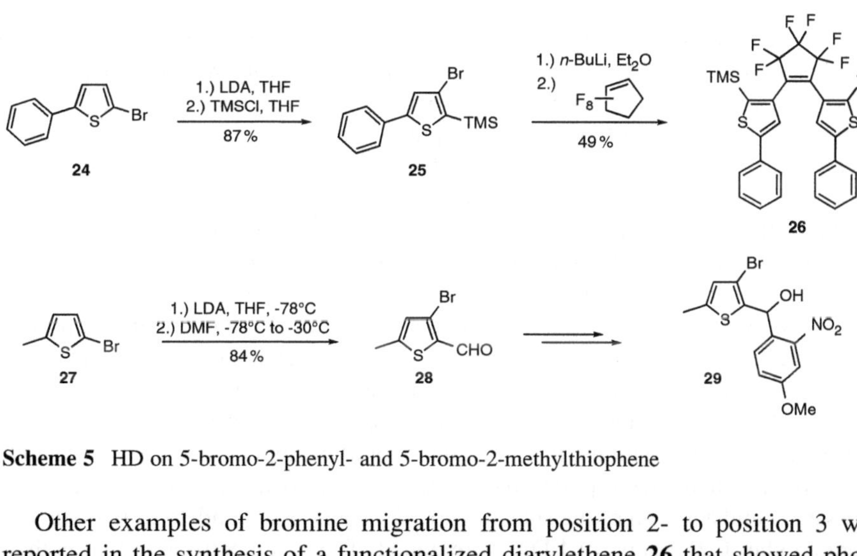

Scheme 5 HD on 5-bromo-2-phenyl- and 5-bromo-2-methylthiophene

Other examples of bromine migration from position 2- to position 3 were reported in the synthesis of a functionalized diarylethene **26** that showed photoreversible photochromism in a polymer film (Scheme 5 top) [36] or in the large scale synthesis of thienobenzazepine derivatives [37]. The latter example reported a bromine migration that was carried out in a 100 g scale in 84% yield that demonstrates that HD reactions can be moved from the laboratory to the production scale (Scheme 5 bottom).

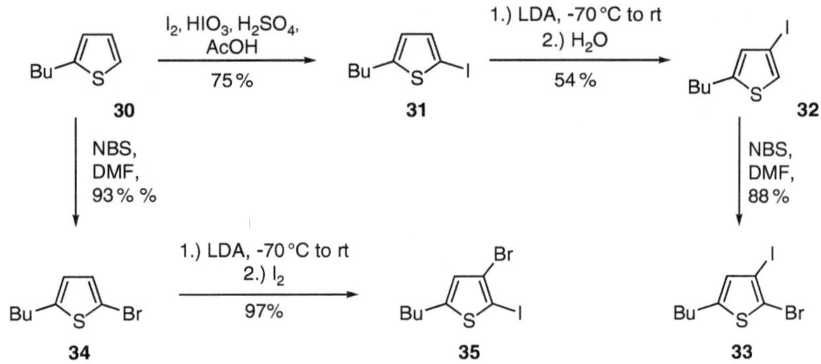

Scheme 6 Synthesis of **33** and **35** via HD reactions

The regioisomeric compounds 2-butyl-4-bromo-5-iodothiophene **35** and 2-butyl-5-bromo-4-iodothiophene **33** were both prepared starting from 2-butylthiophene **30** taking advantage of either an iodine or bromine migration [38]. The iodine migration proved to be less efficient resulting in only 54% yield, whereas the corresponding bromine dance gave an excellent yield of 97% with concomitant introduction of the required iodine (Scheme 6). These two building blocks were then further converted to e.g., thieno-fused dehydro[14]-annulenes, which should have interesting electrochemical and emissive properties.

When 5,5′-dibromo-2,2′-bithiophene **36** is used as starting material in an HD reaction, the question arises whether the lithiation of one ring influences the reactivity of the other. If the reactivity of the second ring remains unaffected, it has to be expected that mixtures of mono- and bis-migrated products will be obtained. However, according to recently published investigations highly selective conversion to 4,5′-dibromo-5-substituted-2,2′-bithiophenes **37** occurred after addition of electrophiles when 1.25 equivalents of LDA were applied [39] (Table 3). Increasing the amount of base to 2.5 equiv. resulted in the formation of double migration products **38**. By adapting the order of reagent addition, it was also possible to obtain products with two different electrophiles in positions 5 and 5′ in good yields. It is important to mention that the order of reagent addition played an important role in the aforementioned protocols. For the mono-migration process, slow addition of LDA to a solution of the substrate was required (so called inverse addition). Also for the double-migration process for the introduction of the same electrophile in both positions, this procedure could be applied. On the other hand, when two different electrophiles had to be introduced a different strategy was necessary: rapid addition of the substrate to 0.95 equiv. LDA and 15 min of stirring before another 0.25 equiv. of LDA were added. The first electrophile that is then introduced has to be inert toward lithiation conditions and either TBDMS or methyl were chosen in this regard.

Several contributions reported on bromine migrations on various fused thiophenes [40–42]. Potential applications of such compounds are as semiconductors in organic field effect transistors, as liquid crystals or more general as organic

Table 3 Single and double HD on 5,5′-dibromo-2,2′-bithiophene

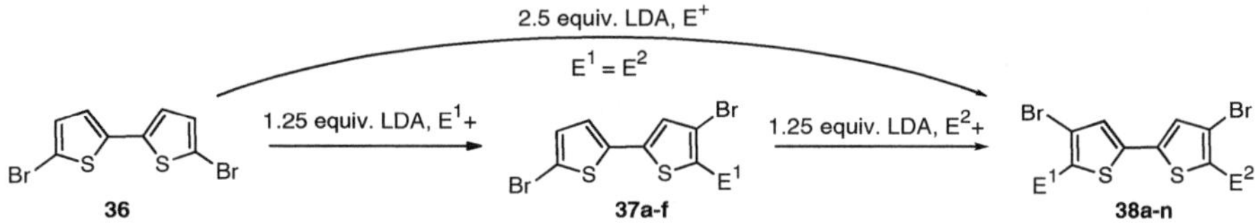

Entry	E^1+	E^2+	E^1	E^2	Compd	Protocol	Percentage (%)
1	MeOH	–	H	–	37a	A	68
2	MeOH	MeOH	H	H	38a	B	74
3	TMSCl	–	TMS	–	37b	A	89
4	TMSCl	TMSCl	TMS	TMS	38b	B	79
5	TBDMSCl	–	TBDMS	–	37c	A	91
6	TBDMSCl	TBDMSCl	TBDMS	TBDMS	38c	B	71
7	MeI	–	Me–	–	37d	A	76
8	MeI	MeI	Mc–	Mc	38d	D	83
9	Me_2S_2	–	MeS–	–	37e	A	51
10	Me_2S_2	Me_2S_2	MeS–	MeS–	38e	B	61
11	DMF	–	–CHO	–	37f	A	63
12	DMF	DMF	–CHO	CHO	38f	B	67
13	–	TMSCl	TBDMS	TMS	38g	C	53
14	–	MeI	TBDMS	Me	38h	C	45
15	–	Me_2S_2	TBDMS	MeS–	38i	C	80
16	–	DMF	TBDMS	CHO	38j	C	53
17	–	MeOH	Me	H	38k	D	81
18	–	TMSCl	Me	TMS	38l	D	64
19	–	Me_2S_2	Me	MeS–	38m	D	70
20	–	DMF	Me	CHO	38n	D	70

Protocol A: Starting from **36**: LDA (1.25 equiv.), E+
Protocol B: Starting from **36**: LDA (2.5 equiv.), E+
Protocol C: Starting from **37c**: LDA (1.05 equiv.), E+
Protocol D: Starting from **37d**: LDA (1.26 equiv.), E+

functional materials. Although the type of the fused ring system varied the principal is again the same, LDA induced migration of bromine from 2- into 3-position, one example using **39** as substrate is given in Scheme 7.

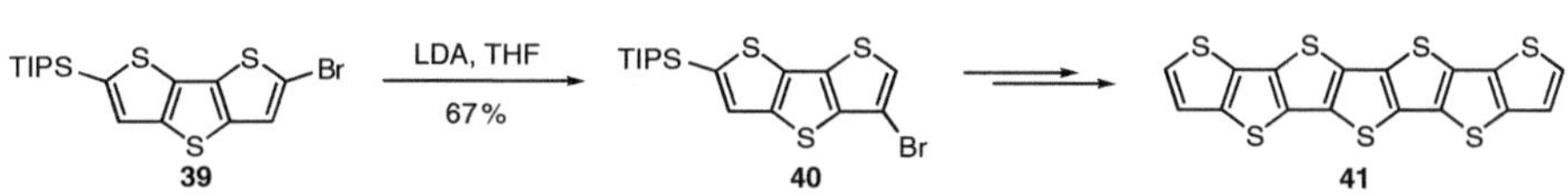

Scheme 7 Application of HD reactions for the synthesis of fused thiophenes

3.2 *Thiazole*

Thiazole was only added in recent years to the portfolio of heterocycles on which a HD reaction has been reported. The acidity of protons on thiazole follows the order H2 > H5 > H4. So principally, halogen migrations from position 2 into positions 5 and 4 as well as from position 5 in position 4 are possible. So far, an HD reaction from position 2 into position 5 was not reported, only from position 2 into position 4. The reason for that is the fact that 2-lithio-thiazole species are often very unstable even at low temperatures. Only in two cases, where position 5 was blocked with an alkyl-silyl group, bromine could be migrated from position 2 into position 4 and the 2-lithiated intermediate trapped with different electrophiles [43, 44]. One of the examples is given in Table 4.

Table 4 Bromine migration from position 2 to 4 on a thiazole system

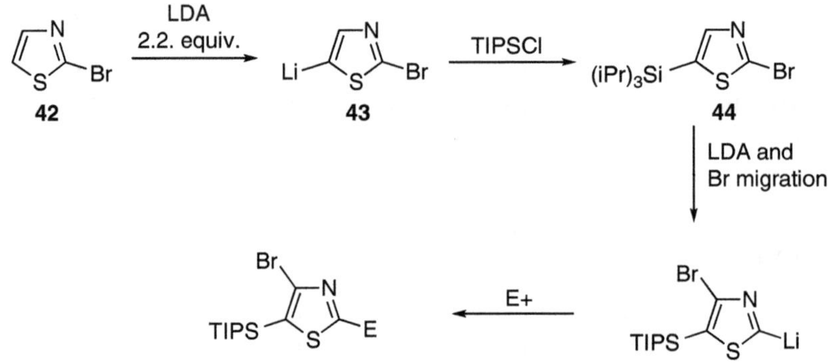

Entry	E+	E	Yield (%)	Compd
1	H_2O	H	85	46a
2	MeI	Me	82	46b
3	Allylbromide	CH_2CHCH_2–	20	46c
4	EtI	Et	<10	46d
5	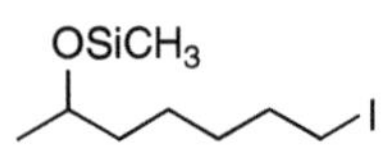	–	0	46e
6	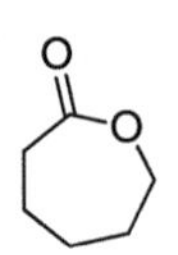	$HO(CH_2)_5CO$–	57	46f
7	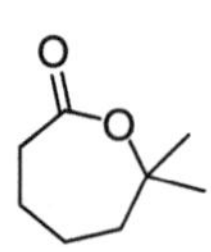	–	0	46g
8	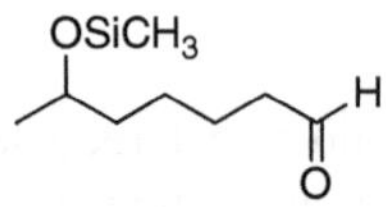	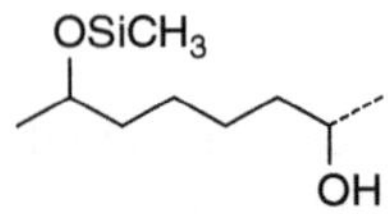	60	46h

On the other hand, bromine migrations from position 5 to position 4 usually work very efficiently and have been reported in a number of cases. The first report of such a reaction was within the total synthesis of the natural product WS75624 B where 5-bromo-2-TIPS-thiazole **47** was converted to 4-bromo-2-TIPS-thiazole **48** in a good yield of 86% [42]. The product was used as reactant in a Negishi cross-coupling reaction to connect the thiazole and pyridine moiety found in the aforementioned natural product (Scheme 8).

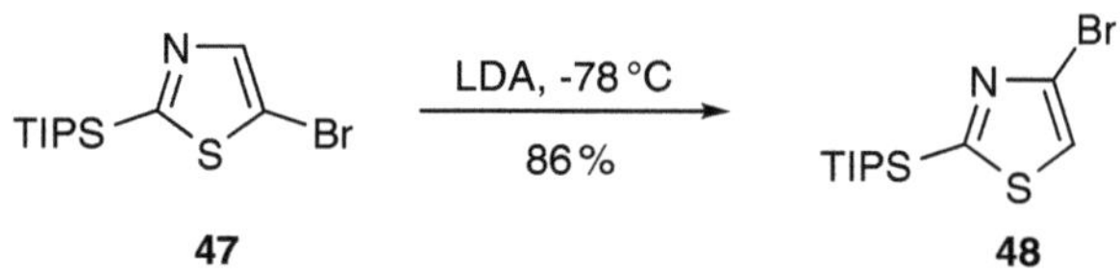

Scheme 8 Bromine migration from position 5 to 4 on a thiazole system

Soon thereafter further studies on thiazole systems were published using different 2-substituted 5-bromothiazoles as starting materials. Here, also the introduction of different electrophiles was investigated. For example, *N*-protected 2-thiazoleamines **49–51** gave good results with carbonyl-containing electrophiles and also TMS and halides (Br, I) were successfully introduced [32] (Table 5, products **52–54**). Here, it was also attempted to suppress the HD reaction; however, this proved to be very difficult. Only in one example, the formation of **55**, could the migration be avoided. It proved to be crucial that the electrophile is already present when the base (LDA) was added so that the initially 4-lithiated thiazole species is immediately quenched before a rearrangement can take place. Hence, only an electrophile that is stable under these conditions, in this case TMSCl, can be successfully applied in this process.

Table 5 HD and HD-prevention on various 2-aminothiazoles

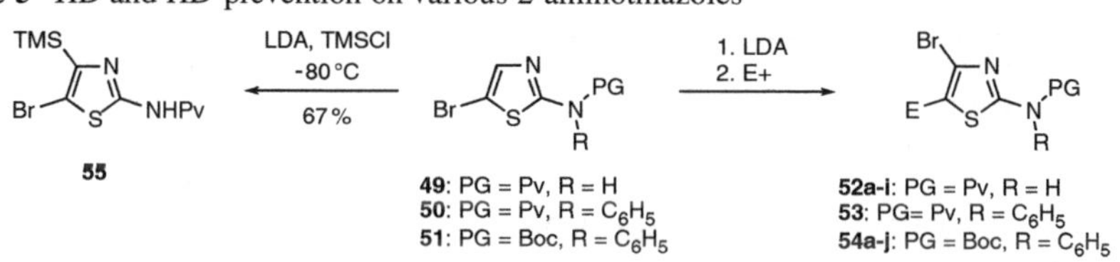

Entry	PG	R	E+	E	Compd	Yield (%)
1	Pv	H	DMF	CHO	52a	92
2	Pv	H	Cyclohexanone	$C_6H_{10}(OH)$	52b	73
3	Pv	H	Br_2	Br	52c	93
4	Pv	H	$BrCH_2CH_2Br$	Br	52d	20
5	Pv	H	I_2	I	52e	76
6	Pv	H	C_6H_5CHO	$C_6H_5CH(OH)$	52f	86
7	Pv	H	TMSCl	TMS	52g	74
8	Pv	H	H_2O	H	52h	99
9	Pv	H	$(C_6H_5)_2CO$	$(C_6H_5)_2COH$	52i	46

(continued)

Table 5 (continued)

Entry	PG	R	E+	E	Compd	Yield (%)
10	Pv	C_6H_5	TMSCl	TMS	53	69
11	Boc	C_6H_5	TMSCl	TMS	54a	97
12[a]	Boc	C_6H_5	Br_2	Br	54b	50
13	Boc	C_6H_5	$C_2Br_2Cl_4$	Br	54c	30
14	Boc	C_6H_5	I_2	I	54d	70
15	Boc	C_6H_5	C_2Cl_6	Cl	54e	70
16	Boc	C_6H_5	Bu_3SnCl	$SnBu_3$	54f	65
17	Boc	C_6H_5	$B(OiPr)_3$	$B(OH)_2$	54g	0
18	Boc	C_6H_5	DMF	CHO	54h	97
19	Boc	C_6H_5	$(C_6H_5)_2CO$	$(C_6H_5)_2COH$	54i	43
20	Boc	C_6H_5	H_2O	H	54j	97

[a]Reported in reference [45]

Another comprehensive study was carried out on 2-chlorothiazole **56** [45] and 2-phenylthiazole **57** [46, 47] (Table 6) towards products **58** and **59**. In some cases, the HD reaction was exploited to get suitable thiazole halides for subsequent cross-coupling reactions, most importantly, Suzuki-Miyaura or Stille [48, 49].

Table 6 HD on 5-bromo-2-chloro and 5-bromo-2-phenylthiazole

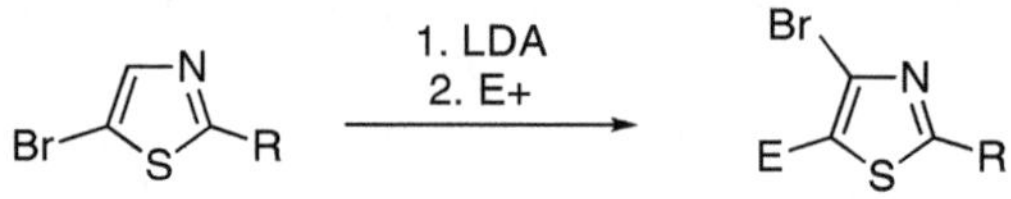

56: R = Cl
57: R = C_6H_5

58a-h: R = Cl
59: R = C_6H_5

Entry	E+	E	R	Compd	Yield
1	EtOH	H	Cl	58a	75
2	C_2Cl_6	Cl	Cl	58b	61
3	$C_2Br_2Cl_4$	Br	Cl	58c	69
4	I_2	I	Cl	58d	76
5	C_6H_5CHO	$C_6H_5CH(OH)$	Cl	58e	77
6	Cyclohexanone	$C_6H_{10}OH$	Cl	58f	77
7	DMF	CHO	Cl	58g	56
8	TMSCl	TMS	Cl	58h	71
9	H_2O	H	C_6H_5	59	78

Additionally, the synthesis of potential SGLT2 inhibitors profited from the application of the HD methodology [50]. It is noteworthy that so far no iodine migrations on thiazole have been reported. Also in our hands such efforts were never successful.

3.3 *Oxazole*

Oxazole reacts under HD conditions very similar to thiazole. The first study investigated the migration process on 5-bromo-2-phenyloxazole **60** and demonstrated that a series of electrophiles can be introduced in position 5 leading to 4-bromo-2-phenyl-5-substituted oxazoles **61a–i** [51] (Table 7).

Table 7 HD on 5-bromo-2-phenylthiazole

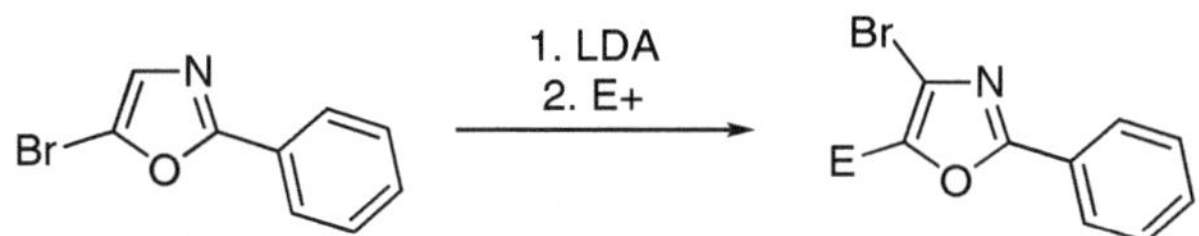

60 **61a-i**

Entry	E+	E	Compd	Yield (%)
1	H_2O	H	61a	60
2	C_6H_5CHO	$C_6H_5(OH)$	61b	78
3	TMSCl	TMS	61c	68
4	C_2Cl_6	Cl	61d	68
5	Br_2	Br	61e	30
6	$BrCH_2CH_2Br$	Br	61e	11
7	$C_2Br_2Cl_4$	Br	61e	76
8	I_2	I	61f	66
9	DMF	CHO	61g	58
10	CO_2	COOH	61h	63
11	Cyclohexanone	$C_6H_{10}OH$	61i	69

Also, 5-bromo-2-(phenylthio)oxazole **62** was a suitable starting material and the resulting 5-lithiated species was trapped with a number of electrophiles in good yields [52] (Table 8, **63a–m**).

Table 8 HD on 5-bromo-2-(phenylthio)oxazole

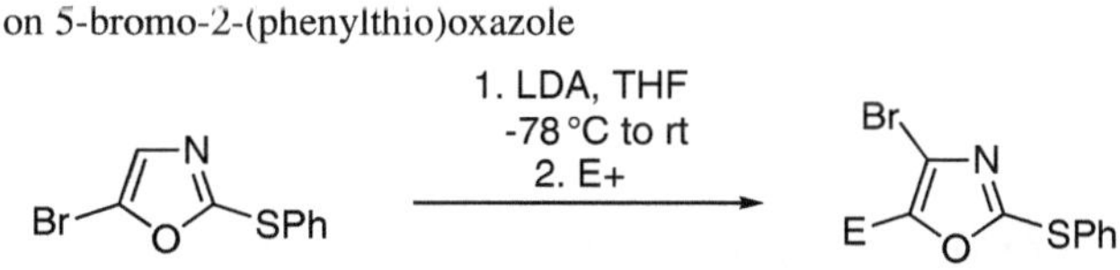

62 **63a-m**

Entry	E+	Product/E	Conditions	Compd	Yield (%)
1	H_2O	H	A	63a	88
2	NIS	I	A	63b	87
3	Bu_3SnCl	$SnBu_3$	A	63c	83
4	TIPSOTf	TIPS	A	63d	89

(continued)

Table 8 (continued)

Entry	E+	Product/E	Conditions	Compd	Yield (%)
5	Furfural	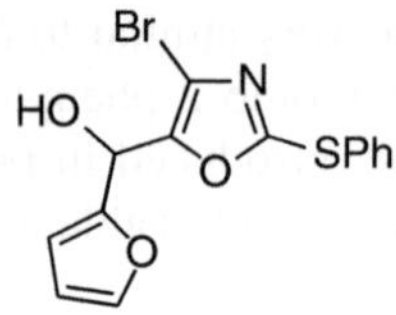	A	63e	82
6	CH_2O	CH_2OH	A	63f	72
7	Cyclopentanone	C_5H_8OH	A	63g	79
8	Acetone	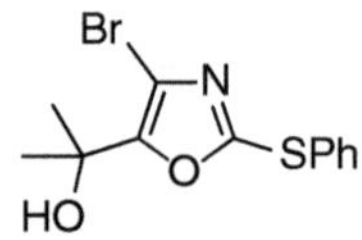	A	63h	74
9	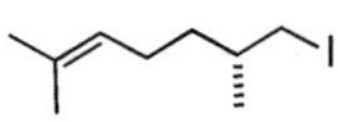	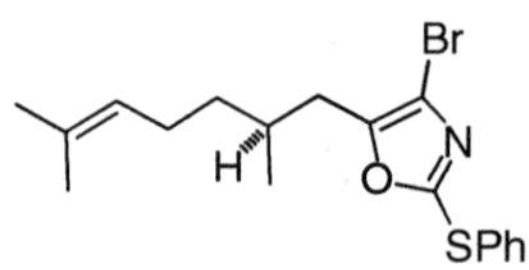	B	63i	87
10	C_6H_5I	C_6H_5	C	63j	80
11	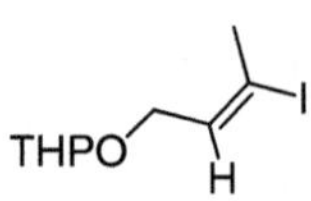	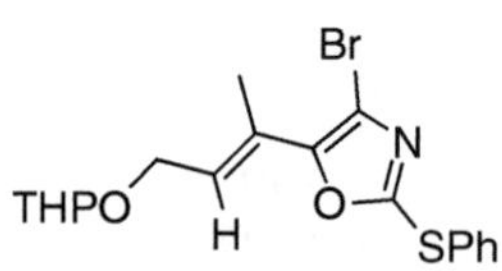	C	63k	67
12	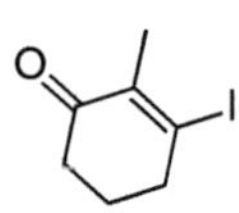	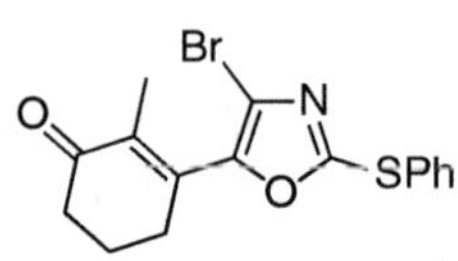	C	63l	72
13	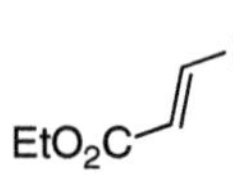	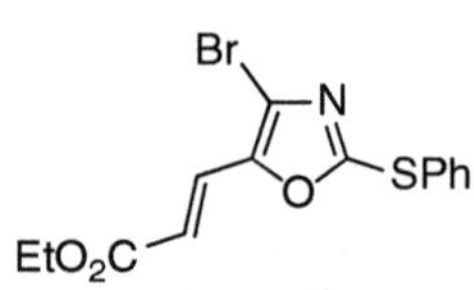	C	63m	77

Conditions A: (1) **62** (1.0 equiv.), THF, LDA (1.5 equiv.), –78°C to 0°C. (2) electrophile (1.5 equiv.), THF, –78°C to rt
Conditions B: (1) **62** (1.0 equiv.), THF, LDA (1.5 equiv.), –78°C to 0°C. (2) anhyd. HMPA (10 equiv.),–78°C. (3) Alkyl iodide (1.5 equiv.), THF, –78°C to rt
Conditions C: (1) **62** (1.0 equiv.), THF, LDA (1.5 equiv.), –78°C to 0°C. (2) anhyd. $ZnBr_2$ (1.2 equiv.), THF, 0°C to rt. (3) aryl or alkenyl iodide (1.5 equiv.), $Pd(PPh_3)_4$ (10 mol%), dry DMF, rt

Again, the HD reaction proved to be very valuable for the synthesis of biologically active compounds such as fatty acid amide hydrolase inhibitors [53] or for the synthesis of substrates for cross-coupling reactions [49].

3.4 Indole

In their continuing quest to make only on first sight simple building blocks easily accessible, the group of Schlosser exploited the HD reaction also on indole in the syntheses of several fluoroindolecarbobylic acids [54]. However, only halogen migrations within the benzo ring were reported. Either 5-bromo-6-fluoro-indole **64a** or 6-fluoro-5-iodo-indole **64b** was converted to 4-bromo-6-fluoro-indole **65** and 6-fluoro-4-iodo-indole **66**, respectively (Scheme 9). In this case, the iodine migration occurred more readily and clean conversion was obtained. The bromine migration gave a 5:1 mixture of danced product and starting material.

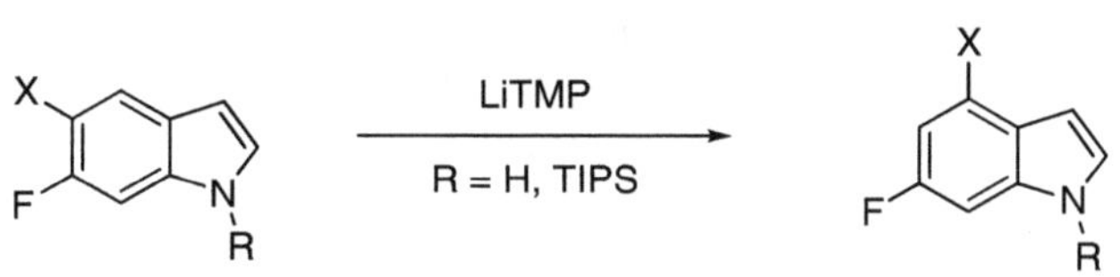

64a: X = Br
64b: X = I

65: X = Br, R = TIPS: 79 %
66: X = I, R = I: 88 %

Scheme 9 HD on indole derivatives

3.5 Pyrazole

An unusual example of a HD reaction was observed on 3,4,5-tribromo-2-PMB-pyrazole-*N*-oxide **67** [55]. Since all three carbons are substituted with bromine already, it is not obvious how an HD should take place on such a substrate. However, it was observed that metal–halogen exchange with *i*-PrMgCl occurred at C3 to give **68** but upon warming from −78°C to room temperature, this intermediate rearranges to the C5-metalated intermediate **69** that can be trapped with electrophiles (Scheme 10). Initial metalation at C3 is unexpected since metalation at C5 would give a less basic, thermodynamically more stable, and sterically less-demanding species. It was speculated that the observed regioselectivity is caused by the low solubility of the C3 metalated species that precipitated at −78°C. Only by warming to room temperature, did the rearrangement occur leading to the readily soluble C5-metalated species **69** (which remains soluble also when cooled again to −78°C). Using an excess of 2.1 equiv. *i*-PrMgCl C3 and C5 could be metalated concomitantly to give intermediate **70**.

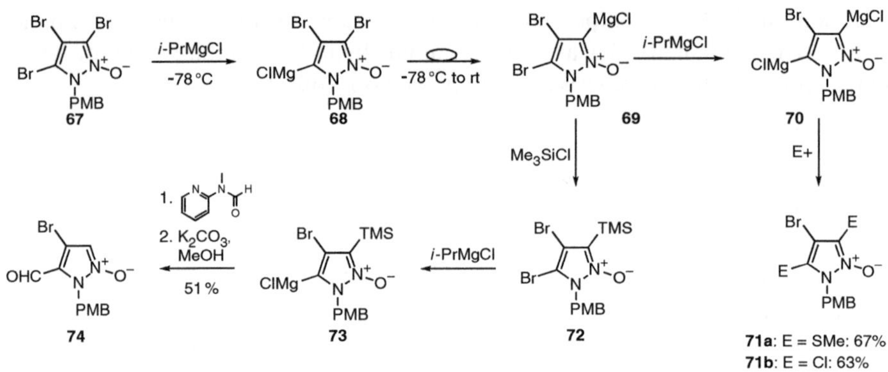

Scheme 10 HD on 3,4,5-tribromo-2-PMB-pyrazole-*N*-oxide

3.6 Selenophene

Also, 2-bromo-5-trimethylsilyselenophene **75** can be efficiently rearranged to 3-bromo-5-trimethylsilyselenophene **76** that was then oxidatively cross-coupled to bis-selenophene product **77** [43] (Scheme 11). The HD product **76** was isolated only in small amounts and no yield was reported.

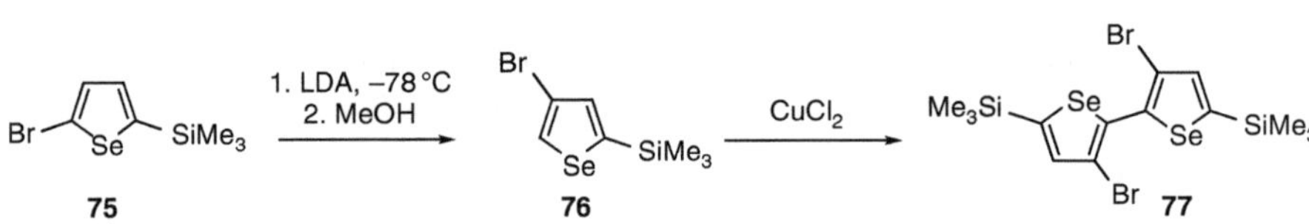

Scheme 11 HD on 2-bromo-5-(trimethylsilyl)selenophene

4 Recent Progress on the Halogen Dance Reaction on Six-Membered Heterocycles

4.1 Pyridine

Besides thiophene, pyridine is the ring system that has been investigated in most detail in HD reactions and many examples have been reported. After the first report by Pieters in 1962 [56], it was mainly Queguiner [57–60] and later Schlosser [61, 62] who delivered paramount contributions in this field. Also in the last decade, fundamental work has been published on relatively simple pyridine halides. However, in recent years more and more examples were reported where the HD methodology was applied in the synthesis of, e.g., natural products, materials, or other more complex compounds of interest.

4.1.1 Fundamental Investigations on Pyridinehalides

Dichloropyridines proved to be a very versatile playground for the exploration of HD reactions although chlorine itself does not migrate. The group of Schlosser investigated all six different dichloropyridines regarding their behavior towards lithium bases [63]. They found that in the case of 3,4- and 2,5-dichloropyridines **78** and **81** the position of initial lithiation depends on the chosen lithiation conditions most importantly the base applied (Scheme 12).

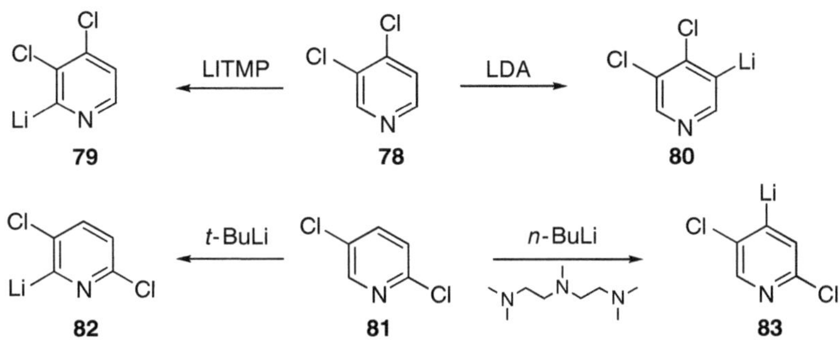

Scheme 12 Direction of lithiation on dichloropyridines

Generally, initial lithiation takes place at the most acidic position, which is always adjacent to a chlorine atom. Additionally, the directing effects of chlorine are cumulative. Hence, if the chlorines are in a *meta*-relationship to each other, lithiation occurs in between. When then another halide (Br or I) is introduced followed by a second lithiation step, the most stable lithiated species is again next to a halide, preferably in between two halides. In light of this general consideration, the lithiation of 2,4-dichloropyridine **84** naturally takes place in position 3 due to the directing effects of the two chlorine atoms. Quenching with iodine gave 2,4-dichloro-3-iodopyridine **85** in good yield. When **85** is submitted to lithiation conditions, lithium is directed in 5-position again due to the adjacent chlorine to give **86**. However, this species quickly rearranges to the corresponding 2,4-dichloro-5-iodo-3-lithiopyridine that was either hydrolyzed with water to give 2,4-dichloro-5-iodo-pyridine **87** (26%) or trapped with dry ice to get to the corresponding carboxylic acid (25%). The low yields might be attributed to steric hindrance. The same reactions were also carried out on 3,5- and 2,6-dichloropyridine **88** and **92** (Scheme 13). The sterically least demanding substrate, 2,6-dichloropyridine **92** also gave the highest yield in the migration process.

In the case of 2,6-difluoro-3-iodopyridine **96**, it is possible to either deprotonate in 5- or 4-position [64]. Due to the stronger directing effect of fluorine, lithiation occurs initially (after 5–15 min) in 5-position (**100**), which gives after iodine quench 2,6-difluoro-3,5-diiodopyridine **101b**. However, when the reaction time with LDA is significantly prolonged to 30 h reversible return to the substrate **96** allows intermittent lithiation in position 4 giving rise to **97**. This intermediate spontaneously gives a basicity gradient-driven heavy halogen migration [65, 66]

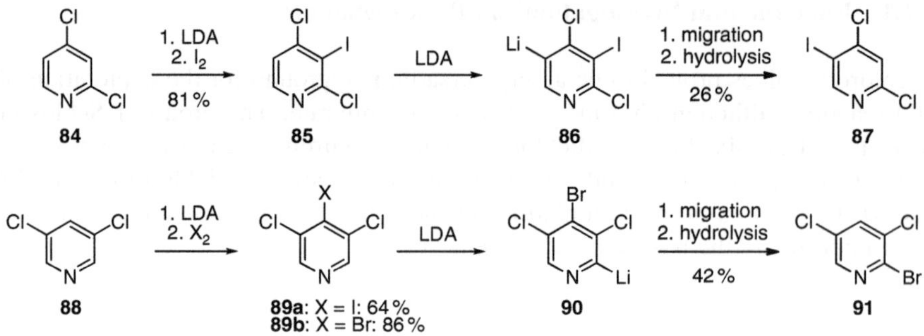

Scheme 13 Iodine dance on various dichloro-iodo-pyridines

giving after hydrolysis 2,6-difluoro-4-iodopyridine **99a** in a good yield of 81% accompanied by 16% starting material (Scheme 14). By quenching **98** with dry ice or I_2, **99b** and **99c** were obtained in good yield. In the case of 3-chloro-2,6-difluoropyridine, no such complementary conditions were identified but the selectivity of lithiation was decreased. Quenching with iodine, gave 77% of 3-chloro-2,6-difluoro-4-iodopyridine accompanied by 15% of 3-chloro-2,6-difluoro-5-iodopyridine. Intrinsically, the 4-position is more acidic but on the other hand fluorine provides stronger neighboring group assistance compared to chlorine [67]. These effects compensate each other to some extent leading to the observed mixture of products.

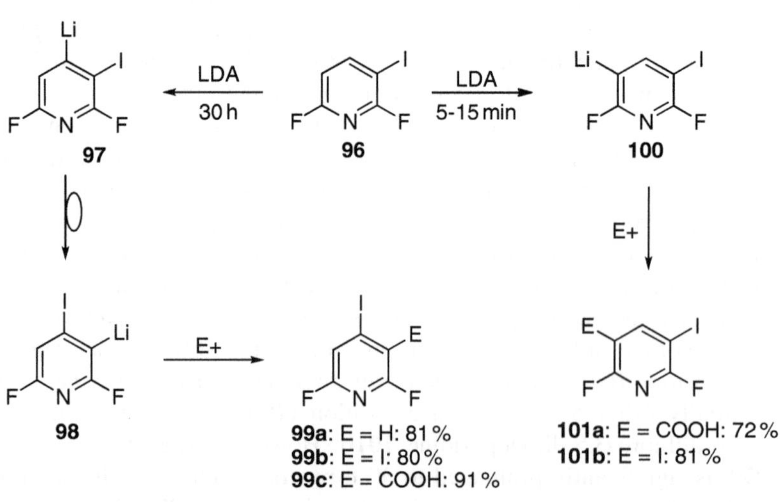

Scheme 14 HD and HD prevention on 2,6-difluoro-3-iodopyridine

The same group also reported an HD reaction on 3-fluoro-4-iodo-2-trimethylsilylpyridine **102**, which was prepared from 3-fluoropyridine via two subsequent directed *ortho*-metalation steps [68]. Iodine-directed deprotonation in position 5 led to migration of iodine in that position leading ultimately to 3-fluoro-5-iodo-2-trimethylsilylpyridine **104** (Scheme 15), which was further elaborated to 5-fluoropyridine-3-carboxylic acid, a compound very difficult to access via alternative routes [69, 70].

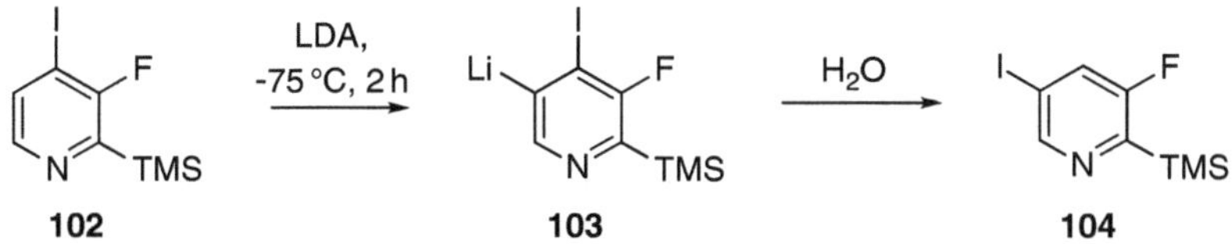

Scheme 15 Iodine migration from position 4 to 5 in a pyridine system

HD reactions have also been reported on tri- and tetrahalogenated pyridines [71]. For example, 3-chloro-5-fluoro-4-iodopyridine was exploited in the syntheses of 5-chloro-3-fluoro-2-iodopyridine and 5-chloro-3-fluoro-2-iodopyridine-4-carboxylic acid. In both cases, the products were not obtained exclusively but were accompanied by the corresponding compounds where the iodine was missing.

In the case of 2,3,4,5-tetrahalogenated pyridines, only position 6 is left for deprotonation. Starting from 2-halo-5-chloro-3-fluoro-4-iodopyridines **105**, the lithiated intermediate **106** rearranges to the corresponding 2-halo-5-chloro-3-fluoro-6-iodo-4-lithiopyridine **107** (Scheme 16). Even if the halide in position 2 is iodine, the iodine in 4-position migrates since this position is in between two halides, which make this the most favorable position for Li. The so obtained intermediates **107a–c** were converted to pyridinecarboxylic acids.

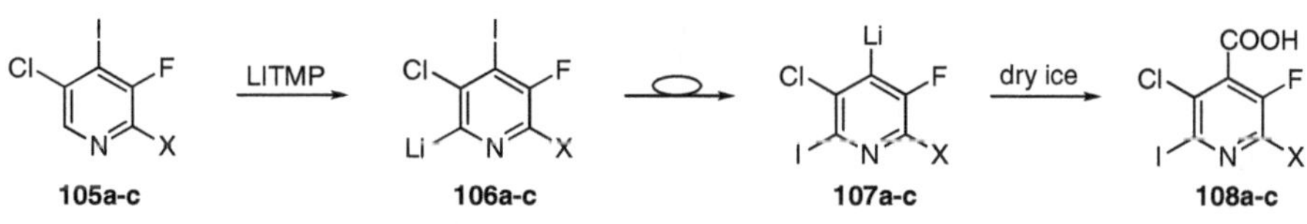

Scheme 16 Tetrahalopyridines in HD reactions

Further examples investigated HD reactions on pyridine halides containing a trifluoromethyl (**109**) [62] or carboxylic acid group (**113**) [72]. It can be expected that also these electron-withdrawing groups influence the direction of lithiation and subsequently also the migration process. However in the reported examples, iodine proved to have a stronger directing effect compared to trifluoromethyl (Scheme 17 top). Also, on 5-bromo-4-iodopyridine-3-carboxylic acid **113** lithiation took place adjacent to bromine in position 6 and subsequent migration of iodine in this position occurred (Scheme 17 bottom).

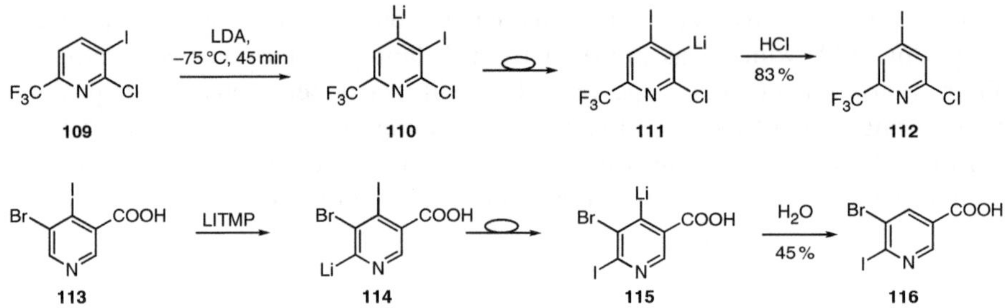

Scheme 17 Iodine migrations on pyridine systems

Also, *O*-carbamates have been applied as directing groups for lithiation in combination with subsequent HD reactions [73]. Three different iodopyridines **117**, **120**, and **123** carrying an *O*-carbamate in 2-, 3-, or 4-position were investigated and gave HD reactions in good yields and different electrophiles were introduced in the former iodine position (Scheme 18). In substrates **120** and **123** where initial lithiation can take place either adjacent to iodine or the *O*-carbamate, the latter group proved to have the stronger directing effect and the initial lithiation took place adjacent to the carbamate.

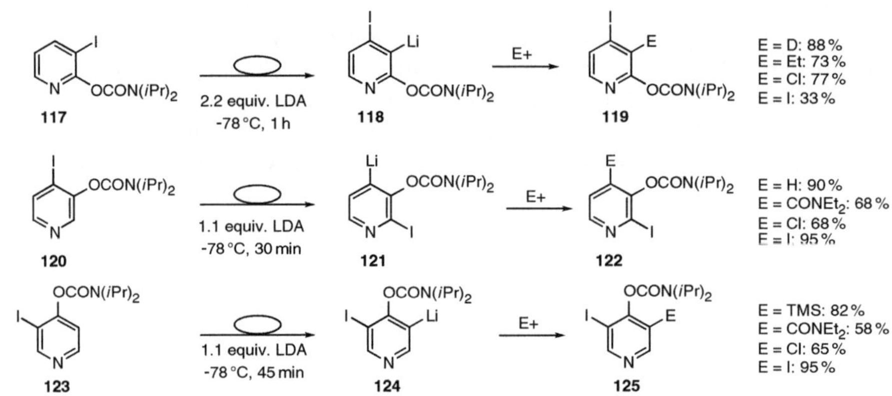

Scheme 18 Iodine migrations on pyridine systems

In one case, an unexpected result was obtained. Performing the lithiation on **126** and subsequent quenching with TMSCl, gave product **130** where iodine was not migrated into 5- but into 6-position (Scheme 19). A possible mechanism for this unusual migration behavior was proposed. Initial iodine migration into 5-position followed by electrophilic quenching with TMSCl leads to **128** as intermediate, which undergoes a second lithiation and halogen migration very quickly to give **129**. This is only possible since TMSCl and LDA are compatible and do not react with each other at low temperatures. It was hypothesized that the driving force of the second migration is steric hindrance between the TMS group and the iodine, which is of course avoided in the rearranged product. By using an excess of TMSCl, TMS groups could be introduced in 4 and 5 position. The structure of **130** was confirmed by X-ray crystallography [67].

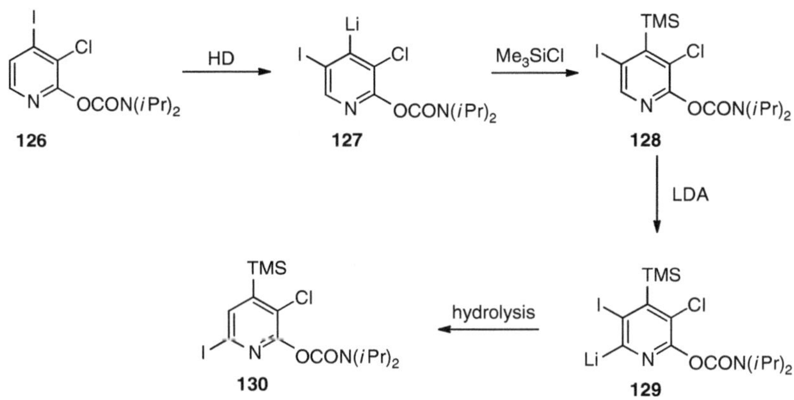

Scheme 19 Iodine migrations on pyridine systems

In one example, pyridine is involved in an HD reaction; however, not as the scaffold on which the HD takes place but as directing group for lithiation [74]. It was reported that in 2-(3-chloro/fluorophenyl)pyridines **131**, lithiation takes place adjacent to the halogen in 2-position of the phenyl ring and e.g. F or Cl can be introduced in that position (Scheme 20). If a second equivalent of base is added, the 4-position of the phenyl ring is deprotonated and the halide then migrates into a 4-position. The same sequence is of course also possible when the 2-phenylpyridine derivative is substituted by the corresponding 2-phenylisoquinoline.

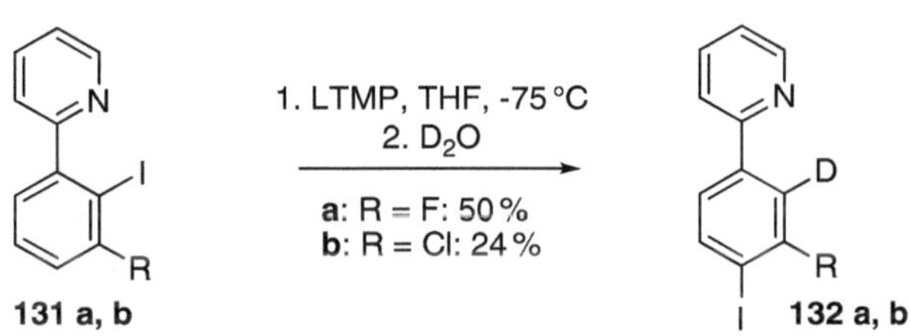

Scheme 20 Pyridine as directing lithiation group in HD reactions

4.1.2 Application of HD Reactions on Pyridine in Target Oriented Synthesis

A well-known HD process is the conversion of 2,3-dihalopyridines into the corresponding 2,4-dihalo derivatives. One such example is depicted in Scheme 21. This transformation was exploited in the synthesis of 2-halo-pyridine-4-boronic acids **135**, which can be further converted to the 4-aryl-2-halopyridines **136** [75]. The same class of compounds was directly prepared also from a 2,4-dihalo compound, in this case 2-bromo-4-iodopyridine via Kumada-Corriu-Tamao coupling [76]. Via this method, ligands for the formation of ruthenium complexes were synthesized that have interesting photoelectrochemical properties [77] or can be applied as colorimetric Hg_2^+ sensor [78].

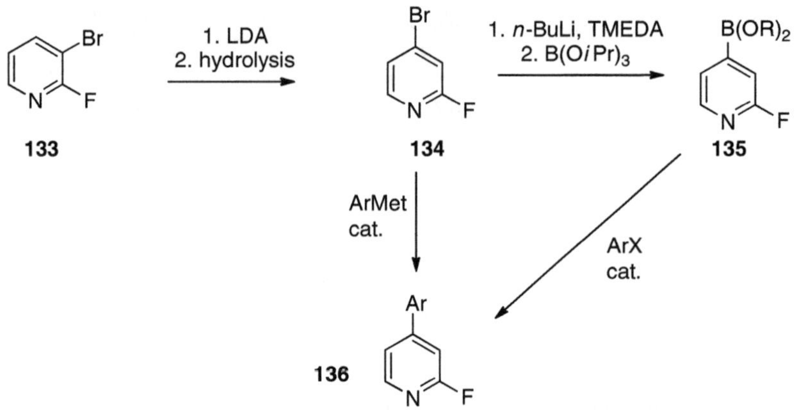

Scheme 21 Synthesis of 4-bromo-2-fluoropyridine via HD and its subsequent application in cross-coupling reactions

Naturally, also 2,3,4-triarylated compounds **140** are accessible via HD reactions on 2,3-dihalopyridines since 2,3,4-trihalogenated pyridines are easily accessible. This synthesis was realized starting from 2-chloropyridine **137**, which was converted into 2-chloro-3,4-diiodopyridine **139** in a one pot protocol and subsequently submitted to iterative cross-coupling reactions [79] (Scheme 22). Although **139** was obtained in only 28% yield in the one pot process compared to 71% in a stepwise reaction sequence, a one pot reaction is superior regarding time efficiency and also less waste is produced since one purification step can be avoided.

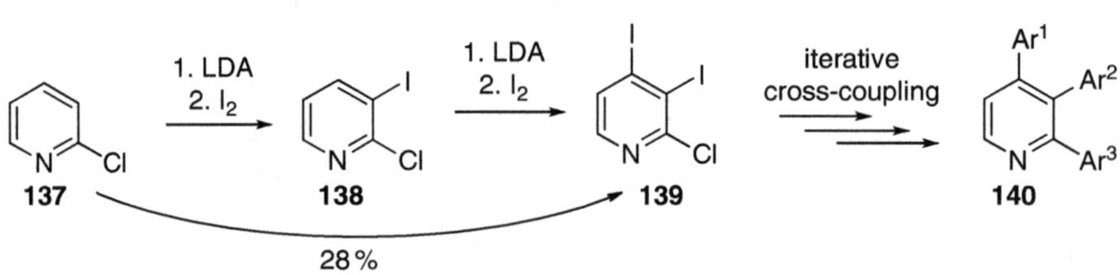

Scheme 22 HD reactions in the synthesis of 2,3,4-triarylated pyridines

In attempts to synthesize tetrahalogenated 4,4′-bipyridines, an HD reaction was observed as unwanted side reaction that gave by the authors unexpected tetrahalogenated 4,4′-bipyridines derivatives. However, the by-product formation can be easily explained taking into account that the conditions applied for the dimerization process are basically often used as conditions to induce HD reactions [80].

Bicyclic pyridinones, interesting scaffolds in medicinal chemistry, have also been accessed taking advantage of a HD reaction to form 2,6-difluoro-4-iodopyridine [81].

In the total synthesis of the natural product caerulomycin C **146**, two groups took advantage of HD processes [82–84]. Sammakia and co-workers elegantly used two HD reactions in their synthesis (Scheme 23). The first HD consisted of an LDA-induced iodine migration from 3- to 5-position (**141–142**), which enabled the introduction of the second methoxy group in position 5 via nucleophilic substitution with CH_3ONa/CuI (**143**). The subsequently introduced bromine in 3-position was then (again LDA induced) migrated in 6-position (**144–145**) and used as leaving group for a Negishi cross coupling reaction with 2-pyridine zinc chloride to give the desired bi-pyridine scaffold. Reduction of the amide to the aldehyde and oxime formation completed the total synthesis of **146**. While Sammakias synthesis took ten steps starting from 2-pyridinecarboxylic acid and gave 11.8% overall yield, the formerly published protocol by the group of Queguiner required seven steps but with a lower overall yield (4.6%).

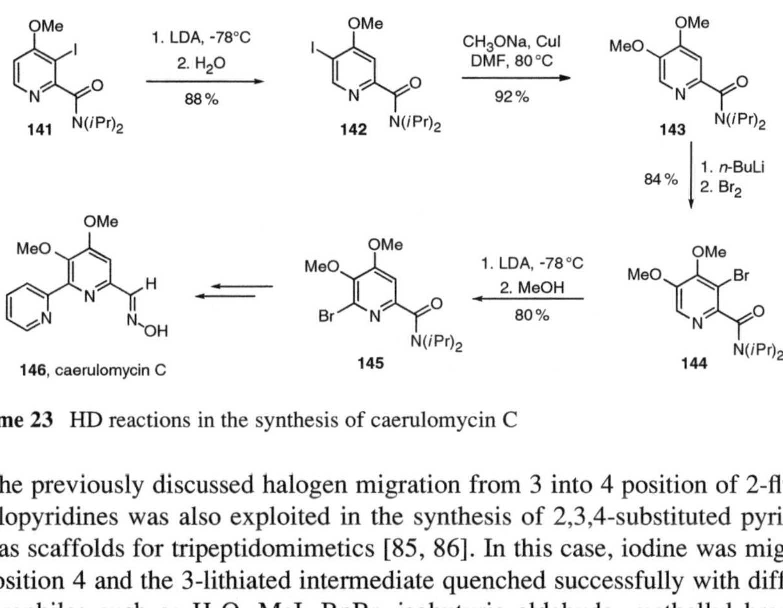

Scheme 23 HD reactions in the synthesis of caerulomycin C

The previously discussed halogen migration from 3 into 4 position of 2-fluoro-3-halopyridines was also exploited in the synthesis of 2,3,4-substituted pyridines **148** as scaffolds for tripeptidomimetics [85, 86]. In this case, iodine was migrated in position 4 and the 3-lithiated intermediate quenched successfully with different electrophiles such as H_2O, MeI, BnBr, isobutyric aldehyde, methallyl bromide, acetaldehyde, and allyl bromide (Table 9). Other electrophiles such as isobutyl halides and ethyl acrylate gave no product formation and ethyl bromoacetate gave only a low yield (13%).

Also an iodine dance starting from 5-iodo-6-chloronicotine **149** was attempted [87]. Although several conditions were explored, mixtures of products were obtained. Besides starting material depending on the choice of base either migration of iodine in 4-position (**150**) or 2-position (**151**) was observed. Additionally, 6-chloro-2,5-diiodo nicotine **152** was observed (Table 10).

Table 9 2-Fluoro-4-iodo-3-substituted pyridines via HD reactions

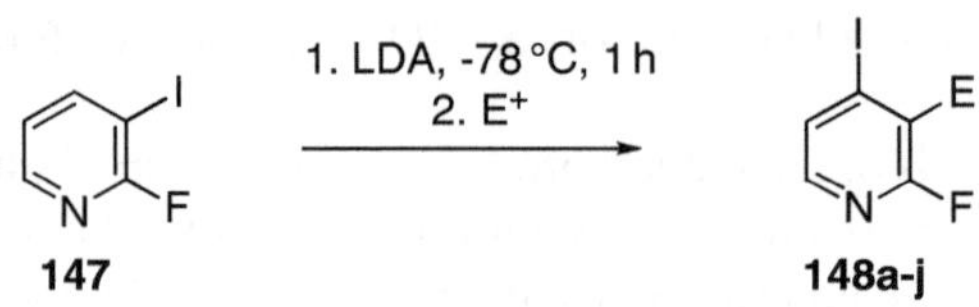

Entry	E+	E	Compd	Yield (%)
1	H_2O	H	148a	80
2	MeI	Me	148b	89
3	BnBr	Bn	148c	71
4	Isobutyl halide	Isobutyl	148d	0[a]
5	Isobutyric aldehyde	$Me_2CHCH(OH)$	148e	96
6	Methallyl bromide	Isobutylene	148f	93
7	CH_3CHO	MeCH(OH)	148g	51
8	Ethyl bromoacetate	CH_2COOEt	148h	13
9	Allyl bromide	Allyl	148i	87
10	Ethyl acrylate	CH_2CH_2COOEt	148j	0

[a]Isobutyl bromide or isobutyl iodide with or without TMEDA at −78°C to −30°C

Table 10 Iodine migrations on 6-chloro-5-iodonicotine

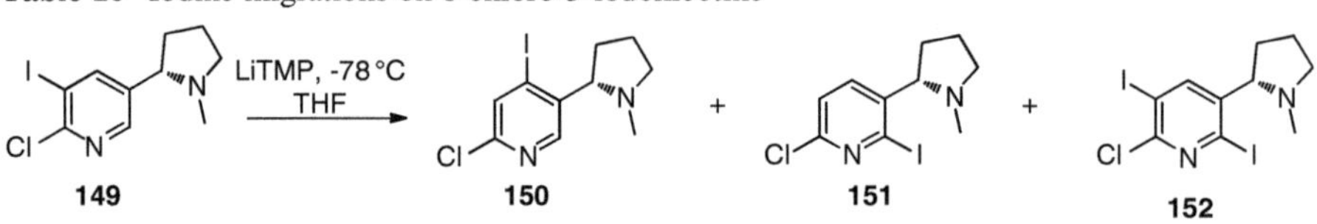

Entry	Conditions	Compd	Ratio
1	1 equiv. LDA, THF, −78°C, 2 h	152	a, b
2	1 equiv. LDA, THF, −78°C, 30 min	149:150	2:1
3	2 equiv. LDA, THF, −78°C, 2 h	149:150:152	2:1:1
4	1.2 equiv. LTMP, THF, −78°C, 1 h	149:151:152	1:2:2

a: 52% isolated yield
b: Traces of x

Another application in the synthesis of a natural product was the total synthesis of atpenin A5 **155** [88]. LDA-induced iodine migration from position 5 to 6 (**153–154**) allowed the introduction of a hydroxyl group after borylation and subsequent oxidation (Scheme 24).

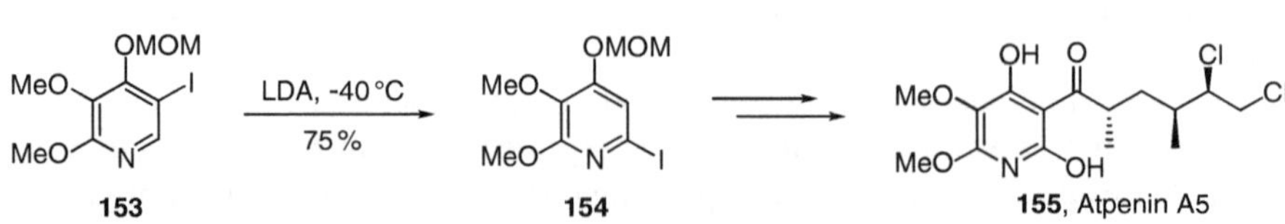

Scheme 24 HD in the synthesis of atpenin A5

4.2 Quinoline

Since HD reactions on pyridine are possible and well investigated, it is not surprising that HD sequences have been reported also on quinolines. For example, the migration of iodine from 4- into 2-position of 3-fluoro-4-iodoquinoline **156** proceeds efficiently with high yield and high selectivity [89]. Several electrophiles could be introduced in excellent yield (Table 11). When the lithiation was carried out at −78°C in the presence of TMSCl, the halogen migration could be omitted and 3-fluoro-4-iodo-2-trimethylsilyquionline was obtained in 95% yield.

Table 11 Iodine migration from the 4- to the 2-position on 3-fluoro-4-iodoquinoline

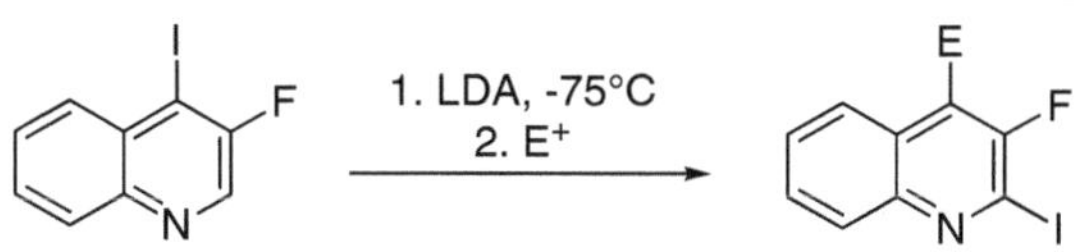

156 **157**

Entry	E+	E	Compd	Yield (%)
1	H_2O	H	157a	95
2	D_2O	D	157b	95
3	I_2	I	157c	98
4	C_2Cl_6	Cl	157d	74
5	MeI	Me	157e	65
6	MeCHO	MeCH(OH)	157f	79
7	C_6H_5CHO	$C_6H_5CH(OH)$	157g	75
8	HCO_2Et	CHO	157h	95

In an anelated ring system such as quinolines, it is also possible that the halides migrate from one ring system to the other. Such a case was observed when 4-bromo-8-iodo-2-(trifluoromethyl)quinoline **158** or 4,8-dibromo-2-(trifluoromethyl) quinoline was treated with *n*-BuLi (Scheme 25). Initially, metal-halogen exchange occurred in position 8 to give 4-bromo-8-lithio-2-(trifluoromethyl)quinolines **159** that upon reaction with starting material **158** induces an HD cascade [90]. The so-obtained 4-lithio-8-bromo-(trifluoromethyl)quinolines were either hydrolyzed or quenched with CO_2 to give the corresponding carboxylic acid **160**. When i-PrMgCl was used for the metal-halogen exchange instead of *n*-BuLi, no migration occurred and the carboxylic acid was introduced in position 8.

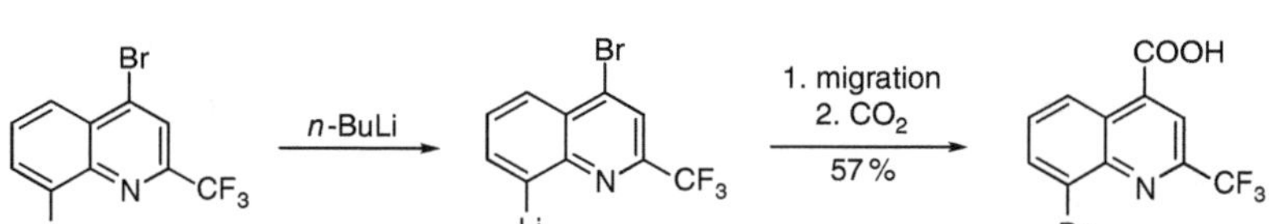

158 **159** **160**

Scheme 25 Bromine migration on a quinoline system from the pyridine to the phenyl ring

On various halogenated 2,2-difluoro-1,3-benzodioxoles, halogen migration in the benzo-ring was observed [91]. For example, iodine migrations were reported and e.g., starting from 2,2-difluoro-4,7-diiodo-1,3-benzodioxole **161** 2,2-difluoro-4,6-diiodo-1,3-benzodioxole **163** was obtained (Scheme 26).

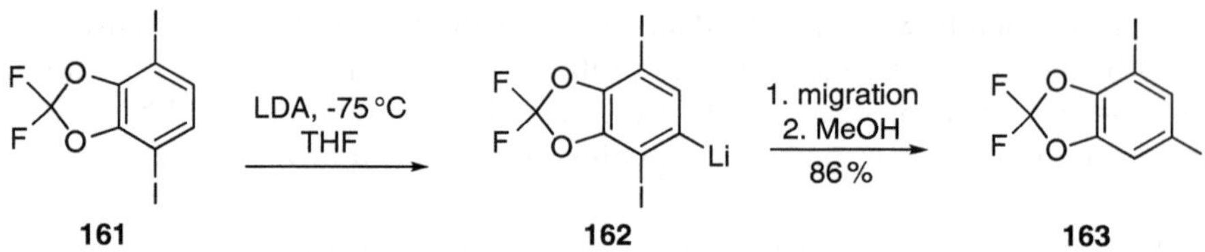

Scheme 26 HD on 2,2-difluoro-1,3-benzodioxole systems

5 Unusual Halogen Dance Reactions

Besides "classical" base induced HD reactions, several unusual halogen migrations have also been reported in recent years, which shall be discussed at this point.

One interesting example was reported on an HD reaction under neutral conditions in the total syntheses of telomerase inhibitors Dictyodendrin B, C, and E. In this case, a very complex *N*-containing anelated heterocyclic system could be selectively brominated at a specific position using NBS to give **164**. Upon standing in $CDCl_3$ at room temperature, the bromine migrated into a more electron rich position forming **165**, a process that could be monitored via NMR spectroscopy [92] (Scheme 27).

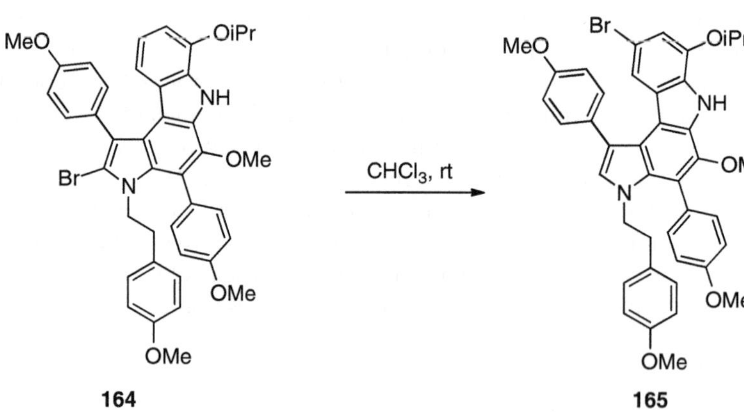

Scheme 27 HD under neutral conditions

Such a neutral halogen migration was also reported on much simpler systems. However, in this case no defined product but complex mixtures were obtained [93].

Starting from 2-bromo-3-methyl-4-nitro-3-thiolene-1,1-dioxide **166** or 4-bromo-3-methyl-4-nitro-2-thiolene-1,1-dioxide **167** (or mixtures thereof) four different products (**168–171**) were obtained in various ratios. It was hypothesized that no base is necessary due to the facile formation of a stabilized anion that then induces the migration process. For example, deprotonation of 2-bromo-3-methyl-4-nitro-3-thiolene-1,1-dioxide **66** leads to a resonance stabilized anion **172** that can react with remaining starting material either via proton exchange or bromine exchange to give **168** and **172** or **169** and **173**, respectively. The newly formed anion **173** deriving from the bromine abstraction can again react with starting material leading to **170** and **171** (Scheme 28). This explains the formation of all observe products.

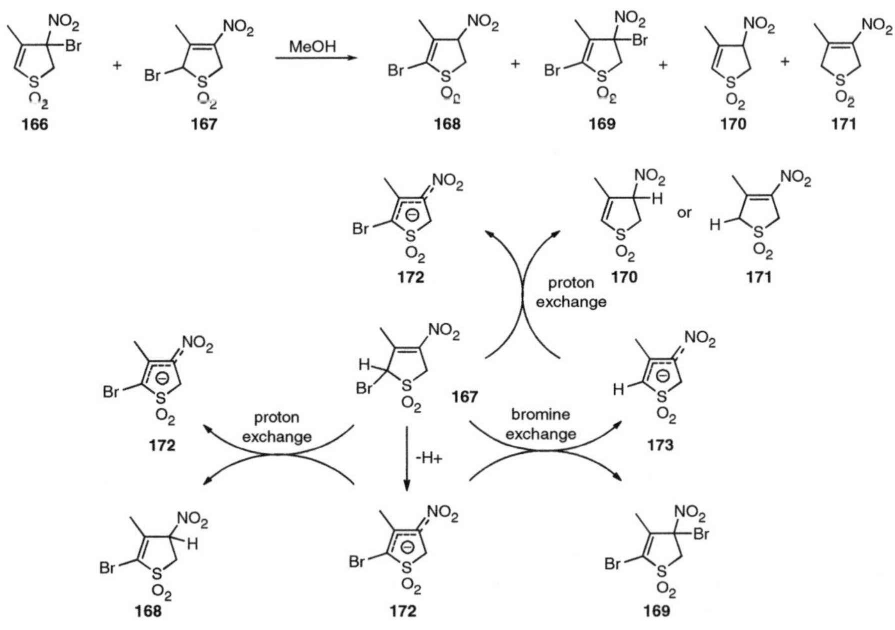

Scheme 28 HD under neutral conditions on thiolene dioxides

The report of an acid-catalyzed HD on deactivated pyrroles is especially remarkable since these are basically the opposite conditions usually required for an HD reaction [94]. However, the reported reactions proceed with low preference for a defined product and mixtures of difficult to separate compounds are obtained (Scheme 29). Substrates **174** or **175** were reacted under strongly acidic conditions and led to four different compounds **174–177**. The transformation of **174** is shown in Scheme 29. A mechanism was suggested that explains the formation of all these products. Initially pyrrole is protonated to **178** that allows subsequent formation of a cyclic bromonium intermediate **179**. This cyclic intermediate can then re-open either to the compound bearing bromine in 4- or 5-position. Additionally, it can be attacked by Br^{-} to give **176** or debrominated to **177**.

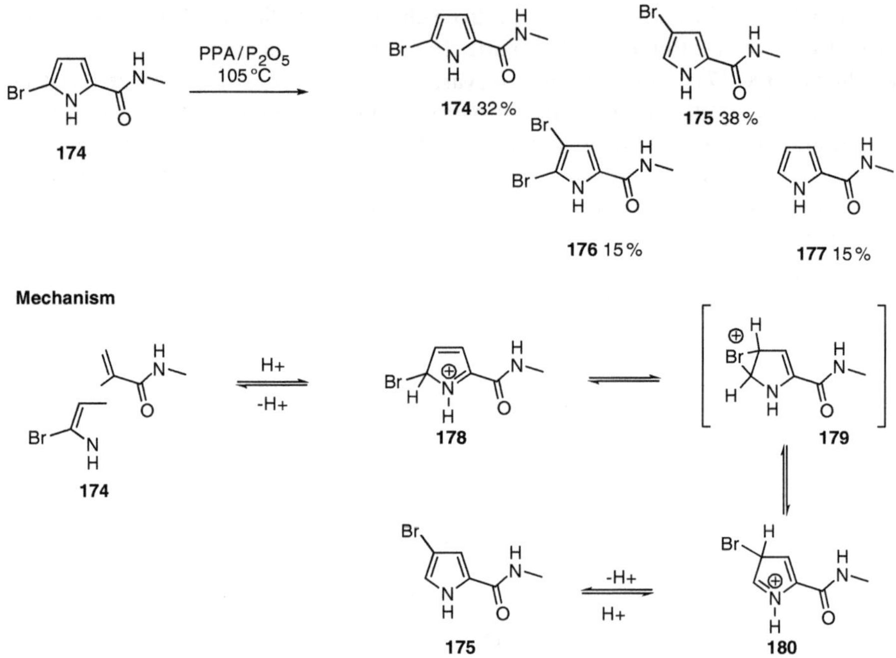

Scheme 29 Acid catalyzed halogen migration on a bromopyrrole system

References

1. Pozharsky AF, Soldatenkov AT, Katritzky AR (1997) Heterocycles in life and society. Wiley, Chichester
2. Eicher T, Hauptmann S (2003) The chemistry of heterocycles. Wiley-VCH, Weinheim
3. Joule JA, Mills K (2010) Heterocyclic chemistry. Wiley-Blackwell, Chichester
4. Negishi E, de Meijere A (eds) (2002) Handbook of organopalladium chemistry for organic synthesis, vol 1 and 2. Wiley, New York
5. de Meijere A, Diederich F (eds) (2004) Metal-catalyzed cross-coupling reactions, vol 1 and 2, 2nd edn. Wiley-VCH, Weinheim
6. Li JJ, Gribble W (eds) (2007) Palladium in heterocyclic chemistry: a guide for the synthetic chemist. Elsevier, Amsterdam
7. Ritleng V, Sirlin C, Pfeffer M (2002) Chem Rev 102:1731–1769
8. Godula K, Sames D (2006) Science 312:67–72
9. Murai S (ed) (1999) Activation of unreactive bonds and organic synthesis. Springer, Berlin
10. Dyker G (ed) (2005) Handbook of C-H transformations: applications in organic synthesis, vol 1 & 2. Wiley-VCH, Weinheim
11. Kakiuchi F, Chatani N (2003) Adv Synth Catal 345:1077–1101
12. Vaitiekunas A, Nord FF (1951) Nature 168:875–876
13. Vaitiekunas A, Nord FF (1953) J Am Chem Soc 75:1764–1768
14. Schnürch M, Spina M, Khan AF, Mihovilovic MD, Stanetty P (2007) Chem Soc Rev 36: 1046–1057
15. Bunnett JF (1972) Acc Chem Res 5:139–147
16. Fröhlich J (1994) Prog Heterocycl Chem 6:1–35

17. Duan X-F, Zhang Z-B (2005) Heterocycles 65:2005–2012
18. Nora de Souza MV (2007) Curr Org Chem 11:637–646
19. Bunnett JF, Moyer CE (1971) J Am Chem Soc 93:1183–1190
20. Bunnett JF, Scorrano G (1971) J Am Chem Soc 93:1190–1198
21. McLennan DJ, Bunnett JF (1971) J Am Chem Soc 93:1198–1201
22. Bunnett JF, Feit I (1971) J Am Chem Soc 93:1201–1205
23. Wotiz JH, Huba F (1959) J Org Chem 24:595–598
24. Moyer CE, Bunnett JF (1963) J Am Chem Soc 85:1891–1893
25. Roberts JD, Vaughan CW, Carlsmith LA, Semenov DA (1956) J Am Chem Soc 78:611–614
26. de Graaff GBR, den Hertog HJ, Melger WC (1965) Tetrahedron Lett 963–968
27. Moses P, Gronowitz S (1961) Ark Kemi 18:119–132
28. Mach MH, Bunnett JF (1980) J Org Chem 45:4660–4666
29. de Bie DA, van der Plas HC (1968) Tetrahedron Lett 36:3905–3908
30. Mallet M, Queguiner G (1979) Tetrahedron 35:1625–1631
31. Peyron C, Navarra J-M, Van Craynest N, Benhida R (2005) Tetrahedron Lett 46:3315–3318
32. Stanetty P, Schnürch M, Mereiter K, Mihovilovic MD (2005) J Org Chem 70:567–574
33. Heiss C, Rausis T, Schlosser M (2005) Synthesis 617–621
34. Lukevics E, Arsenyan P, Belyakov S, Popelis J, Pudova O (2001) Tetrahedron Lett 42: 2039–2041
35. Peyron C, Navarra J-M, Dubreuil D, Vierling P, Benhida R (2008) Tetrahedron Lett 49: 6171–6174
36. Kobatake S, Imagawa H, Nakatani H, Nakashima S (2009) New J Chem 33:1362–1367
37. Holloway WG, Takeuchi K, Prunier ML, Navarro A (2009) Chem Ind 123:61–68
38. O'Connor MJ, Haley MM (2008) Org Lett 10:3973–3976
39. Bobrovsky R, Hametner C, Kalt W, Fröhlich J (2008) Heterocycles 76:1249–1259
40. Zhang X, Cote AP, Matzger AJ (2005) J Am Chem Soc 127:10502–10503
41. Machara A, Pojarova M, Svoboda J (2007) Collect Czechosolv Chem Commun 72:952–964
42. Li C, Shi J, Xu L, Wang Y, Cheng Y, Wang H (2009) J Org Chem 74:408–411
43. Stangeland EL, Sammakia T (2004) J Org Chem 69:2381–2385
44. Getmanenko YA, Tongwa P, Timofeeva TV, Marder SR (2010) Org Lett 12:2136–2139
45. Holzweber M, Schnürch M, Stanetty P (2007) Synlett 3016–3018
46. Schnürch M, Khan AF, Mihovilovic MD, Stanetty P (2009) Eur J Org Chem 3228–3236
47. Khan AF, Schnürch M, Mihovilovic MD, Stanetty P (2009) Lett Org Chem 6:171–174
48. Hämmerle J, Schnürch M, Iqbal N, Mihovilovic MD, Stanetty P (2010) Tetrahedron 66: 8051–8059
49. Hämmerle J, Spina M, Schnürch M, Mihovilovic MD, Stanetty P (2008) Synthesis 3099–3107
50. Kang SY, Song K-S, Lee J, Lee S-H, Lee J (2010) Bioorg Med Chem Lett 18:6069–6079
51. Stanetty P, Spina M, Mihovilovic MD (2006) Synlett 1433–1434
52. Williams DR, Fu L (2010) Synlett 591–594
53. DeMartino JK, Garfunkle J, Hochstatter DG, Cravatt BF, Boger DL (2008) Bioorg Med Chem Lett 18:5842–5846
54. Schlosser M, Ginanneschi A, Leroux F (2006) Eur J Org Chem 2956–2969
55. Eskildsen J, Ostergaard N, Vedso P, Begtrup M (2002) Tetrahedron 58:7635–7644
56. Pieterse MJ, den Hertog HJ (1962) Recl Trav Chim Pays Bas 81:855–863
57. Mallet M, Marsais F, Queguiner G, Pastour PCR (1972) Seances Acad Sci C 275:1535–1538
58. Mallet M, Queguiner G (1982) Tetrahedron 38:3035–3042
59. Mallet M, Branger G, Marsais F, Queguiner G (1990) J Organomet Chem 328:319–332
60. Rocca P, Cochennec C, Marsais F, Thomas-dit-Dummont L, Mallet M, Godard A, Queguiner G (1993) J Org Chem 58:7832–7838
61. Takagishi S, Schlosser M (1991) Synlett 119–121
62. Cottet F, Schlosser M (2004) Eur J Org Chem 3793–3798
63. Marzi E, Bigi A, Schlosser M (2001) Eur J Org Chem 1371–1376
64. Schlosser M, Rausis T (2004) Eur J Org Chem 1018–1024

65. Mongin F, Desponds O, Schlosser M (1996) Tetrahedron Lett 37:2767–2770
66. Rausis T, Schlosser M (2002) Eur J Org Chem 3351–3358
67. Mongin F, Schlosser M (1996) Tetrahedron Lett 37:6551–6554
68. Marzi E, Bobbio C, Cottet F, Schlosser M (2005) Eur J Org Chem 2116–2123
69. Hawkins GF, Roe A (1949) J Org Chem 14:328–331
70. Kyba EP, Liu S-T, Chackalingam K, Reddy BR (1988) J Org Chem 53:3513–3521
71. Schlosser M, Bobbio C (2002) Eur J Org Chem 4174–4180
72. Lazaar J, Rebstock A-S, Mongin F, Godard A, Trecourt F, Marsais F, Queguiner G (2002) Tetrahedron 58:6723–6728
73. Miller RE, Rantanen T, Ogilvie KA, Groth U, Snieckus V (2010) Org Lett 12:2198–2201
74. Mongin F, Rebstock A-S, Trecourt F, Queguiner G, Marsais F (2004) J Org Chem 69: 6766–6771
75. Bouillon A, Lancelot J-C, Collot V, Bovy PR, Rault S (2002) Tetrahedron 58:4369–4373
76. Duan X-F, Li X-H, Li F-Y, Huang C-H (2004) Synthesis 2614–2616
77. Li X, Hou K, Duan X, Li F, Huang C (2006) Inorg Chem Commun 9:394–396
78. Li X-H, Liu Z-Q, Li F-Y, Duan X-F, Huang C-H (2007) Chin J Chem 25:186–189
79. Daykin LM, Siddle JS, Ankers AL, Batsanov AS, Bryce MR (2010) Tetrahedron 66:668–675
80. Abboud M, Mamane V, Auber E, Lecomte C, Fort Y (2010) J Org Chem 75:3224–3231
81. Cheng D, Croft L, Abdi M, Lightfoot A, Gallagher T (2007) Org Lett 9:5175–5178
82. Trecourt F, Gervais B, Mallet M, Queguiner G (1996) J Org Chem 61:1673–1676
83. Mongin F, Trecourt F, Gervais B, Mongin O, Queguiner G (2002) J Org Chem 67:3272–3276
84. Sammakia T, Stangeland EL, Whitcomb MC (2002) Org Lett 4:2385–2388
85. Saitton S, Kihlberg J, Luthman K (2004) Tetrahedron 60:6113–6120
86. Saitton S, Del Tredici AL, Mohell N, Vollinga RC, Boström D, Kihlberg J, Luthman K (2004) J Med Chem 47:6595–6602
87. Wagner FF, Comins DL (2006) Eur J Org Chem 3562–3565
88. Ohtawa M, Ogihara S, Sugiyama K, Shiomi K, Harigaya Y, Nagamitsu T, Omura S (2009) J Antibiot 62:289–294
89. Arzel E, Rocca P, Grellier P, Labaeid M, Frappier F, Gueritte F, Gaspard C, Marsais F, Godard A, Queguiner G (2001) J Med Chem 44:949–960
90. Marull M, Schlosser M (2003) Eur J Org Chem 1576–1588
91. Gorecka J, Leroux F, Schlosser M (2004) Eur J Org Chem 64–68
92. Fürstner A, Domostoj MM, Scheiper B (2006) J Am Chem Soc 128:8087–8094
93. Efremova IE, Abzianidze VV, Elbert JE, Berestovitskaya VM (2004) J Sulfur Chemistry 25: 89–94
94. Tutino F, Papeo G, Quarteri F (2010) J Heterocyclic Chem 47:112–117

Top Heterocycl Chem (2012) 27: 219–246
DOI: 10.1007/7081_2011_61

Published online: 3 November 2011

Halogenated Heterocycles as Pharmaceuticals

Tina Kosjek and Ester Heath

Abstract Pharmaceuticals, which involve halogenated heterocyclic skeleton, are classified into various therapeutic groups such as antipsychotics, benzodiazepine tranquilizers, tricyclic antidepressants, antiallergics, thiazide and quinazolinone diuretics, antineoplastics, antiviral, antibacterial and antifungal agents, and antimalarials. A halogenated heterocycle involved in these drugs' structures either directly impacts drug's intrinsic activity or is responsible for other features such as lipid solubility, pharmacokinetics, or adverse effects. This chapter discusses the chemistry and structure–activity relationships, mechanisms of action, and the clinical use of pharmaceuticals involving the halogenated heterocyclic structure.

Keywords Antimetabolite · Antipsychotic · Fluoroquinolone · Halogenated heterocyclic pharmaceutical · Structure–activity relationship

Contents

T. Kosjek (✉) and E. Heath
Department of Environmental Sciences, Jožef Stefan Institute, Jamova 39, 1000 Ljubljana, Slovenia
e-mail: tina.kosjek@ijs.si

Abbreviations

5-FU	5-Fluorouracil
5-HT	5-Hydroxytryptamine or serotonin
AIDS	Acquired immune deficiency syndrome
ALK	Alkylamino
ATC	Anatomical Therapeutic Chemical Classification
BBB	Blood–brain barrier
CNS	Central nervous system
DNA	Deoxyribonucleic acid
FDA	US Food and Drug Agency
GABA	γ-Aminobutyric acid
HEP	Hydroxyethylpiperazines
HH	Halogenated heterocycles
HHPh	Halogenated heterocyclic pharmaceuticals
MIC	Minimum inhibitory concentration
OROS	Osmotic-controlled release oral delivery system
PHENO	Phenothiazine
PIP	Piperazine
QSAR	Quantitative structure–activity relationship
RNA	Ribonucleic acid
SAR	Structure–activity relationship
THIO	Thioxanthene

1 Introduction

Heterocyclic compounds play a vital role in synthetic chemistry and are widely distributed throughout the world. They are as pyrimidine and purine bases included in deoxyribonucleic acid (DNA), and are therefore considered vital to life. Further, the indole ring appears in a neurotransmitter serotonin and in an essential amino acid tryptophan, whereas imidazole and pyrrolidine rings are structural elements of neurotransmitter histamine and an amino acid proline, respectively.

Halogenated heterocycles (HH) find their use in different fields, and some also possess a pharmacological activity. A HH moiety involved in a drug's structure is not necessarily a part of a drug's pharmacophore; that is, a spatial arrangement of chemical groups or features in a molecule responsible for a particular pharmacological interaction. Instead, the HH moiety may impact other features such as lipid solubility, pharmacokinetics or adverse effects of a given agent. In this view, its role in structure–activity relationship (SAR) of different therapeutic groups is discussed. SAR is the relationship between the chemical structure or three-dimensional arrangement of a molecule and its pharmacological activity. The analysis of SAR enables the determination of drug's pharmacophore and allows the modification of its pharmacological effect or the potency by changing its chemical structure. When drug's activity, physicochemical properties, or structure are expressed quantitatively, the SAR may cast in a form of mathematic equation, or QSAR – quantitative structure–activity relationship [1]. Generally, QSAR analyses attempt to establish linear relationships between selected structural features in a series of related molecules and their known level of activity. If successful, models derived from training sets can be applied to predict molecules with higher potency, which is a basis of the rational drug design.

In this chapter, we attempt to provide a comprehensive review of halogenated heterocyclic pharmaceuticals (HHPh) approved by regulatory organs or under evaluation in clinical trials and discuss advances in their development considering the outcomes of SAR and QSAR analyses. Included herein are HHPh used in treatment of mental disorders (neuroleptics, antidepressants, and anxiolytics), antiallergic pharmaceuticals, diuretics, anticancer, antiviral and antibacterial agents, antimycotics, and antimalarials, as organized by their therapeutic and chemical characteristics.

2 Neuroleptics

The drugs primarily affecting mental processes can be divided into three major categories depending on their clinical usefulness: neuroleptics, antidepressants or mood-stabilizers, and tranquilants [2]. The neuroleptics, called antipsychotic agents, are used in the treatment of psychoses, such as schizophrenia, mania, and psychotic depression. Clinically, these agents counteract or minimize hallucinations and delusions, alleviate psychomotor excitement, and facilitate social adjustment [3]. Pharmacologically, the essential effect of these agents is the reduction of dopaminergic activity in the brain.

One crucial difference between the antipsychotic agents and most other pharmaceuticals is that their site of action is the central nervous system (CNS), which means they need to cross the blood–brain barrier (BBB) to reach it. The BBB separates the CNS from the circulating system (blood). It is formed from special brain capillaries, where the endothelial cells have particularly tight junctions. In order to cross the BBB, a compound either needs a specific transport mechanism present in the endothelial cells, or it has to diffuse into and across the lipid membranes of the BBB. For most endogenous compounds, passive diffusion over the membranes of the BBB is the only way to reach the CNS. Various factors influence the diffusion rate by which this passive transport into the CNS takes place, for example, pH partitioning (p*K*a), octanol/water partitioning (log P), and binding to other compartments (i.e. serum protein binding) [3]. Besides, when given orally, an antipsychotic agent also needs resistance to acid to survive stomach passage, and resistance to hydrolysis and degradation by other intestinal enzymes [2].

2.1 First Generation Antipsychotic Drugs

All typical antipsychotics have a high binding affinity for the dopamine receptor (particularly at receptors known as D_2 dopamine receptors) and their therapeutic effectiveness can be correlated with this affinity [4]. By antagonizing the binding of dopamine, these compounds reduce positive psychotic symptoms including hallucinations, delusions, disorganized speech, and behavior.

Among the typical antipsychotics, the most important chemical classes are phenothiazine (PHENO), thioxanthene (THIO), and butyrophenone derivatives. As illustrated in Fig. 1, the PHENO and THIO derivatives frequently involve halogenated heterocycles in their chemical structures. The halogen substituent is chlorine, which is located at carbon no. 2 of the ring A (Fig. 1).

Since the ionized molecules do not readily penetrate the BBB, the degree of ionization depends on the p*K*a and the pH, which is 7.4 for the blood. For example,

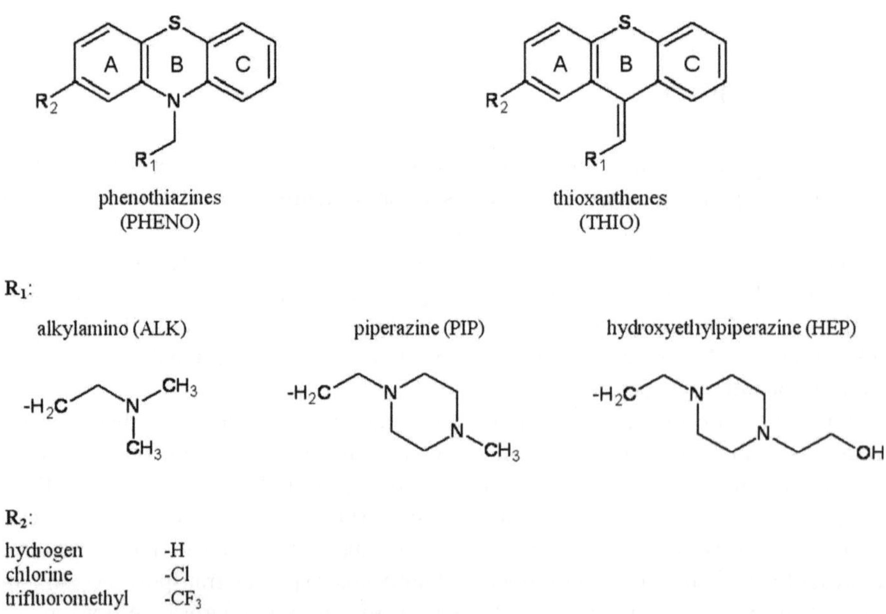

Fig. 1 Structure of PHENO and THIO drugs (5)

Table 1 PHENO and THIO drug structures and effects on dopamine-sensitive adenylate cyclase

Generic name	Ring	R_1	R_2	Relative clinical potency[a]
cis-Flupenthixol	THIO	HEP	CF3	1
cis-trans-Flupenthixol	THIO	HEP	CF3	2
Fluphenazine	PHENO	HEP	CF3	3
Trifluperazine	PHENO	PIP	CF3	4
Triflupromazine	PHENO	ALK	CF3	5
cis-Clopenthixol	THIO	HEP	Cl	6
Perphenazine	PHENO	HEP	Cl	7
cis-Chlorprothixene	THIO	ALK	Cl	8
Prochlorperazine	PHENO	PIP	Cl	9
Chlorpromazine	PHENO	ALK	Cl	10
Promazine	PHENO	ALK	H	11
trans-Chlorprothixene	THIO	ALK	Cl	12
trans Clopenthixol	THIO	HEP	Cl	13
trans-Flupenthixol	THIO	HEP	CF3	14

[a]Drugs are listed in descending order of milligram potency in treating schizophrenia and eliciting extrapyramidal side effects; adopted from [5]

PHENO derivative chlorpromazine, which contains two benzene rings, can be predicted to be sufficiently lipid soluble to penetrate the brain. Chlorpromazine is a weak base (p*K*a 9.2), and at pH 7.4, 98.4% is in the charged form. However, the chlorine substituent on the benzene ring further enhances its lipid solubility, which makes it to partition to a greater extent in lipids than into aqueous media [2]. Accordingly, the portion of chlorpromazine passing the BBB is higher and thus the drug shows higher potency than its nonchlorinated analogue promazine, as evident from Table 1.

The chlorinated PHENO and THIO derivatives differ in side chain substituents (R_1, Fig. 1) giving the following PHENO compounds: perphenazine, chlorpromazine and prochlorperazine, or THIO derivatives: chlorprothixene and clopenthixol.

Feinberg and Snyder [5] studied effect of R_1 and R_2 functional groups on the activity of dopamine-sensitive adenylate cyclase, which correlates directly and is a measure for the dopamine receptor antagonistic activity [6]. The results of their research proposed: (1) greater potency of drugs with trifluoromethyl rather than chlorine as the R_2 substituent; (2) enhanced activity of PHENO substituted by piperazine (PIP) compared to alkylamino (ALK) side chains; (3) increased potency associated with hydroxyethylpiperazines (HEP) as contrasted to piperazine side chains; (4) greater potency of *cis* rather than *trans* THIO analogues; and (5) crucial location of the ring A substituent (chlorine or trifluoromethyl group) at carbon no. 2 [5]. The results of this research are gathered in Table 1.

2.2 Dibenzepines

Another group of tricyclic antipsychotic agents are the dibenzepines, containing a seven-membered dibenzazepine central ring substituted with oxygen, nitrogen,

Table 2 Antipsychotic 11-piperazinyldibenzazepines (adopted from [2])

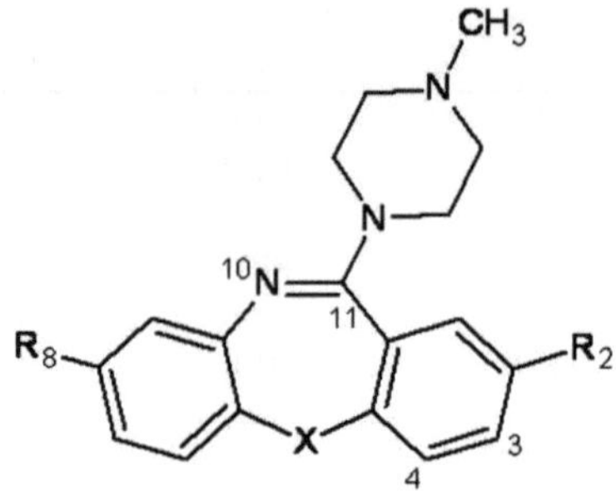

Generic name/trade name	Chemical type	X	R_8	R_2
Clozapine/Leponex®	Dibenzodiazepine	NH	Cl	H
Fluperlapine	Dibenzazepine	C	F	H
Loxapine/Loxitane®	Dibenzo-oxazepine	O	H	Cl
Clothiapine	Dibenzothiazepine	S	H	Cl

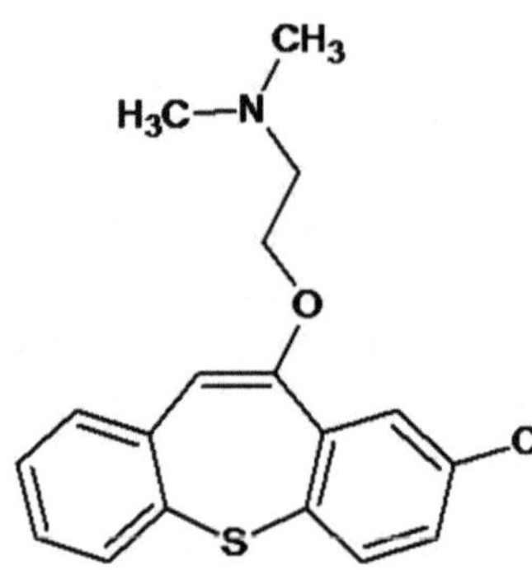

Fig. 2 Chemical structure of dibenzothiepin zotepine

sulfur, or carbon [7]. Similarly to PHENO and THIO derivatives, several dibenzepines contain a piperazine ring attached to the central ring (Table 2).

Loxapine and clozapine exemplify two growing families of agents. The loxapine-like family includes typical antipsychotics with prominent antidopaminergic activity (clothiapine, loxapine, zotepine). They contain an electron-withdrawing moiety (chlorine) at the position 2, relatively close to the side chain of nitrogen atoms (Table 2) [8]. As illustrated in Fig. 2, zotepine is the clothiapine analogue, holding a different (oxy-N,N-dimethylethanamine) side chain and lacking the nitrogen at position 10.

In contrast to the loxapine-like family, the clozapine-like family (clozapine, fluperlapine) has an electronegative substituent at position 8, away from the side-chain nitrogen atoms (Table 2). Clozapine has been shown the most promising among the dibenzazepine-type antipsychotics. The exact basis for the activity of clozapine is unknown, but importantly, it has low affinity at most dopamine receptors and it acts at a range of receptors and their subtypes such as muscarinic, serotonergic (5-HT_2), α-adrenergic, H_1 histamine, and others [8]. This fact suggests that a complex blend of interactions is required for its antipsychotic drug efficacy [7]. A consequence of clozapine's affinity to such a large variety of receptors is the manifestation of several clinically limiting side effects such as sedation, hypersalivation, and tachycardia [7]. Regardless of its severe adverse effects, clozapine was shown to be more effective than classical neuroleptic drugs and is therefore used particularly in case of treatment-resistant schizophrenia [9].

2.3 Atypical Antipsychotics

Among the first generation antipsychotics as well as dibenzazepines, clozapine is an exception, since it produces virtually no extrapyramidal side effects [2]. Extrapyramidal side effects are a group of movement disorders associated with physical disability and subjective discomfort and distress, including parkinsonism, dystonia (twisting and repetitive movements) and tardive dyskinesia (involuntary, repetitive body movements). The absence or the reduced risk for extrapyramidal symptoms as adverse effects classifies clozapine into the group of the so-called "Atypical antipsychotics," together with few other commercially available halogenated heterocyclic neuroleptics: risperidone, zipraspidone [10], and paliperidone [11].

Many theories have been developed to explain the atypical pharmacological profile of these compounds. It was suggested that a combination of serotonin (5-hydroxytryptamine, 5-HT) and dopamine receptor affinity in a proper ratio is crucial for atypicality [12], which served as the basis for the design of a series of potential atypical antipsychotic agents [13]. As illustrated in Fig. 3, risperidone

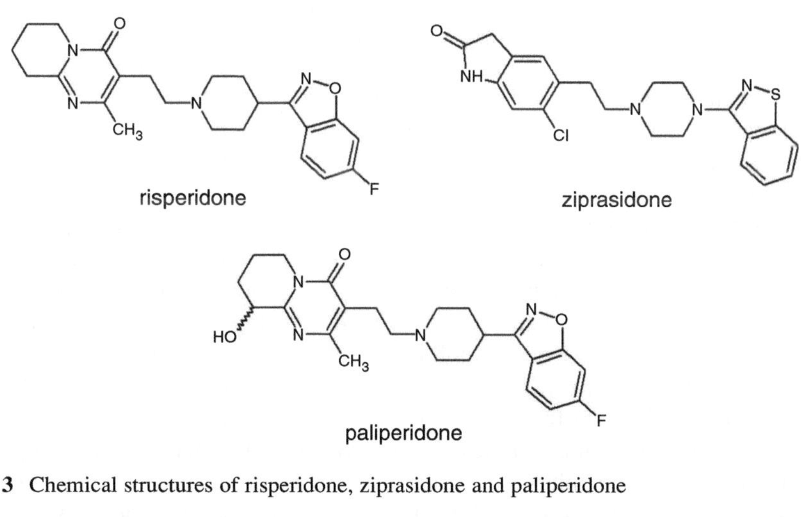

Fig. 3 Chemical structures of risperidone, ziprasidone and paliperidone

(Risperdal®), ziprasidone (Zolrix®), and paliperidone (Invega®) may be considered as mimicking the 5-HT structure, since they involve the tryptamine indole ring analogue structures: 6,5-fused heterocyclic rings such as benzisoxazole or benzisothiazole [14].

Paliperidone (9-hydroxyrisperidone) is an atypical antipsychotic, a primary active metabolite of risperidone, which in comparison shows improved pharmacokinetic behavior and reduced drug–drug interactions [11]. Paliperidone is marketed as Invega® or Invega Sustenna®. The first drug is an extended release formulation of paliperidone that uses the “OROS” (Osmotic-controlled Release Oral delivery System) extended release system to allow for once-daily dosing. Invega Sustenna (palperidone palmitate) is a long-acting injectable formulation of paliperidone indicated for once-monthly injection.

3 Tricyclic Antidepressants

The treatment of depression relies on a varied group of antidepressant therapeutic agents, in part because the clinical depression is a complex syndrome of widely varying severity. The first agents used successfully were tricyclic antidepressants, which elicit a wide range of neuropharmacological effects, but their primary pharmacological activity is inhibiting norepinephrine and serotonin uptake into nerve endings, which leads to sustained facilitation of noradrenergic and serotonergic function in the brain [8].

Because of the three-ring molecular core and therapeutic responses in most patients with major depression the trivial name “tricyclic antidepressants” is used for this group. SAR in the classical tricyclic antidepressant series shows a relative lack of specificities: tricyclic antidepressants have a nucleus consisting of two phenyl rings and a third, seven-membered central ring, which may have one, several, or no heteroatoms, and it may or may not be saturated [15]. In contrast to the structurally similar antipsychotics (see Sect. 2.1), ring substitution has little positive impact on antidepressant activity [16]. This might justify the fact that only two halogenated representatives of the series classical tricyclic antidepressant series are found: clomipramine and amoxapine (Fig. 4). Another important difference is the lack of coplanarity in the tricyclic structure of antidepressants as compared with relative planarity of the two aromatic rings in antipsychotics [16].

As illustrated in Fig. 4, tianeptine (Coaxil®) differs from classical tricyclic antidepressants as it involves a 3-chlorodibenzothiazepin nucleus and an aminoheptanoic side chain. In contrary with the lack of specific requirements that characterize the classical tricyclic series, the tianeptine series is sharply characterized with the following features: an aminocarboxylic chain (with an optimal length of six methylene links), a tricyclic system with an electron-donor heteroatom in position 5, and an aromatic substitution with a moderate electron-acceptor atom in position 3 [15].

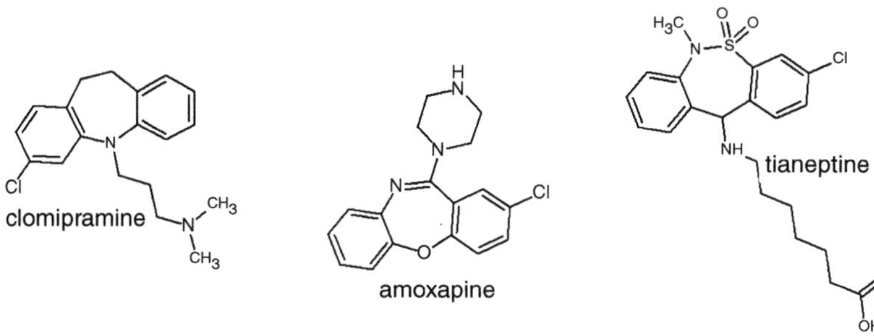

Fig. 4 Chemical structures of halogenated heterocyclic antidepressants: clomipramine, amoxapine and tianeptine

4 Benzodiazepines

Benzodiazepines are the drugs of choice in the pharmacotherapy of anxiety and related emotional disorders, sleep disorders, status epilepticus, and other convulsive states; they are used as centrally acting muscle relaxants, for premedication and as inducing agents in anesthesiology [2]. Potential shortcomings of benzodiazepines include tolerance and withdrawal symptoms [17]. When used alone, benzodiazepines carry low risk of acute toxicity. However, benzodiazepines often are used with other types of medications, including other drugs with abuse potential and these drugs can enhance the toxic effects of benzodiazepines. The latter interact synergistically with other central nervous system depressants, including other hypnotics, sedating antidepressants, neuroleptics, anticonvulsants, antihistamines, and alcohol. Fatal overdose with benzodiazepines is rare. When it does occur, the combination of benzodiazepines and alcohol and/or opiates, is often the cause [18].

Their anxiolytic action probably involves the neuronal effects of γ-aminobutyric acid (GABA), the most prominent inhibitory neurotransmitter in the central nervous system [2]. Benzodiazepine receptors are linked predominantly to GABA receptors. Binding of benzodiazepines enhances the affinity of the recognition site for GABA by inducing conformational changes that make GABA binding more efficacious. Activation of the benzodiazepine–GABA–chloride ionophore complex is responsible for producing the therapeutic anxiolytic effects of benzodiazepines and for mediating many of the side effects and, possibly, dependence and withdrawal from these drugs [19].

Sternbach and coworkers described the fundamental structure–activity relationships (SAR) of this classic benzodiazepine derivatives' series [20]. They recognized that the presence of the seven-membered imino-lactam ring (ring B) is essential and that substitution is advantageous only in positions 1, 3, 7, and 2′ (Fig. 5). In particular, electron-withdrawing groups, e.g. halogens, NO_2, and CF_3 increase the biological reactivity of benzodiazepines, whereas electron-releasing groups such as CH_3, and OCH_3 show the opposite effect. Further,

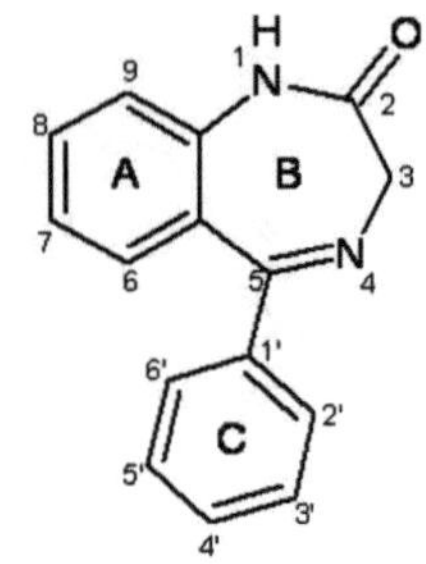

Fig. 5 5-phenyl-1,4-benzodiazepin-2-one: ring system of "classic" benzodiazepines [2]

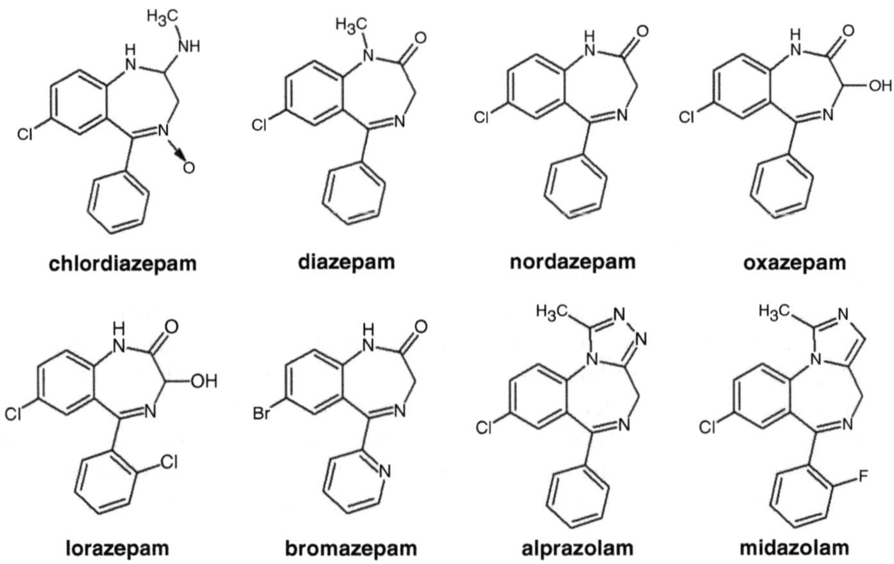

Fig. 6 Structural formulae and generic names of few commercially available halogenated benzodiazepines

a methyl group at position 1 of ring B increases the intrinsic activity, but it is decreased by larger substituents with *tert*-butyl derivative being completely inactive. On the ring C, the intrinsic activity is increased by halogen substitution at the 2′ position (e.g. Cl and F), and decreased or abolished by a substitution at the 4′ position [20].

Numerous 5-phenyl-1,4-benzodiazepin-2-one derivatives have been synthesized all over the world, and more than a dozen have found their way into the current therapy [2]. As evident in Fig. 6, the classical 5-phenyl-1,4-benzodiazepin-2-one nucleus has been modified by condensation of an additional ring on the 1,2-bond, giving triazolo- and imidazo- benzodiazepines. Figure 6 illustrates the halogenated benzodiazepine derivatives. Chlordiazepoxide/chlorodiazepam (Librium®) was the first benzodiazepine introduced to the marked in 1960. With structural modifications, the compound was replaced by 5–10 times more potent diazepam marketed as Valium®.

The pharmacokinetic parameters for benzodiazepines can often be misleading because the active metabolites with long half-lives can markedly alter the duration of effects [8]. For example, diazepam is in the human organism metabolized into N-desmethyldiazepam or nordazepam, which has a half-life of up to 100 h. Nordazepam subsequently is 3-hydroxylated into oxazepam. The hydroxyl group of oxazepam enables its rapid glucuronidation and excretion with the urine, which explains a significantly shorter half-life time of oxazepam as compared to diazepam or nordazepam [2]. Both, nordazepam and oxazepam are marketed also as individual drugs.

5 Antiallergic Agents

Allergy is a term comprising different hypersensitivity reactions, which are manifested as chronic or acute symptoms, such as hay fever, pruritus, contact dermatitis, drug rashes, utricaria, atopic dermatitis, and anaphylactic shock [21]. A mediator associated with manifestations of many allergic reactions is histamine (1-H-imidazole-4-ethanamine). Activation of H_1-histaminic receptors stimulates the contraction of smooth muscles in gut, uterus, and bronchi, and causes relaxation of capillaries, resulting in increased permeability leading to edema [21]. Besides histamine, other allergic mediators such as eicasonoids (prostaglandins, thromboxanes, leukotrienes) and platelet-activating factor, can be involved in allergic reactions and may surpass the role of histamine [3].

"Antihistamines" or histamine H_1-antagonists are commonly used terms for antiallergic agents that bind to but do not activate histamine H_1-receptors, thereby blocking the actions of endogenous histamine. H_1-antagonists can be classified into two major groups: classical and nonclassical H_1-antagonists. One major disadvantage of the classical H_1-antagonists is the induction of sedation, possibly attributed to occupation of cerebral H_1-receptors. CNS effects are to be anticipated because most classical antihistamines have the ability to cross the BBB [21]. Due to the unfavorable therapeutic profile of classical H_1-antagonists, new, second generation H_1-antagonists have been developed, which exhibit higher selectivity and less sedative potential due to their decreased ability to penetrate the BBB [22]. In his Thesis, ter Laak [3] in-depth discussed optimal physicochemical properties and conformational requirements for the rational design of potent and selective nonsedating H_1-antagonists. The halogenated heterocyclic representatives of the second generation H_1-antihistamines are illustrated in Fig. 7.

Classical H_1-receptor antagonists share a common pharmacophore consisting of two aromatic rings and a basic nitrogen atom [3]. The second-generation H_1-antagonists, among which some few halogenated heterocyclic antihistamines can be found, either partly resemble the molecular structure of classical antihistamines (loratadine and desloratadine), or these structural features are absent (temelastine, Fig. 7. In the case of antihistamines, the halogenated heterocycle is not a part of their pharmacophore, but more likely controls the pharmacokinetics, bioavailability, or side-effects and is thus not a common structural element for antihistamines.

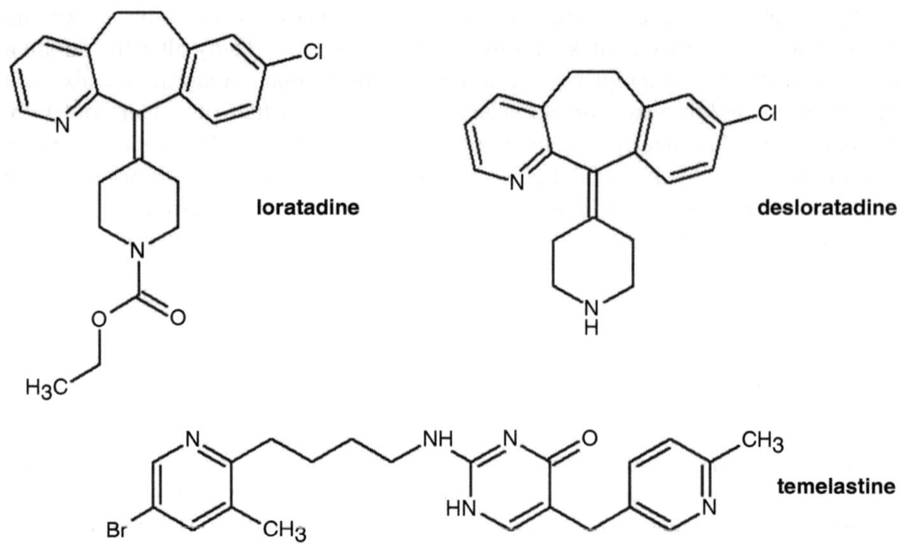

Fig. 7 Non-sedative H_1-antagonists containing halogenated heterocycles

6 Diuretics

Diuretics are chemicals that increase the rate of urine formation leading to the increased excretion of electrolytes (especially sodium and chloride ions) and water from the body. These pharmaceuticals are used in various edematous conditions resulting from congestive heart failure, nephrotic syndrome, chronic liver disease, and in the management of the hypertension, and as a sole agent or adjunctive therapy in treatment of hypercalcemia, acute mountain sickness, glaucoma and others [23]. Diuretics are classified by their chemical class, mechanism, and site of action, and effects on urine contents. One major class of diuretics, which involves the halogenated heterocyclic compounds, is thiazide diuretics. Alternatively, their structural analogues quinazolinone diuretics show similar pharmacologic profile.

6.1 Thiazide Diuretics

The members of thiazide diuretics are derived from benzothiadiazine, with the 2 *H*-1,2,4-benzothiadiazine nucleus illustrated in Fig. 8, and chlorothiazide, the simplest member of this series of diuretics (Fig. 8). Thiazide diuretics inhibit the sodium-chloride transporter protein in the distal tubule, which results in reduced reabsorption of sodium and chloride ions from the distal convoluted tubules in the kidneys. Thiazides also cause loss of potassium and may thus cause hypokalemia, and lower urinary calcium excretion. This effect is associated with

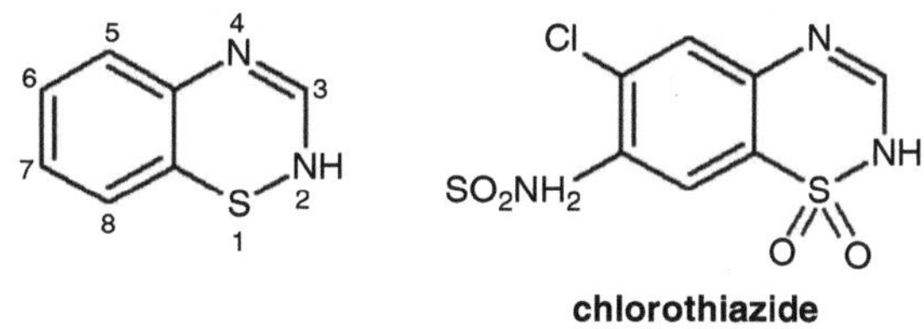

Fig. 8 Thiazide diuretics: their nucleus and the most recognized representative (chlorothiazide)

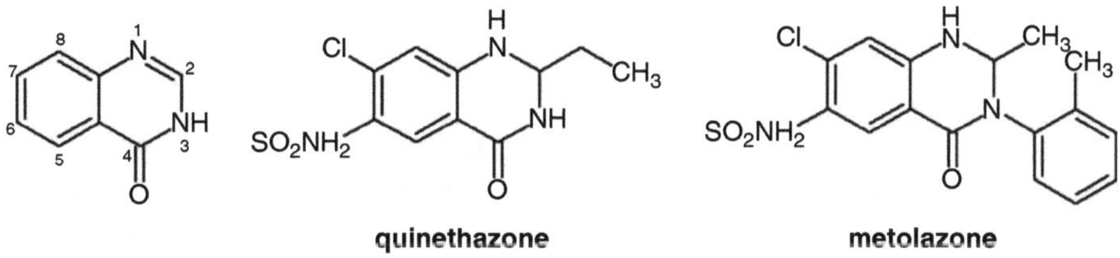

Fig. 9 Quinazolinone diuretics: the "pharmacophore" and two representatives (quinethazone and metolazone)

positive calcium balance and consequently with an increase in bone mineral density and reductions in fracture rates attributable to osteoporosis.

Thiazide diuretics are weakly acidic, due to acidity of N-2 hydrogen atom affected by electron withdrawing effects of the neighboring sulfone group. The sulphonamide group that is substituted at C-7 is the additional point of acidity. These acidic protons make possible the formation of a water-soluble sodium salt that can be used for intravenous administration of the diuretics. An electron withdrawing group at C-6, such as Cl or trifluoromethyl increases the intrinsic activity of the thiazide diuretics. The trifluoromethyl-substituted diuretics are more lipid-soluble resulting in a longer half-life time as compared to their chlorinated analogues. On the contrary, the electron-releasing groups, such as methyl or methoxyl, or no substitution at C-6 markedly reduce the diuretic activity of thiazides. Replacement or removal of sulphonamide group at C-7 yields compounds with little or no diuretic activity, whereas saturation of the 3,4-double bond produces a 10-times more effective analogue [23].

6.2 *Quinazolinone Diuretics*

The quinazolin-4-one molecule has been structurally modified from the thiazide diuretics by replacement of the sulfone group by carbonyl group. With this structural analogy it is not surprising that the quinazolinones have the similar diuretic effect as well as the adverse effects. The representatives of this class are quinethazone and metolazone, illustrated in Fig. 9.

7 Halogenated Heterocyclic Compounds in Cancer Chemotherapy

Cancer represents one of the major causes of death in the developed world. The medical term for cancer or tumor is "neoplasm." Tumour is a general term indicating any abnormal mass or growth of tissue, not necessarily life threatening. A "cancerous tumor" is a malignant neoplasm. A critical difference between benign and malignant neoplasms is that malignant neoplasms are subjected to uncontrolled proliferation, loss of function due to lack of capacity to differentiate, invasiveness, and the ability to metastasize [24].

Different and often combined approaches to treat cancer are used today including surgery, radiation therapy, immunotherapy, and chemotherapy. In contrast to surgery and radiotherapy, the chemotherapy is not so much limited by metastasis as by the total mass of tumors [24]. Anticancer drugs, named also antineoplastic drugs are according to the Anatomical Therapeutic Chemical (ATC) Classification System classified into the group L01, and can be divided into the following subgroups:

- L01A – Alkylating agents
- L01B – Antimetabolites
- L01C – Plant alkaloids and other natural products
- L01D – Cytotoxic antibiotics and related substances
- L01X – Other antineoplastic agents [25]

In recent years, the discovery of new agents extended to entirely new treatment approaches, including genetic therapies, manipulations of the immune system, stimulation of normal hematopoietic elements, induction of differentiation in tumor tissues, and inhibition of angiogenesis [26]. However, it is unlikely that these new approaches will completely replace the conventional therapy, since the conventional drugs have become increasingly effective and their toxicity manageable and predictable [26].

Among the antineoplastic agents there are several representatives of halogenated heterocyclic compounds, which are all classified into the subgroup L01B antimetabolites. These agents inhibit a metabolic pathway essential for survival or reproduction of cancer cells through inhibition of folate, purine, pyrimidine, and pyrimidine nucleoside pathways required for DNA synthesis [24].

DNA stores genetic information, and is in the cell located in the chromosomes in its nucleus. When a cell divides, the genetic information is transferred and maintained in daughter cells, unless a change is induced by a mutation or evolution [24]. The DNA carries the information necessary for its exact duplication in the precise sequence of purine (adenine and guanine) and pyrimidine (thymine and cytosine) bases along its sugar phosphodiester backbone. The purine and pyrimidine bases are in DNA bound to deoxyribose sugar via a β-glycosidic linkage and form nucleosides. Nucleosides are phosphorylated on

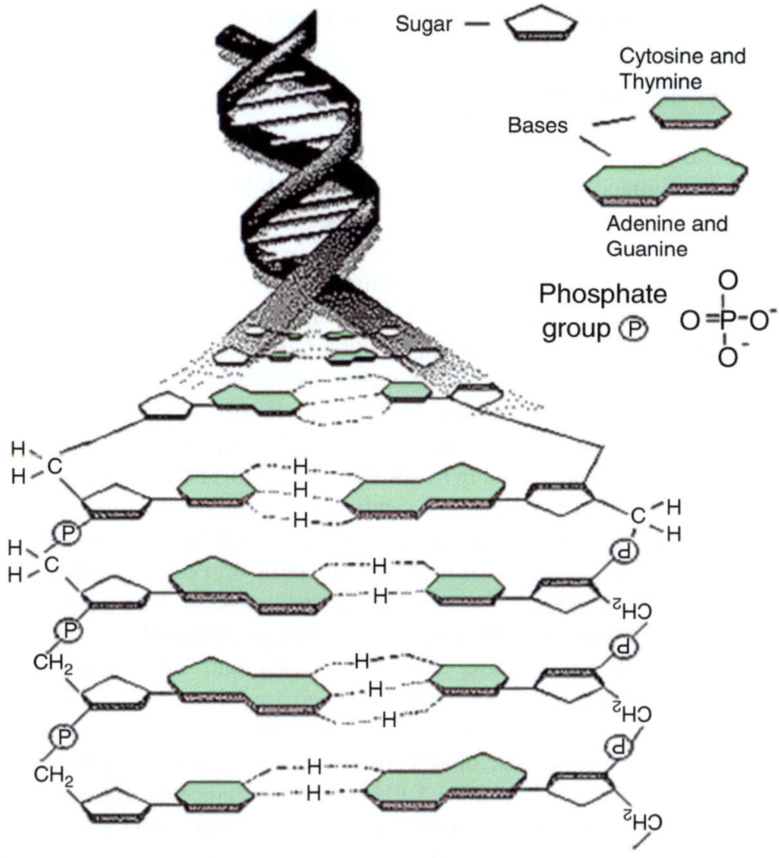

Fig. 10 DNA structure (adopted from [27])

the sugar's primary alcohol group (–CH_2-OH), producing deoxyribonucleotides, which are the molecular building-blocks of DNA. As illustrated in Fig. 10, DNA is a double-stranded helix of two individual molecular chains held together by hydrogen bonding between pairs of bases in opposite positions in the two chains.

By interfering with the processes of DNA replication and consequently cell division and the growth of tumors, halogenated heterocyclic antimetabolites are used in cancer treatment. These antimetabolites mimic the structure of the purine or pyrimidine bases or their nucleosides in DNA, but hold one crucial difference: the base and/or the sugar heterocycles involve a halogen substituent. In this view, the halogenated heterocyclic metabolites can be classified into purine analogues, pyrimidine analogues, and their nucleoside analogues.

7.1 Antimetabolites: Pyrimidine Analogues

7.1.1 Fluoropyrimidines

The most renowned representative of halogenated heterocyclic pyrimidine antimetabolites is 5-fluorouracil (5-FU), which was developed by Roche in the United States in the 1950s and is still being used widely today for systemic and local cancer treatment [28]. In 5-FU, a fluorine atom is substituted for hydrogen at the 5 position of the pyrimidine of uracil (Fig. 11). This carbon fluorine is extremely stabile and precludes the addition of a methyl group at the 5 position, thus preventing the formation of thymidine. To be active, 5-FU must be phosphorylated to the nucleotide, and, as such inhibits thymidilate synthase, a key enzyme in biosynthesis of the DNA [24].

A number of reviews on the mechanism of thymidilate synthase inhibition by 5-FU and derivatives have been published [29, 30] and an understanding of 5-FU mechanism of action has resulted in major therapeutic advances in the past 15 years. As a result, the inhibition of thymidilate synthase by antimetabolites remains a classic approach and a key strategy for suppressing cell division in cancerous tissue [30, 31].

5-FU has been extensively used in the treatment of skin cancers and a variety of solid tumors such as breast, colorectal, and gastric cancers. It is usually administered by intravenous bolus or by continuous infusion. Although the latter route is generally more efficient and less toxic, it is costly and inconvenient [30, 32]. Unfortunately the treatment with 5-FU has been found to cause neurotoxic and cardiotoxic side effects, and can show devastating bone marrow and gastrointestinal toxicity deriving from the lack of selectivity of a drug toward tumors [24, 30], or development of resistance toward 5-FU [33]. To enhance the therapeutic index of 5-FU, different combination therapies have been used, which improved the efficiency of the treatment [30, 34]. Furthermore, in response to the need for a more convenient agent with an improved safety profile and equivalent/superior efficacy compared with intravenous 5-FU, several new oral fluoropyrimidines have been studied and developed [35]. These 5-FU prodrugs are being used clinically, either alone or in combination therapies in several countries. The examples involve furtulon (doxifluridine), camofur, ftorafur (tegafur), galocitabine, capecitabine, and emitefur [30, 36, 37], which are converted into the bioactive compounds in vivo. Their chemical structures are shown in Fig. 12.

The most prominent of oral fluoropyrimidines is Capecitabine (Xeloda®), an oral tumor-selective fluoropyrimidine carbamate approved in the treatment of

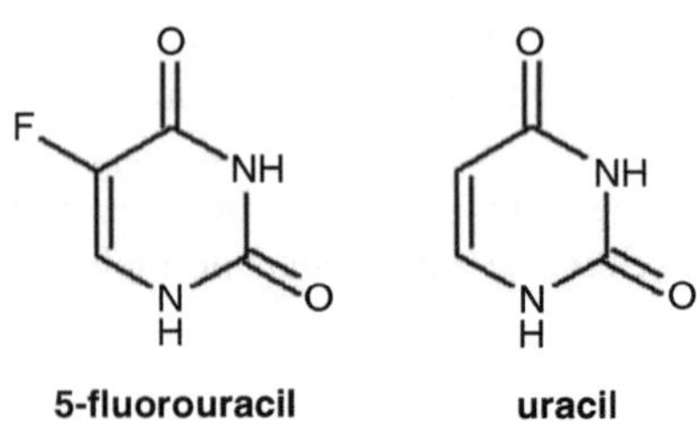

Fig. 11 Chemical structures of 5-FU and uracil

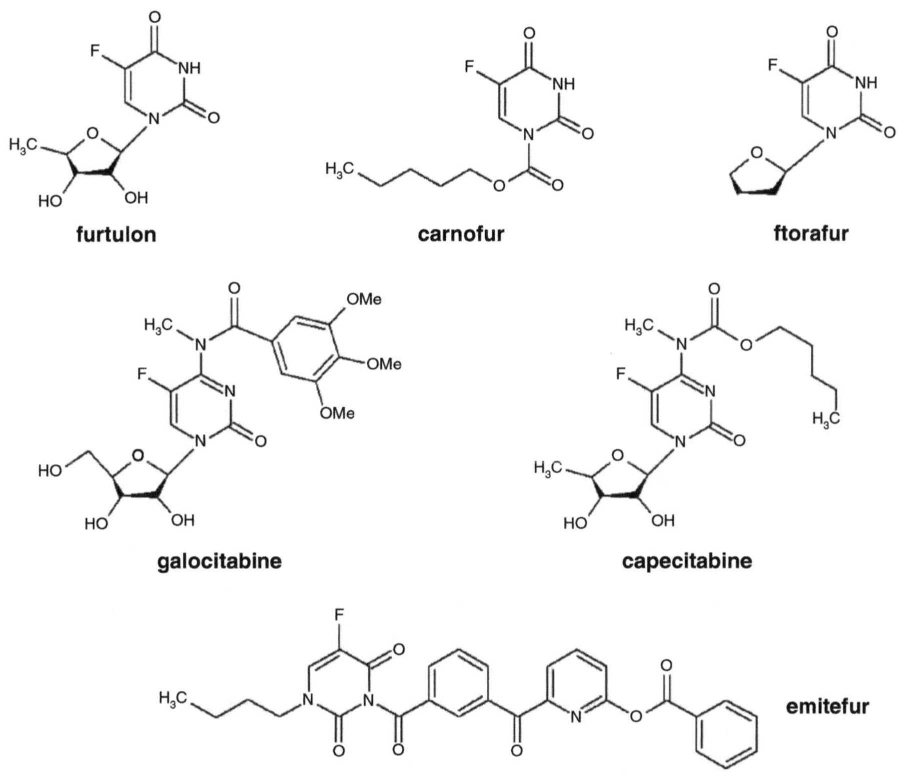

Fig. 12 Chemical structures of fluoropyrimidines

colorectal and breast cancer. It is itself a prodrug of another 5-FU prodrug (furtulon or doxifluridine) and was designed to minimize the substantial local gastrointestinal toxicity of doxifluridine without compromising its antitumor efficacy [28, 38, 39]. The pharmacologically inactive capecitabine is reliably absorbed unchanged from the gastrointestinal tract and then converted through three enzymatic reactions to 5-FU.

In summary, the unifying theme for all of the oral fluoropyrimidine therapies is that they are pharmacologically rational renderings of oral 5-FU. Oral 5-FU alone has erratic absorption and nonlinear pharmacokinetics, whereas most of the new oral fluoropyrimidine compounds release 5-FU that has linear pharmacokinetics. Finally, oral fluoropyrimidines have the potential to provide the more patient-friendly oral administration therapy and resulting improved quality of life, a more favorable toxicity profile, the economic advantage and the equivalent antitumor efficacy compared to standard regimens [38, 40].

7.1.2 Pyrimidine Fluoronucleosides

Halogenated pyrimidine nucleosides are well documented and explored as potential anticancer agents. They mimic natural nucleosides in terms of uptake and

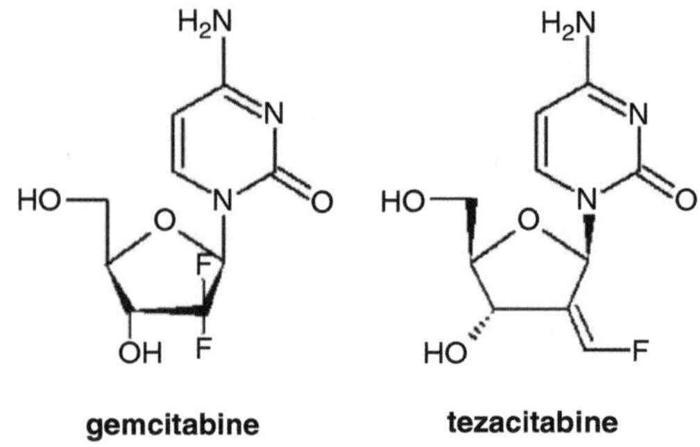

Fig. 13 Chemical structures of pyrimidine fluoronucleosides

metabolism, and can incorporate into newly synthesized DNA resulting in chain termination. Some of these drugs also inhibit key enzymes involved in the biosynthesis of purine and pyrimidine nucleotides and (ribonucleic acid) RNA synthesis as well as directly activate the apoptosis as anticancer agents [41].

Halogenated pyrimidine nucleosides are nucleoside analogues with the halogen positioned on the deoxyribose sugar (gemicitabine, tezacitabine, Fig. 13). Chemically, gemicitabine and tezacitabine have the hydrogen atoms on the 2′ carbon of deoxycytidine replaced by two fluorine atoms and a fluoromethylene group, respectively. These nucleoside analogues are potent irreversible inhibitors of an enzyme present at elevated levels in tumor tissue that catalyzes the biosynthesis of deoxyribosenucleotides, key building blocks for DNA replication [42]. Gemicitarabine (Gemzar®) and tezacitabine are prodrugs that are metabolized intracellularly to the active diphosphate and triphosphate nucleosides. Their cytotoxic effects are exerted through incorporation of the triphosphate nucleosides into DNA, resulting in inhibition of DNA synthesis and induction of apoptosis [43]. Gemicitabine also shows promising antiviral activity [44].

In many cases, the stability of the nucleoside analogue, particularly the stability of the glycosyl bond, is an important factor determining the biological activity as well as the therapeutic usefulness of nucleosides as drug candidates. Fluorine substitution at the 2- or 3-position of a sugar is known to increase the chemical stability of nucleoside analogues, particularly in an acidic environment [45]. Fluorine substitution also has a favorable effect of increasing the metabolic stability [46–48]. These findings suggest the importance of a fluorine moiety in the nucleoside as therapeutic agents.

7.2 Antimetabolites: Purine Analogues

Fludarabine (Fludara®), cladribine (Litak®), and clofarabine (Evoltra®) are synthetic purine analogues in which a hydrogen is replaced by a halogen, such as fluorine (fludarabine) or chlorine (cladribine and clofarabine) atom (Fig. 14). Such substitution allows the halogenated purines to be incorporated into DNA but further replication of the DNA is disrupted. Besides the halogenated purine moiety,

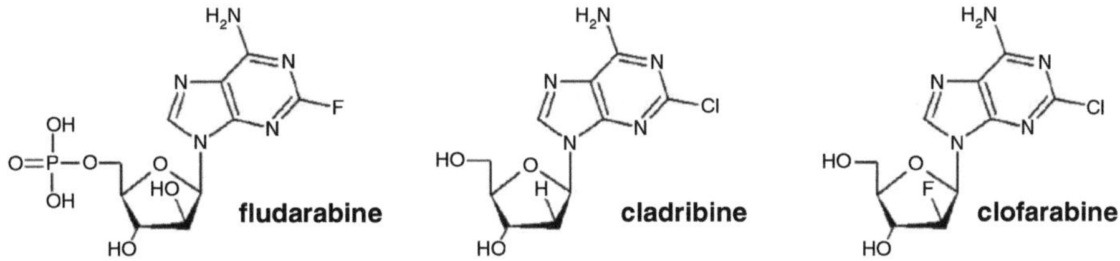

Fig. 14 Halogenated purine antimetabolites

clofarabine also contains a fluorine on the 2′ carbon of the deoxyribose, analogously to pyrimidine fluoronucleosides discussed in the previous section. Clofarabine is a newer generation nucleoside antineoplastic agent, which like other nucleoside analogues, exerts its anticancer effect by inducing "apoptosis" by blocking DNA synthesis as well as through inhibition of DNA repair [49]. In comparison to fludarabine [30], clofarabine has a higher potency in treatment of acute myeloid leukemia, and is appropriate for a combination therapy with other anticancer agents and in combination with radiotherapy. The current experimental evidence showed that a moderate dose of clofarabine along with radiotherapy led to a significant inhibition of tumor growth [50]. This result suggests the potential of clofarabine as a significant radiosensitizer [49].

8 Halogenated Heterocyclic Antiviral Agents

Viruses are microscopic organisms that can infect all living cells. They are parasitic and multiply at the expense of host's metabolic system [51]. Viruses are among smallest microorganisms, varying in size from 0.02 to 0.4 μm, and consist of nucleic acid core that contain either DNA or RNA. The nucleic acid core is surrounded by a protein coat known as capsid. The entire structure is called nucleocapsid [51].

Because viruses are obligate intracellular parasites, their replication depends on host's cellular processes. Ideally, an antiviral drug is considered most effective, if it interferes with the viral replication without affecting normal cellular metabolic processes. Unfortunately, this objective has not been achieved with many antiviral compounds, with many antiviral drugs being proved toxic to humans at therapeutic levels or having a limited spectrum of activity [51].

Antiviral agents discussed in this chapter show similar mode of action and are also structurally strongly related to nucleoside analogues used in the therapy of antineoplastic diseases (see Sect. 7.1.2). Nucleoside analogues have been the cornerstone of antiviral therapy over the past thirty years. In the effort to discover effective antiviral agents against acquired immune deficiency syndrome (AIDS) and viral hepatitis, a large number of nucleoside analogues have been synthesized and evaluated. Although SAR studies have not led to a pharmacophore model for the antiviral activities of nucleosides, some structural features have been particularly successful [46]. For example, all six of the nucleoside reverse transcriptase inhibitors approved by the US Food and Drug Agency (FDA) for the treatment of AIDS (zidovudine, didanosine, zalcitabine, stavudine, lamivudine, and abacavir) can be considered as 2′,3′-dideoxynucleosides. Among their ring substituents, electron withdrawing groups such as azido [52] and fluorine [53–55] have often produced potent antiviral activity. Another structural feature that is often beneficial for antiviral activity is a 2′,3′-unsaturated bond [46]. The chemical structure involving the above described features is given in Fig. 15.

A number of fluorine-substituted nucleoside analogues have demonstrated their potent antiviral activity as well as favorable chemical and pharmacological properties by virtue of the small but highly electronegative nature of the fluorine atom, which is also capable of participating in hydrogen bonding [47]. Thus, fluorination at the 5-position of the pyrimidine base as well as at the 2′- and/or 3′-position of the sugar moiety of the nucleoside analogue have been extensively studied in the pursuit of safe, effective, and chemically stable antiviral agents [47]. It has also been found that the introduction of a fluorine atom into the 2′-position of nucleosides, and of purine nucleosides in particular, resulted in a stabilized glycosyl bond with good antiviral activity [48, 56, 57]. Watanabe et al. [56] compared the antiherpes virus activity in different halogen-substituted arabinonucleosides, finding that the 2′-fluoro moiety exhibits, generally, more potent antiviral activity than do their 2′-chloro or 2′-bromo analogues. However, myelosuppression and neurotoxicity may limit the usefulness of these nucleosides, and it was found that the unnatural L configuration gives selective antiviral agents having lower toxicities than the corresponding D isomers [57]. Figure 16 shows examples of halogenated nucleosides potentially useful in treatment of various virus infections [45, 48, 58].

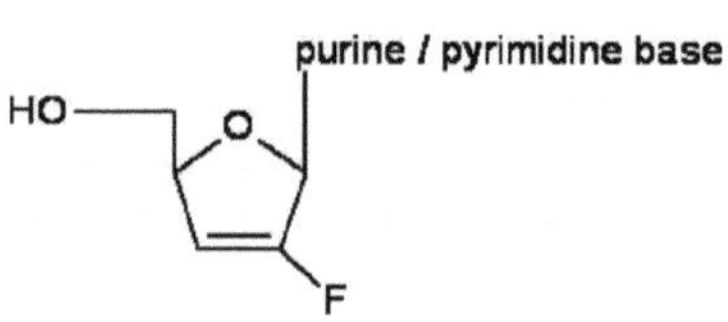

Fig. 15 Structure of 2′,3′-didehydro-2′,3′-dideoxy-2′-fluoro nucleosides

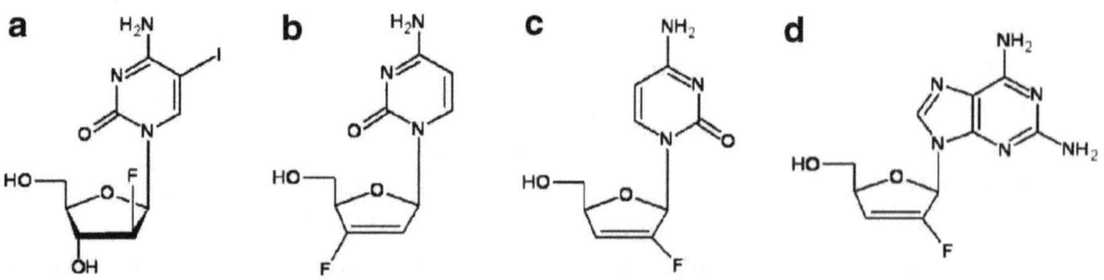

Fig. 16 Halogenated heterocyclic antiviral agents: (**a**) fluoroiodoaracytosine [51], (**b**) pyrimidine nucleoside with 3′-fluoro position on sugar, (**c**) pyrimidine nucleoside with 2′-fluoro position on sugar, (**d**) purine nucleoside with 2′-fluoro position on sugar

9 Antibacterial Agents: Fluoroquinolones

Antibiotics are substances produced by various species of microorganisms (bacteria, fungi, actinomycetes) that suppress the growth of other microorganisms and eventually destroy them. However, common usage often extends the term antibiotics to include synthetic antibacterial agents, such as the sulphonamides and quinolones, which are not the products of microbes [59]. This section will focus on the quinolones, since they are the only class of antibacterials holding the halogenated heterocycles in their structures.

Quinolones are chemotherapeutic bactericidal drugs, eradicating bacteria by inhibiting the bacterial DNA gyrase or the topoisomerase II enzyme, thereby inhibiting DNA replication and transcription. The quinolone antimicrobials comprise a group of synthetic substances possessing in common an N-alkylated 3-carboxypyrid-4-one ring fused to a second aromatic ring that carries other substituents [60]. Figure 17 shows the basic fluoroquinolone "pharmacore" [61]. Two major groups have been developed from this basic structure: quinolones and naphthyridones. The presence of nitrogen at position 8 identifies the naphthyridones, whereas a carbon and associated group at position 8 identifies the quinolone (Fig. 17). A major breakthrough in the study of the quinolone antibacterial agents, originally derived from quinine, was the introduction of fluorine at position 6, which resulted in up to 100-fold improvement in minimum inhibitory concentration (MIC) [61].

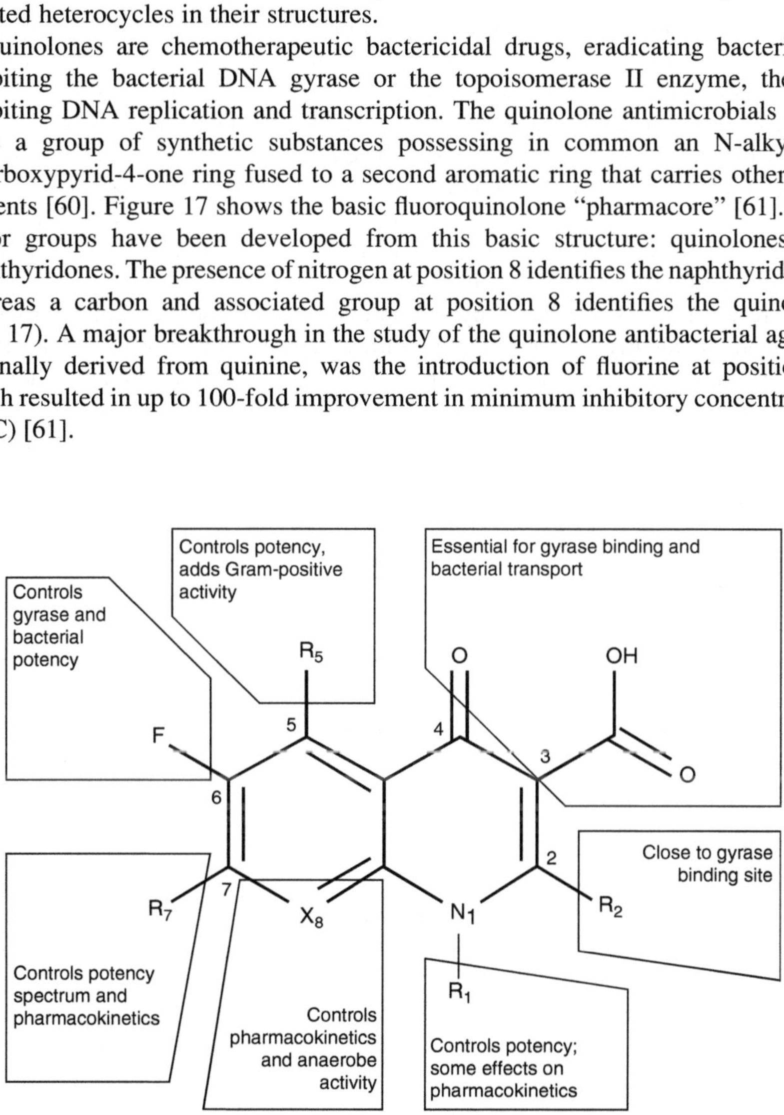

Fig. 17 Structure of quinolone and napthyridone molecule (adapted from [61] and [62])

The potency of quinolones and napththyridones were further enhanced by the addition of groups to N-1, C-5, and C-7. The substituents at position C-7 are associated with a number of key attributes, such as antibacterial spectrum, bioavailability, and side effects. The most common are cyclic amino groups, for example piperazine and pyrrolidine rings; other groups have been less successful. Piperazine rings are particularly common (e.g. norfloxacin, ciprofloxacin, Fig. 18) conferring the activity against Gram-negative organisms. Further, pyrrolidine ring is associated with improved activity against Gram-positive bacteria (e.g. clinafloxacin, Fig. 18); however, this group decreases water solubility and oral bioavailability. Introduction of methyl group on the pyrrolidone ring helps to overcome some of these physical properties [61]. The potency of a drug is further increased by introduction of cyclopropyl group to the N-1 position, which is found in ciprofloxacin, clinafloxacin, gatifloxacin, moxifloxacin, and balofloxacin (Fig. 18).

Considering the structural changes to the quinolone molecule and their correlation with adverse effects, the photoreactivity is probably most influenced by substitution on the position 8, with chlorine of fluorine producing most phototoxic potential (e.g. clinfloxacin, Fig. 18) and methoxy groups the least (e.g. moxifloxacin, balofloxacin, gatifloxacin, Fig. 18).

There are several reports of hybrid antibiotics constructed from fluoroquinolones and covalently connected penicillins, cephalosporins, and carbapenems. Such examples are derivatives of 6-(3-ethyl-4-methylanilino)uracil, shown to be potent inhibitors of a topoisomerase subtype, a DNA-dependent DNA polymerase that is required for the DNA synthesis in Gram + bacteria. A series of hybrid DNA polymerase-topoisomerase inhibitors were prepared by linking 6-(3-ethyl-4-methylanilino)uracil and fluoroquinolones [63]. A potent member of

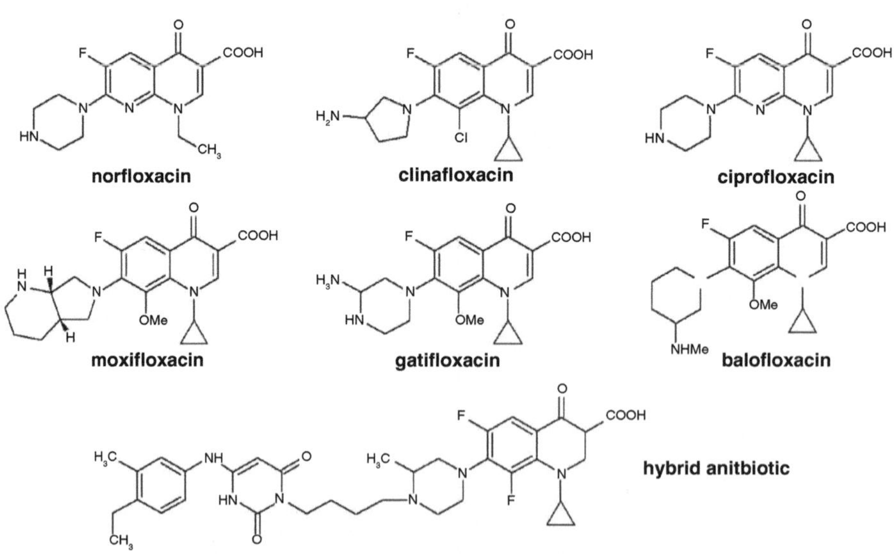

Fig. 18 Fluoroquinolone antibiotics

this series is illustrated in Fig. 18. The compound at nontoxic doses completely protected mice from lethal infections of *S. aureus*. However, problems with water solubility are being addressed.

10 Halogenated Heterocyclic Antifungal Agents

Fungi are plant-like, nonphotosynthetic eukaryotes growing either in colonies of single cells (yeasts) or in filamentous multicellular aggregates (molds). Some fungi species are parasites on terrestrial plants, a smaller number produce disease in humans and animals. Mycotic illnesses in humans are divided into three groups: contagious skin and hair infections; noncontagious soilborne infecions or airborne systemic infections; and noncontagious foodborne toxemias [64]. The responsible organisms and methods of prevention differ with each group. A high percentage of humans are infected by various soilborne or airborne fungi, but few healthy persons develop serious diseases or die. Individuals who do become seriously ill often have been weakened by immunosuppresive treatment for other clinical conditions (chemotherapy or organ transplants) or have reduced cellular immune response due to, for example, AIDS [64]. The particularly high susceptibility of these individuals to life-threatening infections caused by pathogenic fungi has increased the urgency for improved antifungal therapies.

A representative of halogenated heterocyclic antifungal agents is 5-fluorocytosine or flucytosine, a fluorinated pyrimidine, which was originally developed as a potential antileukemic agent [64]. Susceptible fungi are capable of deaminating assimilated 5-fluorocytosine to 5-FU, a potent antimetabolite (see Sect. 7.1.1) [65]. 5-FU is then converted to 5-fluoro-2′-deoxyuridylic acid, an inhibitor of thymidylate synthetase. DNA synthesis is impaired as a result of this reaction (Fig. 19). Alternatively, 5-FU can be incorporated into RNA to result in inhibition of protein synthesis [64]. Mammalian cells do not convert flucytosine to 5-FU, which is crucial for the selective action of this compound on fungi [65]. Still, serious adverse effects may occur, including depression of the bone marrow function and the resulting development of trombocytopenia or leukopenia. Other adverse effects,

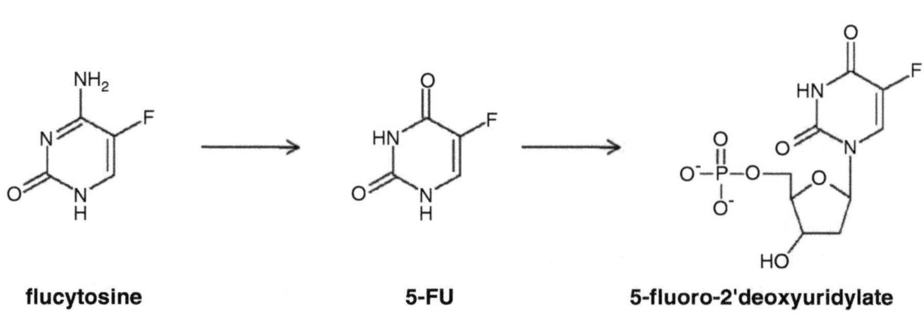

Fig. 19 Conversion of 5-fluorocytosine to 5-fluoro-2′-deoxyuridylic acid via 5-FU

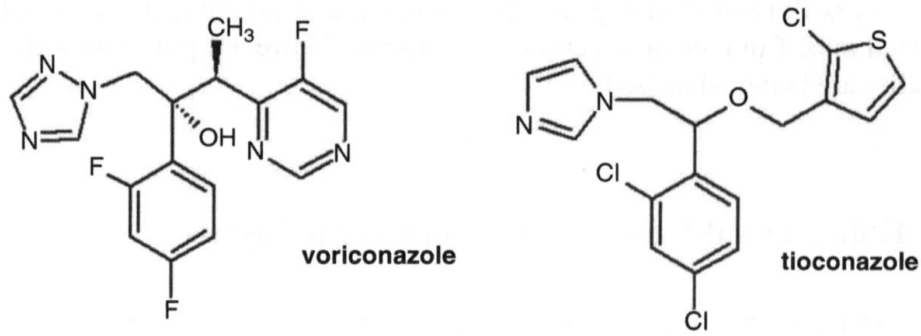

Fig. 20 Azole antifungal agents

including rash, nausea, vomiting, diarrhea, and severe enterokolitis, have been noted. Toxicity of flucytosine may be the result of its conversion to 5-FU by the microbial flora in the gastrointestinal tract of the host [65].

The azole antifungals include two broad classes, imidazoles and triazoles. Both classes share the same antifungal spectrum and mechanism of action. They are used topically for the treatment of cutaneous dermatophytic and yeast infections. Several azoles are also available for oral administration to treat systemic mycoses and fluconazole can also be administered intravenously [64]. The antifungal azoles have a five-membered N-1 atom linked to other aromatic rings via an aliphatic carbon atom(s). They inhibit a cytochrome P-450, which catalyses the 14 α-demethylation of lanosterol. The N-3 atom of the azole ring binds to the ferric iron atom in the heme prosthetic group to prevent the activation of oxygen for insertion into lanosterol [64]. This is a key step in the biosythesis of ergosterol, an essential sterol of fungal membrane [4].

The imidazole ring is crucial for the antifungal activity, but it is associated with a rapid metabolism and high lipophilicity. In order to bolster resistance to metabolism, the imidazole moiety was replaced by triazole [4]. Figure 20 illustrates a triazole (voriconazole) and diazole (tioconazole) antifungals. In general, the azole rings of antifungal agents are not halogenated; instead the halogen is positioned on the pyrimidine and thiophene ring of voriconazole and tioconazole, respectively. Further, good activity was observed in compounds with a halogen in the 2 and/or 4 positions of benzene ring, which can be noted in both azole antifungals illustrated in Fig. 20 [4, 64].

11 Parasite Chemotherapy

Parasitic diseases are widely distributed around the world and have a heavy impact on social and economic development, especially in Third World countries. The incidence of some parasitic diseases reaches as high as 80% in developing countries, where some of the extremely serious parasitic diseases are endemic [66]. Parasites vary in size from single-celled protozoa that cause malarial infections and

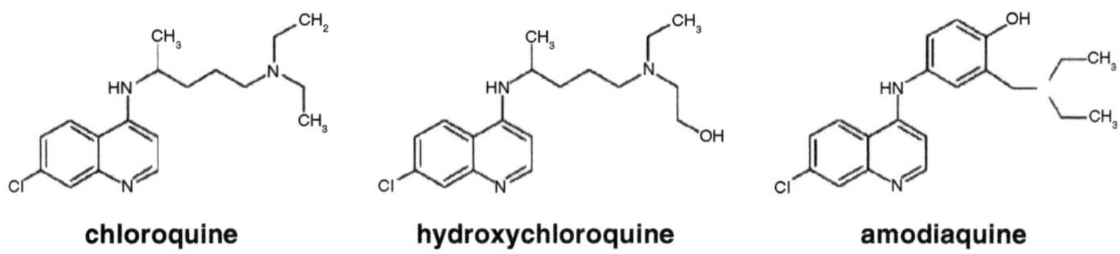

Fig. 21 Halogenated heterocyclic antimalarials

amoebic dysentery, to larger, more complex helmiths or worms. Protozoans cause more human parasitic infections than any other type of organism, with malaria being one of the most common causes of death in the world [66]. Malaria is transmitted by an infected female Anopheles mosquito. The specific protozoan organisms causing malaria belong to a genus *Plasmodium.*

Among the antimalarial agents the first known is quinine, an alkaloid derived from a bark of an evergreen cinchona tree. Quinine is a quinolinemethanol derivative bearing a substituted bicyclic quinuclidine ring. Quinine treated millions of malaria patients, but it is a very bitter substance, which may lead to serious unwanted effects such as premature contractions in late state of pregnancy, hemolysis after parenteral administration and visual disturbances after overdose [66]. These issues motivated further research for other antimalarials basing on quinoline pharmacophore, which included halogenated quinoline structures, among others. An example of a halogenated (chlorinated) quinoline is chloroquine, which is considered one of the most important antimalarial drugs for suppression and prophylaxixs in most regions where malaria is endemic [66]. Chloroquine involves 7-chloro-4-aminoquinoline structure, which is common to other two halogenated heterocyclic antimalarials, amodiaquine and hydroxychloroquine. The chemical structures of all three antimalarials are illustrated in Fig. 21.

12 Conclusions

Halogenated heterocycles are incorporated in various pharmaceuticals' structures, among which their role in antipsychotics, antineoplastics, and antibacterial agents is particularly notable. The introduction of a halogen can increase the metabolic stability of a heterocyclic compound or add specific effects that enhance its binding to target macromolecules.

Specific benefits include the following:

- In drugs affecting the central nervous system, the halogen increases permeability of blood–brain barrier due to changes in lipophilicity or pK_a.
- The fluoro substituent in pharmacologically active heterocycles may be considered as an isosteric replacement of hydrogen, and frequently leads to a dramatic change in pharmacological activity.

- Fluorine is incorporated into most antineoplastic antimetabolites, either into base (pyrimidine or purine) or into sugar heterocycle.
- Fluorinated nucleosides are antiviral agents exhibiting analogous inhibitory mechanisms as antineoplastic antimetabolites.
- The introduction of fluorine into the quinolone heterocycle resulted in important improvement in their antibacterial potency.

Halogenated heterocycles are also present in other therapeutic groups such as antimalarials, benzodiazepine tranquilizers, tricyclic antidepressants, antiallergics, thiazide and quinazolinone diuretics, azole antifungal agents, and in individual halogenated heterocyclic pharmaceuticals such as nonsteroidal anti-inflammatory agent carprofen, or potassium-sparing diuretic amiloride. Drugs targeting novel mechanisms are constantly under development and introduction of halogenated heterocycles has become an important strategy in the design and discovery of novel drug candidates.

Acknowledgments The research leading to these results has received funding from the Ministry of Higher Education, Science and Technology of Slovenia, Project Z1-3677: "Psychoactive pharmaceuticals and their transformation products in water treatment processes," from the Slovenian Research Agency under Research Programme P1-0143 – "Cycling of nutrients and contaminants in the environment, mass balances and modeling of environmental processes and risk analysis," and from the European Community's Seventh Framework Programme (FP7/2007-2013) under grant agreement n. 265264: "CytoThreat": "Fate and effects of cytostatic pharmaceuticals in the environment and the identification of biomarkers for improved risk assessment on environmental exposure."

References

1. Tute MS, Kier LB (1995) In: Lemke TL, Williams DA, Foye WO (eds) Principles of medicinal chemistry, 4th edn. Williams & Wilkins, Philadelphia
2. Neumeyer JL, Booth RG (1995) In: Lemke TL, Williams DA, Foye WO (eds) Principles of medicinal chemistry, 4th edn. Williams & Wilkins, Philadelphia
3. ter Laak Am (1995) Doctoral thesis, Vrije Universiteit, Amsterdam. http://dare.ubvu.vu.nl/bitstream/1871/12721/1/tekst.pdf. Accessed 10th April 2011
4. Kirk KL (2006) J Fluor Chem 127:1013–1029
5. Feinberg AP, Snyder SH (1975) Proc Nat Acad Sci 72:1899–1903
6. Kebabian JW, Petzold GL, Greengard P (1972) Proc Nat Acad Sci 69:2145–2149
7. Tehan BG, Lloyd EJ, Wong MG (2001) J Mol Graph Model 19:417–426
8. Baldessarini RJ (1996) In: Limbird LE, Hardman JG (eds) Goodman & Gilman's the pharmacological basis of therapeutics, 9th edn. McGraw-Hill, New York
9. Meltzer HY (1997) Curr Med Res Opin 14:1–20
10. Farah A (2005) Prim Care Companion J Clin Psychiatry 7:268–274
11. Teague SJ (2011) Drug Discov Today 16:398–411
12. Hrib NJ, Jurcak JG, Burgher KL, Conway PG, Hartman HB, Kerman LL, Roehr JE, Woods AT (1994) J Med Chem 37:2308–2314
13. Lowe JA III, Seeger TF, Nagel AA, Howard HR, Seymour PA, Heym JH, Ewing FE, Newman ME, Schmidt AW (1991) J Med Chem 34:1860–1866
14. Prakash C, Kamel A, Chui D (1997) Drug Metab Dispos 25:897–901

15. Labrid C, Moleyre J, Poignant JC, Malen C, Mocaër E, Kamoun A (1988) Clin Neuropharmacol 11:21–31
16. Nieforth KA, Gianutsos G (1995) In: Lemke TL, Williams DA, Foye WO (eds) Principles of medicinal chemistry, 4th edn. Williams & Wilkins, Philadelphia
17. Riss J, Cloyd J, Gates J, Collings S (2008) Acta Neurol Scand 118:69–86
18. Longo LP, Johnson B (2000) Am Fam Physician 61:2121–2128
19. Arana GW, Hyman SE (1991) Handbook of psychiatric drug therapy, 2nd edn. Little Brown, Boston
20. Sternbach LH (1979) J Med Chem 22:1–7
21. Di Bella M, Braghiroli D, Witiak DT (1995) In: Lemke TL, Williams DA, Foye WO (eds) Principles of medicinal chemistry, 4th edn. Williams & Wilkins, Philadelphia
22. Simons FER, Simons KJ (1999) Clin Pharmacokinet 36:329–352
23. Rankin GO (1995) In: Lemke TL, Wilkins DA, Foye WO (eds) Principles of medicinal chemistry, 4th edn. Williams & Wilkins, Philadelphia
24. Foye WO, Sengupta SK (1995) In: Lemke TL, Williams DA, Foye WO (eds) Principles of medicinal chemistry. Williams & Wilkins, Philadelphia
25. ATC/DDD Index (2011) International language for drug utilization research. WHO Collaborating Centre for Drug Statistics Methodology/Norwegian Institute of Public Health. http://www.whocc.no/. Accessed 15 Apr 2011
26. Calabresi P, Chabner BA (1996) In: Limbird LE, Limbird JG, Hardman JG (eds) Goodman & Gilman's the pharmacological basis of therapeutics, 9th edn. McGraw-Hill, New York
27. DNA Molecule - Two Views. Access Excelence @ the National Health Museum Resource Center. http://www.accessexcellence.org/RC/VL/GG/dna_molecule.php. Accessed 15 Apr 2011
28. Straub JO (2009) Integrated Environ Assess Manag 6:540–566
29. Malet-Martino M, Jolimaitre P, Martino R (2002) Curr Med Chem Anticancer Agents 2: 267–310
30. Isanbor C, O'Hagan D (2006) J Fluor Chem 127:303–319
31. Lehman NL (2002) Expert Opin Investig Drugs 11:1775–1787
32. Levy-Piedbois C, Durand-Zaleski I, Juhel H, Schmitt C, Bellanger A, Piedbois P (2000) Ann Oncol 11:157–161
33. Mader RM, Muller M, Steger GG (1998) Gen Pharmacol 31:661–666
34. Macdonald JS (1999) CA Cancer J Clin 49:202–219
35. Zufıa L, Aldaz A, Giráldez J (2004) J Chromatogr B 809:51–58
36. Danenberg PV (2004) Frontiers Biosci 9:2484–2494
37. Petty RD, Cassidy J (2004) Curr Cancer Drug Targets 4:191–204
38. Lamont EB, Schilsky RL (1999) Clin Cancer Res 5:2289–2296
39. Bajetta E, Colleoni M, Di Bartololmeo M, Buzzoni R, Bozzetti R, Doci R, Somma L, Cappuzzo F, Stampino CG, Guenzi A, Balant LP, Zilembo N, Di Leo A (1995) J Clin Oncol 13:2613–2619
40. Tauxe-Wuersch A, de Alencastro LF, Grandjean D, Tarradellas J (2006) Intern J Environ Anal Chem 86:473–485
41. Liu P, Sharon A, Chu CK (2008) J Fluor Chem 129:743–766
42. Rodriguez GI, Jones RE, Orenberg EK, Stoltz ML, Brooks DJ (2002) Clin Cancer Res 8: 2828–2834
43. Gemcitabine. Cancer Care Ontario. http://www.cancercare.on.ca/common/pages/UserFile.aspx?fileId=10623. Accessed 15 Apr 2011
44. Cerqueira NMFSA, Fernandes PA, Ramos MJ (2007) Chem Eur J 13:8507–8515
45. Lee K, Choi Y, Gumina G, Zhou W, Schinazi RF, Chu CK (2002) J Med Chem 45:1313–1320
46. Gumina G, Schinazi RF, Chu CK (2001) Org Lett 3:4177–4181
47. Chong Y, Gumina G, Mathew JS, Schinazi RF, Chu CK (2003) J Med Chem 46:3245–3256
48. Zhou W, Gumina G, Chong Y, Wang J, Schinazi RF, Chu CK (2004) J Med Chem 47: 3399–3408
49. Faderl S, Gandhi V, Kantarjian HM (2008) Curr Opin Hematol 15:101–107

50. Cariveau MJ, Stackhouse M, Cui XL, Tiwari K, Waud W, Secrist JA III, Xu B (2008) Int J Radiat Oncol Biol Phys 70:213–220
51. Sethi ML (1995) In: Lemke TL, Williams DA, Foye WO (eds) Principles of medicinal chemistry, 4th edn. Williams & Willkins, Philadelphia
52. Mitsuya H, Weinhold KJ, Furman PA, St. Clair MH, Lehrman SN, Gallo RC, Bolognesi D, Barry DW, Broder S (1985) Proc Natl Acad Sci USA 82:7096–7100
53. Owen GR, Verheyden JPH, Moffatt JG (1976) J Org Chem 41:3010–3017
54. Marquez VE, Tseng CKH, Mitsuya H, Aoki S, Kelley JA, Ford H Jr, Roth JS, Broder S, Johns DG, Driscoll JS (1990) J Med Chem 33:978–985
55. Pai SB, Liu SH, Zhu YL, Chu CK, Cheng YC (1996) Antimicrob Agents Chemother 40: 380–386
56. Watanabe KA, Su TL, Klein RS, Chu CK, Matsuda A, Chun MW, Lopez C, Fox J (1983) J Med Chem 26:152–156
57. Chu CK, Ma TW, Shanmuganathan K, Wang CG, Xiang YJ, Pai SB, Yao GQ, Sommadossi JP, Cheng YC (1995) Antimicrob Agents Chemother 39:979–981
58. Clercq E (1984) Eur J Clin Microbiol Infect Dis 3:96–107
59. Chambers HF, Sande MA (1996) In: Limbird LE, Hardman JG (eds) Goodman & Gilman's the pharmacological basis of therapeutics, 9th edn. McGraw-Hill, New York
60. Mitscher LA (1995) In: Lemke TL, Williams DA, Foye WO (eds) Principles of medicinal chemistry, 4th edn. Williams & Wilkins, Philadelphia
61. Andersson MI, MacGowan AP (2003) J Antimicrob Chemother 51:1–11
62. Domagala JM (1994) J Antimicrob Chemother 33:685–706
63. Zhi C, Long Z, Manikowski A, Comstock J, Xu WC, Brown NC, Tarantino PM, Holm KA, Dix EJ, Wright GE, Barnes MH, Butler MM, Foster KA, LaMarr WA, Bachand B, Bethell R, Cadilhac C, Charron S, Lamothe S, Motorina I, Storer R (2006) J Med Chem 49:1455–1465
64. Weinberg ED (1995) In: Lemke TL, Williams DA, Foye WO (eds) Principles of medicinal chemistry, 4th edn. Williams & Wilkins, Philadelphia
65. Bennett JE (1996) In: Limbird LE, Hardman JG (eds) Goodman & Gilman's the phatmacological basis of therapeutics, 9th edn. McGraw-Hill, New York
66. Foye WO (1995) In: Lemke TL, Williams DA, Foye WO (eds) Principles of medicinal chemistry, 4th edn. Williams & Wilkins, Philadelphia

Top Heterocycl Chem (2012) 27: 247–268
DOI: 10.1007/7081_2011_59

Published online: 22 September 2011

Sources, Occurrence and Fate of Halogenated Heterocyclic Pharmaceuticals in the Environment

Ester Heath and Tina Kosjek

Abstract After pharmaceuticals are excreted from target organisms, they enter the environment, where they can be diluted, dissociated, biologically or abiotically degraded or sorbed/accumulated in various environmental compartments, depending on their physicochemical properties. This chapter will review the existing literature on three model halogenated heterocyclic pharmaceuticals belonging to different pharmaceutical classes as environmental contaminants. Their occurrence and fate in the environment will be discussed and conclusions will be drawn regarding their common structure and fate.

Keywords 5-Fluorouracil · Antibacterial · Antineoplastic · Anxiolytics · Ciprofloxacin · Diazepam · Environment · Halogenated heterocyclic pharmaceutical

Contents

Abbreviations

5–FU	5-Fluoroacil
AOP	Advanced oxidation processes

E. Heath (✉) and T. Kosjek
Department of Environmental Sciences, Jožef Stefan Institute, Ljubljana, Slovenia
e-mail: ester.heath@ijs.si

BCF	Bioconcentration factor
CIP	Ciprofloxacin
DOC	Dissolved organic carbon
DZ	Diazepam
HHPh	Halogenated heterocyclic pharmaceuticals
K_d	Distribution coefficient
k_H	Henry's coefficient
K_{oc}	Organic carbon partition coefficient
K_{ow}	Octanol-water partition coefficient
LOD	Limit of detection
˙OH	Hydroxyl radicals
PBT profiler	Persistent, bioaccumulative, and toxic profiles estimated for organic chemicals
pK_a	Dissociation constant
WW	Wastewater
WWTP	Wastewater treatment plants

1 Pharmaceuticals in the Environment

The environment is contaminated by a number of organic micropollutants released from urban, industrial and agricultural activities, and an increase in their presence as a result of human activities has been observed over the last decade. One of the emerging groups of organic pollutants of special concern is pharmaceutical residues, especially in natural water systems. Pharmaceuticals are pharmacologically active substances, whose influence on ecosystems, and their potential threat to human health, is yet to be fully realized. Their main source in the environment is wastewater treatment plants (WWTP), whose effluents, even after classical (biological) treatment, still contain pharmaceutical residues and release them into receiving water bodies. Although environmental monitoring covers more and more types of organic compounds, such as biocides and pesticides, recent efforts have been taken to recognize the importance of pharmaceutical residues bioaccumulation and effects due to chronic exposure from low environmental concentrations and synergistic effects. Diclofenac is one such compound. Research published in the scientific journal Nature in 2004 [1] confirmed that veterinary use of Diclofenac, a halogenated aromatic compound with non-steroidal anti-inflammatory application, is responsible for the recent devastating declines in South Asian vulture populations. Albeit Diclofenac has been used in human medicine for decades, only recently it was introduced for veterinary use in India and Pakistan. Vultures appear to have been exposed to the drug while scavenging livestock carcasses. Its effect on the birds has been rapid, with death following a few days after exposure. These findings show that even if certain pharmaceuticals are safe for humans and mammals, they can have fatal effects for other organisms. This makes

Diclofenac a candidate drug to be included in legislation and regular monitoring programmes.

Pharmaceuticals are an important and indispensable element of modern life. Many of these compounds are used in large quantities in everyday life and their demand is increasing due to a developing health care system, demographic changes and in response to increased exposure to the psychological stresses of modern life. In parallel to increasing consumption, the burden on the environment posed by pharmaceutical residues is higher. Pharmaceutical residues from human and veterinary medicine are mostly eliminated from the body via urine and faeces either in its original or partially transformed form. Alternatively, they can be present in discharges from the pharmaceutical industry or may enter the environment after improper disposal. Once eliminated, they enter water ecosystems via wastewater effluents, and/or in the case of animal waste, via surface run-off, where they can be diluted, dissociated, biologically or abiotically degraded or sorbed/accumulated in various environmental compartments, depending on their physicochemical properties.

Pharmaceuticals are, due to their continuous consumption and discharge into the environment, ubiquitously present in surface waters in ng/L to μg/L levels. Concentrations measured in the aquatic environment are usually at least three to four orders of magnitude lower than those having a recognized pharmacological effect. Risks arising from acute exposure to pharmaceutical residues can therefore be regarded as unlikely. However, possible effects of life-long exposures as well as synergistic effects have still to be determined.

2 Halogenated Heterocyclic Pharmaceuticals

Within Europe there are approximately 3,000 different pharmaceuticals commonly used [2] including painkillers, antibiotics, antidiabetics, antihypertonics, anxiolytics, contraceptives, lipid regulators, antidepressants, impotence drugs and cytostatic agents; however, only about 50 have halogenated heterocyclic structure and most of them are in Table 1 [3].

HHPh are expected, mostly due to the presence of a halogen atom, to be problematic for biodegradation [4] and consequently could show possible persistence in the environment. Albeit by definition they share a halogenated heterocyclic skeleton, they greatly differ in chemical structure, which conditions their mode of action [5] as well as their environmental fate and behaviour.

This chapter will address three potentially environmentally significant HHPh belonging to different pharmaceutical groups. The parameters determining the fate and distribution of organic pollutants including HHPh into the various environmental compartments can, to some extent, be predicted from their chemical structures (Table 2) and physico-chemical properties (Table 3) including the dissociation constant (pK_a), bioconcentration factor (BCF), octanol–water partition coefficient (K_{ow}), organic carbon partition coefficient (K_{oc}),

Table 1 List of halogenated heterocyclic pharmaceuticals and their pharmacological activity

Compound	Pharmacological group
Diazepam	Anxiolytic, anticonvulsant
Clobazam	Anxiolytic, anticonvulsant
Nordazepam	Anxiolytic, anticonvulsant
Oxazepam	Anxiolytic, anticonvulsant
Lorazepam	Anxiolytic, anticonvulsant
Bromazepam	Anxiolytic
Alprazolam	Anxiolytic, hypnotic
Midazolam	Anxiolytic, hypnotic
Clozapine	Atypical antipsychotic
Chlorpromazine	Antipsychotic
Chlorprothixene	Antipsychotic
Clopenthixol	Antipsychotic
Perphenazine	Antipsychotic
Prochlorperazine	Antipsychotic
Loxapine	Antipsychotic
Clothiapine	Antipsychotic
Fluperlapine	Antipsychotic
Zotepine	Antipsychotic
Risperidone	Atypical antipsychotic
Zipraspirone	Atypical antipsychotic
Paliperidone	Atypical antipsychotic
Amoxapine	Tricyclic antidepressant
Clomipramine	Tricyclic antidepressant
Tianeptine	Tricyclic antidepressant
Chlorotiazide	Thiazide diuretic
Trichlormethiazide	Thiazide diuretic
Methyclothiazide	Thiazide diuretic
Polythiazide	Thiazide diuretic
Cyclothiazide	Thiazide diuretic
Amiloride	Potassium sparing diuretic
Metolazone	Quinazolinone diuretic
Quinethazone	Quinazolinone diuretic
Temelastine	Antiallergic
Loratadine	Antiallergic
Desloratadine	Antiallergic
Carprofen	Nonsteroidal anti-inflammatory drug
5-Fluorocytidine	Metabolite antagonist
Fludarabine	Anticancer agent: purine analogue
Cladribine	Anticancer agent: purine analogue
Clofarabine	Anticancer agent: purine analogue
Gemicitarabine	Anticancer agent: pyrimidine fluoronucleoside
Tezacitarabine	Anticancer agent: pyrimidine fluoronucleoside
5-Fluorouracil	Anticancer agent: pyrimidine analogue
Furtulon	Anticancer agent: pyrimidine analogue
Camofur	Anticancer agent: pyrimidine analogue
Ftorafur	Anticancer agent: pyrimidine analogue

(continued)

Table 1 (continued)

Compound	Pharmacological group
Galocitarabine	Anticancer agent: pyrimidine analogue
Capecitarabine	Anticancer agent: pyrimidine analogue
Emitefur	Anticancer agent: pyrimidine analogue
Chloroquine	Antimalarial
Amodiaquine	Antimalarial
Hydroxychloroquine	Antimalarial
Quinacrine	Antimalarial
Norfloxacin	Antibacterial agent
Clinafloxacin	Antibacterial agent
Ciprofloxacin	Antibacterial agent
Moxifloxacin	Antibacterial agent
Gatifloxacin	Antibacterial agent
Balofloxacin	Antibacterial agent
Ofloxacin	Antibacterial agent
Lomefloxacin	Antibacterial agent
Griseofulvin	Antifungal agent
Fluoroiodoaracytosine	Antiviral agent
Tioconazole	Antifungal agent
Voriconazole	Antifungal agent
Flucytosine	Antifungal agent

For details see Chap. 7. Halogenated heterocycles as pharmaceuticals

Table 2 Main characteristics of selected HHPh

Selected HHPh	Ciprofloxacin	Diazepam	5-Fluorouracil
Abbreviation	CIP	DZ	5-FU
Trade name	Ciloxan Ciprobay Cipro	Valium	Adrucil Carac Efudex
Chemical structure	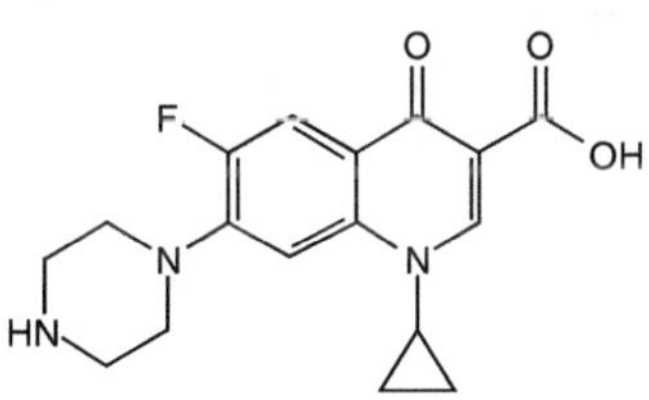	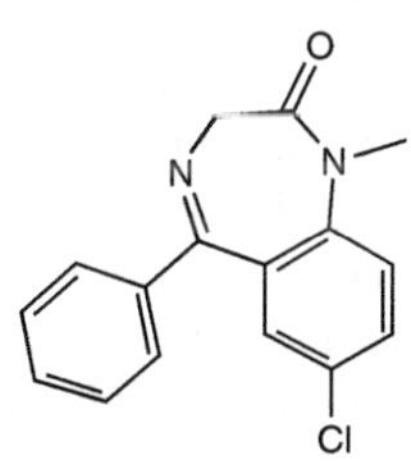	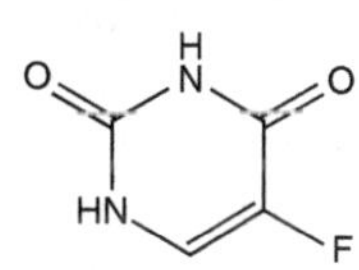
Formula	$C_{17}H_{18}FN_3O_3$	$C_{16}H_{13}ClN_2O$	$C_4H_3FN_2O_2$
CAS number	85721-33-1	439-14-5	51-21-8
Molecular weight	331.133 g/mol	284.072 g/mol	130.018 g/mol

Table 3 Main physico-chemical parameters of selected HHPh

HHPh/physico-chemical parameter	CIP	DZ	5-FU
p*K*a	$pK_{a1} = 6.09$ $pK_{a2} = 8.74$	3.4	$pK_{a1} = 8.0$ $pK_{a2} = 13$
BCF	3.2	34	3.6
log K_{ow}	0.28	2.82	−1.0 to 0.34
K_{OC}	61,000	n.a.	8
Atmospheric ˙OH rate (10^{-11} cm^3/molecule × s)	31.3	0.99	0.583
Solubility (mg L^{-1})	3.00×10^4	50	1.11×10^4
Henry's law constant (atm × m^3/mole)	5.09×10^{-19}	3.64×10^{-09}	1.66×10^{-10}
Vapour pressure (mm Hg)	1.65×10^{-12}	2.78×10^{-8}	2.68×10^{-6}
REF	[6, 7]	[6, 8]	[7, 9]

n.a. not available

atmospheric OH rate, solubility, Henry's coefficient (k_H) and vapour pressure. Data gathered in Table 3 will be used to explain the expected environmental fate of selected HHPh.

2.1 *Antibacterial Agents (Ciprofloxacin)*

Antibiotics are substances produced by various species of microorganisms (bacteria, fungi, actinomycetes) that suppress the growth of other microorganisms and eventually may destroy them. However, common usage often extends the term antibiotics to include synthetic antibacterial agents, such as the sulphonamides and quinolones, which are not the products of microbes [10]. Antibiotics are among the most frequently prescribed medications in modern medicine. Although antibiotics are useful in a wide variety of infections, it is important to realize that antibiotics only treat targeted bacterial infections and once they reach the environment, they can have unwanted consequences. Antibiotics are among the first addressed and most-studied groups of pharmaceuticals in the environment. As an example of widespread fate and behaviour of antibiotics in the environment we will address Ciprofloxacin (CIP), a broad-spectrum antibacterial agent with halogenated heterocyclic structure. The results presented below are not a review of available literature and only representative studies, which give us an idea about environmental occurrence and cycling of this HHPh, are presented.

2.1.1 Sources of Pollution

Widespread production and application of CIP as a broad-spectrum antibacterial agent may result in its release to the environment through various waste streams. Consequently, it has been detected in hospital effluents, municipal waste treatment plant influents and effluents as well as in surface waters.

2.1.2 Main Physico-Chemical Parameters and Their Influence on Environmental Behaviour

Dissociation

pK_a is an equilibrium constant that describes the degree of dissociation of a compound at a particular pH. CIP has pK_a values of 6.09 for the carboxylic acid group and 8.74 for the nitrogen on the piperazinyl ring (Table 3). It is expected that at pHs greater than 6.09, the acid will be primarily dissociated and at pHs less than 8.74, the nitrogen will be primarily protonated. Therefore, CIP is expected to have an ionic charge at almost any environmental pH.

Sorption

Sorption is a process in which one substance takes up or holds another. CIP has an ionic charge at any environmental pH (explained under Dissociation); therefore, its sorption occurs through ionic binding [11]. CIP physico-chemical parameters (Table 3) indicate following:

1. The octanol–water partition coefficient (log K_{ow}) is the ratio of the concentration of a chemical in octanol and in water at equilibrium at a specified temperature. Octanol is an organic solvent that is used as a surrogate for natural organic matter. This parameter is used in many environmental studies to help determine the fate of chemicals in the environment. Polar compounds with log K_{ow} values <1 will distribute into the water phase mostly. CIP has a log K_{ow} of 0.28, which suggests the potential for bioconcentration in aquatic organisms is low.
2. The soil organic carbon–water partitioning coefficient K_{oc} is the ratio of the mass of a chemical that is adsorbed in the soil per unit mass of organic carbon in the soil per the equilibrium chemical concentration in solution. It is the "distribution coefficient" (K_d) normalized to total organic carbon content. K_{oc} values are useful in predicting the mobility of organic soil contaminants; higher K_{oc} values correlate with less mobile organic chemicals, while lower K_{oc} values correlate with more mobile organic chemicals. K_{oc} of 61,000 indicates that CIP is expected to adsorb to suspended solids and sediment.
   CIP (log K_{oc} values of 4.3–4.9 l kg^{-1}, dependent on pH [11]) was shown to adsorb to particles with fine particulate matter [12].
   CIP (K_{oc} of 61,000) was determined to be immobile in soil in a batch sorption studies using soil [11]. High sorption to soil prevents leaching into ground or surface waters [11]. However, it also prevents its possible photodegradation [13] and depends on type of particulate matter. Belden and co-workers [13] compared the sorption coefficients for CIP between two types of organic material. Sorption to fine particulates (log K_d of 4.54 $\pm$ 0.09 l kg^{-1}) was 1.6 orders of magnitude greater than sorption to coarse particulate (log K_d of 2.92 $\pm$ 0.10 l kg^{-1}) potentially resulting in differential toxicity among similar organisms that occupy these

different niches and leading to different estimates of environmental fate and effects.

3. The bioconcentration factor (BCF) is a measure of the ability for water-borne chemical substance to concentrate in fatty tissue of fish and aquatic organisms relative to its surroundings. An estimated BCF of 3.2, in agreement with log K_{ow}, suggests the potential for bioconcentration in aquatic organisms is low.

Volatility

Henry's Law is a chemistry law, which states that the mass of a gas that will dissolve into a solution is directly proportional to the partial pressure of that gas above the solution. CIP's Henry's Law constant is estimated as 5.09×10^{-19} atm $\times$ m^3/mole (Table 3) indicating that CIP is expected to be essentially non-volatile [7]. Based upon an estimated vapour pressure of 1.65×10^{-12} mm Hg (Table 3, [14]), CIP is not expected to volatilize from dry soil surfaces.

2.1.3 Biological Degradation

Biodegradation is a process, where compounds are capable of being decomposed by biological agents, especially bacteria. Biodegradation is unlikely to be an important dissipation route from water for CIP [15]. This was shown also using the OECD Closed Bottle Test protocol and an inoculum from a WWTP accepting hospital wastewater effluent, where CIP (3.5 mg/L) resulted in no significant biodegradation during whole incubation period (28 and 49 days) [16].

2.1.4 Abiotic Degradation

Abiotic degradation is a process in which a substance is converted to simpler products by physical or chemical mechanisms, examples include hydrolysis and photolysis.

A rate coefficient "Atmospheric OH rate" is a physico-chemical property of organic compounds that describes a compound's potential for indirect photolysis, e.g. reaction rate of a compound with hydroxyl radicals ($^{\cdot}$OH) [17, 18]. CIP has the potential to react with $^{\cdot}$OH radicals 3.13×10^{-10} cm^3/molecule $\times$ s, which is approximately 30 and 50 times higher than DZ and 5-FU ($9.90\ 10^{-12}$ cm^3/molecule $\times$ s and $5.83\ 10^{-12}$ cm^3/molecule $\times$ s, respectively). In other words, CIP with higher atmospheric OH rate constant (Table 3), has a higher potential for advanced oxidation processes than compounds with lower atmospheric OH rate constant. This was confirmed by Meylan and co-workers [17, 19] who studied the vapour-phase reaction of CIP with photochemically-generated hydroxyl radicals what resulted in the rate constant of 3.13×10^{-11} cm^3/molecule $\times$ s at 25°C using a structure-estimation method. This corresponds to an atmospheric half-life of about

1.23 h at an atmospheric concentration of 5×10^5 $^{\cdot}$OH per cm^3, which shows a potential of CIP for AOP.

Torniainen and co-workers [20] studied photodegradation of CIP in aqueous solutions in a wide pH range (4.0–10.6). CIP was irradiated with a high-pressure mercury lamp (313 nm) and gave the following initial half-lives in increasing order from: 22 min, 40 min, 55 min, 104 min and 200 min at pH 8.6, 6.0, 4.0, 5 and 10.6, respectively. They explained the variability of results with pH by the ampholytic nature of CIP (p*K*a 6.09 for the carboxylic acid and 8.74 for the nitrogen on the piperazinyl ring). Photolysis half-lives were also examined at constant pH 5 and varying concentrations. Half lives were showed to be in the following order from 1.7 h to 15 h at concentrations from 0.05 to 0.5 mM. Later De Bel et al. [21] confirmed CIP degradation rate being highly pH dependent. They concluded that pH determines the CIP degree of photodegradation. Torniainen et al. [20] also studied photodegradation in natural sunlight and showed it was much slower. Exposure of aqueous solutions (pH 5.0) to daylight for 15 days resulted in CIP losses of less than 16%. The main degradation product of CIP photolysis was later showed to be a compound that replaces the entire piperazinyl ring with an amino (NH_2) group [22].

Vasconcelosa et al. [23] studied photo-induced oxidation, heterogeneous photocatalysis, ozonation and peroxone of CIP in a hospital effluent. As expected, photo-induced oxidation was much slower than the other processes. Ozonation showed the best performance, e.g. total degradation in 30 min treatment, while heterogeneous photocatalysis and peroxone led to almost complete CIP degradation after 60 min treatment, which was not expected since the latter have greater capacity for hydroxyl radical generation. The by-products formed during the application of the processes were found to be very similar, most likely deriving from the oxidation of the piperazine group.

Studies by Belden and coworkers [13] reviewed CIP photodegradation vs adsorption. The half-life of CIP in surface water is expected to be short due to rapid photodegradation [12, 24] with a half-life of approximately 2 h in natural water [12]. However, the fate of CIP within aquatic systems was proved to be dependent on both photodegradation rates and sorption rates. Although these processes tend to immediately reduce the amount of CIP freely available within water, ultimately sorption may prolong the environmental half-life of CIP [13]. Further, type of organic material present plays an important role in the sorption kinetics [25].

2.1.5 Occurrence in Wastewaters and in the Environment

Solubility of CIP as well as of all selected model HHPh (Table 3) is notably higher than its actual concentrations in the aquatic environment (Table 4), which indicates that the solubility does not limit CIP occurrence in aqueous matrices.

Table 4 Environmental fate and occurrence (maximal reported concentration) of the investigated HHPh in WW and environment

HHPh/Environmental fate	CIP	DZ	5-FU
Biodegradation/ Ref	NO [15, 16]	YES [26] NO [27–31]	YES [32, 33] NO [9, 34]
Photodegradation/ Ref	YES [12, 13, 17, 19–21, 23, 24]	NO (direct) [26] YES [26, 27, 31, 35]	NO (direct) [7, 36] YES [9, 37]
Adsorption onto sludge /sediment/ Ref	YES [11–13]	YES [29] NO [27]	NO [33]
Occurence in WW (max detected concentration) Ref	WWTP influent: 1,000 ng/ [38–40] WWTP effluent: 370 ng/L (primary) and 80 ng/L (secondary) [39] Hospital effluents: 124.5 μg/L [38, 40–42]	WWTP influent:1,180 ng/L [31] WWTP effluent: 660 ng/L [27, 31, 43] Hospital effluents: approx 1 μg/L [40]	Hospital effluent: 124 μg/L [33, 44–46]
Occurence in environment (max detected concentration)/ Ref	Rivers: 0.03 μg/L [47]	Rivers: 880 ng/L [31, 43, 48, 49] Drinking water: 23.5 ng/L [48, 49]	n.a.

n.a. not available

Surface Water. Kolpin et al. [47] performed a survey in 115 water samples from a network of US stream sampling sites across 30 states during 1999–2000. CIP was detected at a maximum concentration of 0.03 μg/L and was determined at a 2.6% frequency.

A French study of the aqueous phase of river water under different hydrological conditions at 5 sampling locations in the Seine River inner Estuary showed no significant frequency and presence of CIP [50]. On the contrary, other investigated antibiotics such as 17 representatives of quinolones, sulfonamides, nitroimidazoles and diaminopyrimidines were during six-month survey shown to occur at individual concentrations reaching 544 ng L^{-1} (sulfamethoxazole) clearly indicating wastewaters as a point source. Similar were outcomes of a Chinese study [51] and Mexican study [38].

Wastewater. CIP was detected in concentrations of 250–370 ng/L in primary effluent of a municipal waste treatment plant in Zurich, Switzerland [39]; secondary effluent contained CIP concentrations of 70–80 ng/L indicating that 70–80% is eliminated during the activated sludge treatment. The residual amounts are being discharged to ambient water [39]. Study by Brown and co-workers [38] investigated occurrence of antibiotics in municipal wastewater. Results showed the concentrations of CIP reached up to 1,000 ng/L, whereas effluent levels were all bellow LOD. Verlicchi et al. [40] published a review study on the content of emerging pollutants, including CIP, in urban WW and showed CIP to be present in concentrations from 0.01 to 10 μg/L in urban WW.

Batt et al. [52] studied the occurrence of CIP and other antibiotics in four full-scale WWTPs differing in design and operating conditions. The overall removal of the antibiotics (effluent *vs* influent) ranged from 33% to 97% and clearly depends on a combination of biological and physicochemical treatment processes and operating conditions of the treatment system.

Hospital Effluents. CIP was detected in concentrations of 3–87 μg/L in wastewater samples collected from the main sewer of a hospital in Switzerland [41] and in 24 wastewater samples from the main sewers of five hospitals in southern Germany (0.7–124.5 μg/L, [42]). Study by Brown and co-workers [38] investigated the occurrence of antibiotics in effluent from hospitals and showed that concentrations of CIP reached up to 2,000 ng/L. Verlicchi et al. [40] published a review study on the content of emerging pollutants, including CIP, in hospitals and urban WW and showed CIP to be present in concentrations from 0.1 to 100 μg/L in hospital effluents. The concentrations in urban WW were generally 10 times lower for this compound.

2.1.6 Environmental Fate

CIP is a widely produced and applied broad-spectrum antibacterial agent and may be released to the environment through various waste streams. Consequently, it has been detected in hospital wastewater effluents, municipal waste treatment plants influents and effluents and even in surface waters.

The PBT profiler estimates [8] the half-life of a chemical in each environmental medium (air, water, soil and sediment). These half-lives are for reactivity and do not take into consideration partitioning into or out of the medium, or physical transport. Reported half-lives for CIP in water, soil, air and sediment (60, 120, 540 and 0.05 days, respectively) defines CIP to be moderately persistent in water and soil ($\geq$ 60 days), very persistent in sediment (> 180 days) and not persistent in air (below 2 days). If CIP is exposed to a mixture of water, soil, sediment and air, it will be distributed only between water (48%) and soil (52%) according to PBT profiler estimates [8].

CIP has an estimated vapour pressure of 1.65×10^{-12} mm Hg at 25°C and when released to air, CIP will exist solely in the particulate phase in the ambient atmosphere and will be removed from the atmosphere by wet and dry deposition. If CIP is released to soil, it is expected to have no mobility (K_{oc} of 61,000). Basing on the estimated Henry's Law constant of 5.09×10^{-19} atm $\times$ m^3/mole, volatilization from moist soil surfaces is not expected to be an important fate process.

Biodegradation is unlikely to be an important dissipation route from water for CIP [15, 16]. If released into water, CIP is expected to adsorb to suspended solids and sediment (K_{oc} of 61,000). Also, volatilization from water surfaces is not expected to be an important fate process (Henry's Law constant). An estimated BCF of 3.2 suggests that the potential for bioconcentration in aquatic organisms is low. Aqueous CIP solutions are susceptible to photolysis in sunlight. The summary of CIP environmental fate is presented in Table 4 and was adopted from TOXNET [7].

2.2 Anxiolytics (Diazepam)

Benzodiazepines are the drugs of choice in the pharmacotherapy of anxiety and related emotional disorders, sleep disorders, status epilepticus and other convulsive states; they are used as centrally acting muscle relaxants, for pre-medication and as inducing agents in anaesthesiology.

In general, the consumption of psychoactive drugs is worldwide increasing due to developing health care system, demographic changes and increasing awareness of mental health issues as well as in response to the psychological stresses caused by modern life, which can result in various orders of psychiatric disorders. Diazepam (DZ) is a worldwide applied anxiolytic-sedative drug useful in the symptomatic relief of anxiety and tension states. It has also adjunctive value in the relief of certain neurospastic conditions and in certain cases, its anticonvulsant activity has been found useful in controlling status epilepticus.

2.2.1 Sources of Pollution

Psychoactive pharmaceuticals enter the environment via the same routes as other pharmaceuticals, e.g. mainly through wastewater treatment plants, after they are excreted with urine or faeces. Consequently, DZ has been detected in waste water streams from hospitals and in effluents from municipal wastewater treatment plants. Alternatively, it can be present in discharges from the pharmaceutical industry or may enter the environment after improper disposal. Due to its high consumption and continuous discharge into the environment, it is ubiquitously present in surface waters in ng/L levels.

2.2.2 Main Physico-Chemical Parameters and Their Influence on Environmental Behaviour

Dissociation

Given an average environmental pH of 7 (5–9) and according to the p*K*a value listed in Table 3 (3.4), DZ is not likely to be dissociated at environmental pH.

Sorption

DZ physico-chemical parameters indicate following (Table 3):

1. According to log K_{ow} of 2.82, DZ is not expected to be removed onto suspended soils.
2. For municipal wastewater treatment, only substances with K_d values > 500 L kg_{SS}^{-1} partition significantly (> 10%) onto sludge; hence, DZ with primary sludge K_d 43.9 ± 26.1 L kg_{SS}^{-1} and secondary sludge K_d of 21.1 ± 7.6 L kg_{SS}^{-1} [27], is not expected to.
3. BCF factor for DZ (34, Table 3) indicates the slight potential of compound to bioconcentrate and bioaccumulate. Currently a value above 1,000 is used to indicate bioaccumulation potential. Basing on BCF, DZ is not expected to be removed with suspended soils.

Volatility

Based on an estimated Henry's Law constant of 3.64×10^{-09} atm × m^3/mole and vapour pressure of 2.78×10^{-8} mm Hg (Table 3), volatilization from moist soil surfaces and water is not expected to be an important fate process for DZ.

2.2.3 Biological Degradation

Several psychoactive pharmaceuticals partially or totally resist to wastewater treatment [28, 31], which they pass almost unaffected and thus enter the aquatic environment. For DZ, which is during classical biological treatment removed < 10%, the anaerobic sludge treatment is more efficient (10–50%), but the removal is a result of adsorption to activated sludge, rather than degradation [29]. Other studies [27] presented the results on DZ originating from studies of laboratory and full-scale plants and showed no significant removal in laboratory scale, corresponding to a $k_{biol} < 0.1$ L g_{SS}^{-1} d^{-1}; while compound was not found in full-scale plants.

Several studies report on great variability in DZ biodegradation test results [26, 30]. Redshaw et al. [30] reported a short-term bacterial liquid culture experiment on biodegradability of DZ that gave very variable results and overall showed no statistically significant difference between the concentration of DZ measured in the samples at the start (day 0) and end (day 60) of the experiment that suggests no biodegradation. Kosjek et al. [26] studied the efficiency of DZ removal in laboratory-scale bioreactors under aerobic conditions and observed an increase from 20% to 70% during a one-year operation, which is most likely the result of biomass adaptation. Under anoxic conditions, however, the removal remained unchanged (20%). Sequential anoxic–oxic system showed up to 50% DZ removal, whereas oxic–anoxic resulted in up to 90% removal of DZ.

2.2.4 Abiotic Degradation/Removal

Photodegradation. During photodegradation studies the effect of ferrioxalate or iron nitrate on the photo-Fenton degradation efficiency of the DZ was evaluated. The DZ degradation was improved in the presence of ferrioxalate in relation to $Fe(NO_3)$, either under black-light or solar irradiation and high DOC removal percentages were achieved. The removal of DZ was 80% after 60 min irradiation [35]. In comparison to classical treatment technologies, advanced oxidation methods more efficiently eliminate DZ [31]. Kosjek et al. [26] studied DZ photolysis with monochoromatic ($\lambda = 254$nm) and polychromatic UV light and photocatalytic removal in the presence of H_2O_2 (0.1% and 0.5%). Results showed that H_2O_2 addition successfully degrades persistent DZ (up to 96%), whereas its removal is negligible without the presence of the radical source.

Ozonation. Ozone is a strong oxidant used as a disinfectant in drinking water treatment where ozone and $^{\cdot}OH$ radicals act on model compound. DZ is relatively resistant to ozonation [40]; however, it has some potential to be oxidized with $^{\cdot}OH$ radicals formed during ozone treatment [27].

Chlorination. ClO_2: Comparing to ozone, fewer pharmaceuticals (including DZ) react with ClO_2 and their constants are considerably lower [27]. Chlorine: Generally disinfection with Cl_2 and ClO_2 does not lead to a general oxidation/removal of pharmaceuticals and DZ was shown to be refractory to oxidation by chlorine [27].

Adsorption on activated carbon. Activated carbon is a common process applied in waterworks to eliminate many organic micropollutants. DZ, along with other neutral compounds with rather high K_{ow} value (log K_{ow} between 2.4 and 6.4) was classified as easily adsorbable (99% removal, 0.2 mg/L of activated carbon) [27].

Stripping into air. Air stripping is a common decontamination of environmental matrices containing volatile organics. Generally a Henry's coefficient $>3.10^{-3}$ is required for significant stripping and pharmaceuticals, including DZ, generally have Henry's coefficient lower than 10^{-5} and are therefore rather hydrophilic [27] and do not classify for this kind of removal.

Membrane filtration is an emerging technology in drinking water treatment. Ultrafiltration can be applied in combination with activated carbon. As expected, the removal is achieved by adsorption on activated carbon and ultrafiltration using membranes sized approximately 10 nm has no effect on removal [27]. Nanofiltration was for DZ (μg/L) tested with different types of nanomembranes and reverse osmosis on pilot scale. Ultrafiltration was performed as a pre-treatment step. DZ removal was greater than 98% [27].

2.2.5 Occurrence in Wastewaters and in the Environment

Solubility of DZ (Table 3) is higher than its actual concentrations in the aquatic environment (Table 4), which indicates that the solubility does not limit DZ occurrence in aqueous matrices.

Surface Water. Ternes and co-workers determined DZ in concentrations of up to 33 ng/L in German rivers [43] while Calisto and Esteves [31] in their review list the highest-reported surface water concentration to be 880 ng/L. In another study, DZ was determined in concentrations up to 1.2 ng/L in river water and 23.5 ng/L in drinking water [49]. Stuer-Lauridsen et al. [48] showed DZ to be present in river and drinking water in concentrations around 10 ng/L.

Wastewater. Among the psychoactive pharmaceuticals, most attention has been focused on the prototype of benzodiazepine tranquillizers, DZ. Ternes et al. [43] determined DZ in substantially low concentrations (up to 53 ng/L) in effluents from wastewater treatment plants. In raw water of German and Finnish waterworks, DZ was showed in concentrations between 10 and 40 ng/L [27]. Calisto and Esteves [31] reviewed the occurrence of DZ in different WWTP and showed the highest concentration determined to be in an influent in Belgium 1,180 ng/L and in corresponding effluent 660 ng/L. This is in agreement with Ternes et al. [27] who state that neutral substances such as DZ hardly show any removal during wastewater treatment and remain stable during post-treatment steps as well as in the groundwater.

Hospital Effluents. Verlicchi et al. [40] reported DZ in hospital effluents in concentrations around 1 μg/L.

2.2.6 Environmental Fate

The PBT profiler estimates [8] reported half-lives for Diazepam in water, soil, air and sediment to be 38, 75, 340 and 1.6 days, respectively. These indicate DZ is not persistent in water (< 60 days), moderately persistence in soil (>60 days), very persistent in sediment (>180 days) and not persistent in air (<2 days). If DZ is exposed to a mixture of water, soil, sediment and air, it will be distributed only between water (14%) and soil (86%) according to PBT profiler estimates [8].

Based on K_{ow} of 2.82 (Table 3), DZ is expected to be quite immobile in soil, which also indicates low bioconcentration potential since the compound is expected to be mainly in water phase due to high water solubility. Generally, bioconcentration is expected to play an important environmental role for compounds with K_{ow} greater than 5. According to vapour pressure of 2.78×10^{-8} mm Hg (Table 3), volatilization from moist soil and water surfaces is not expected to be an important fate process, which is also in agreement with an estimated Henry's Law constant of 3.64×10^{-09} atm $\times$ m^3/mole (Table 3). DZ will exist solely in the particulate phase in the ambient atmosphere. Particulate-phase DZ may be removed from the air by wet and dry deposition.

Neutral substances such as DZ hardly show any removal during wastewater treatment and remain stable during post-treatment steps as well as in the groundwater [27].

2.3 *Antineoplastic Pharmaceuticals (5-Fluorouracil)*

Cytostatic drugs (often referred to as antineoplastic and anticancer drugs) are used in the chemotherapy of oncological patients to fight a disease of uncontrolled multiplication of the body's own cells and spread of abnormal forms within the body, commonly referred to as cancer. Chemotherapy works by preventing the growth and proliferation of cancer cells, and is, besides surgical excision and irradiation, the most common approach to treat established cancer.

Cytostatics are designed to damage DNA, inhibit DNA synthesis, and interrupt cell replication. They act unselectively on all growing cells and possess a carcinogenic potency [53]. Due to their mode of action, practically all eukaryotic organisms could be vulnerable to damage, with teratogenicity being of greatest concern at low ng/L levels [4]. Comparing the consumption of cytostatics with other pharmaceutical classes, it is low and the possible threat posed by cytostatics finding their way in the environment is recognized in only limited scientific circles. Despite deficiency of application of sensitive, specific and accurate analytical methods in the area of environmental fate of cytostatics, available literature has shown their presence particularly in hospital discharges, WWTP influents and effluents and even in surface waters [53]. However, to date no one has confidently stated whether cytostatics have an impact on the environment. In order to do so, their environmental occurrence and toxicity studies at chronic exposure and environmental concentrations should be carried out. The lack of these studies has recently been addressed at European level with approved financing (1.1.2011) of two EU FP7 projects "CytoThreat" ("Fate and effects of cytostatic pharmaceuticals in the environment and the identification of biomarkers for improved risk assessment on environmental exposure") and "Pharmas" ("Ecological and human health risk assessment of antibiotics and anticancer drugs found in environment").

5-Fluorouracil is one of the most administered and, from the environmental point of view, addressed anticancer drugs worldwide [4, 46]. Below are presented the outcomes on 5-FU adopted from a review paper by Kosjek and Heath [53].

2.3.1 Sources of Pollution

Fluorouracil is one of the fifty typically used active substances to treat oncology patients in European hospitals. Its worldwide production and consumption may result in its release to the environment through various waste streams. 5-FU is mostly administered intravenously in either a clinic or a hospital. Most oncology patients receive the treatment at the oncology wards and leave for home after the infusion or injection has been administered [4, 33]. Those patients may excrete part of the cytostatics in the hospital, as the treatment takes up to a couple of hours and the pharmacokinetics of some cytostatics is relatively fast. Due to lower production and limited consumption, 5-FU has been detected mostly in hospital effluents only. Due to recent developments in advanced analytical instrumentation with detection limits below ng/L, its occurrence is expected to be revealed in other environmental compartments [53].

2.3.2 Main Physico-Chemical Parameters and Their Influence on Environmental Behaviour

Dissociation

Average environmental pH is estimated to be around 7. According to its p*K*a values of 8 and 13 (Table 3), 5-FU is not likely to be dissociated under common environmental conditions and only environmental $pH > 8$ would be expected to cause 5-FU to have some ionic charge [53].

Sorption

5-FU physico-chemical parameters (Table 3) indicate following:

According to its log K_{ow} between -1.0 and 0.34, 5-FU will be mainly distributed in water phase and its potential for bioconcentration in aquatic organisms is low.

1. Experimentally-derived K_{oc} of 8 suggests that 5-FU will not adsorb to suspended solids and sediment [33] and is expected to have very high mobility in soil. Therefore, if not biodegraded in waste water treatment plant, it will pass unaltered through the WWT to receiving waters [53].
2. BCF for 5-FU is estimated to be 3.6, which, in agreement with log K_{ow}, suggests the low potential for bioconcentration in aquatic organisms.

Volatility

Because of the low values for the Henry's coefficient (1.66×10^{-10} atm $\times$ m^3/mole (Table 3) and vapour pressure (2.68×10^{-6} mm Hg, Table 3), the fraction removed by volatilization is considered negligible from both water and dry soil surfaces [54].

2.3.3 Biological Degradation

Straub [9] reviewed 5-FU OECD 301F, 302B and 301D biodegradability tests performed before 2009 and reported no degradation under the different test conditions and 5-FU concentrations (9 to 854 mg/L). Buerge et al. [54] indicated that the high concentrations of 5-FU may be having cytotoxic effects on the degrading microorganisms potentially leading to false negative results. This was also shown in a study by Kiffmeyer et al. [32], where 5-FU (5 mg/L) was completely removed from a spiked influent in a laboratory sewage plant within days, but the rate appears to depend on the initial concentration of 5-FU (higher initial concentration resulted in slower degradation) that was consistent with Mahnik et al. [33]. Similarly, contradictory results by Kümmerer and Al-Ahmad [34] suggest that 5-FU was unbiodegradable in the closed bottle (OECD 301D) and Zahn–Wellens tests (OECD 302B).

2.3.4 Abiotic Degradation

5-FU is not expected to undergo hydrolysis in the environment due to the lack of functional groups that hydrolyze under environmental conditions [7].

5-FU absorbs light at wavelengths 256–266 nm and not at wavelengths greater than 290 nm. Therefore, 5-FU is not expected to be susceptible to direct photolysis by sunlight [36]. Straub et al. [9] reported that 5-FU was light sensitive in solution with an UV maximum absorbance peak at 266 nm and was rapidly photodegraded when exposed to an Hg medium-pressure lamp. The compound can also be degraded by ozonation [37].

2.3.5 Occurrence in Waste Waters and in the Environment

Solubility of 5-FU (Table 3) is notably higher than its actual concentrations in the aquatic environment (Table 4), which indicates that the solubility does not limit 5-FU occurrence in aqueous matrices and due to its persistence 5-FU is likely to pass through a WWTP in the effluent and into surface waters.

Surface Water. The recalcitrance of 5-FU in the highly biologically active environment of activated sludge indicates that it will also be extremely persistent in river water [4]. The dilution plays an important role by the time 5-FU reaches the surface waters, which may be the reason that 5-FU has not been reported yet in surface waters.

Waste Water Treatment. 5-FU is unlikely to be significantly removed during WW treatment because of its hydrophilicity and not having the potential to readily sorb to the sludge. Also, 5-FU contains halogen atoms, which are known to be problematic for biodegradation and also is reported to be toxic to bacteria in higher concentrations [4].

Hospital Effluents. The cytostatic compounds appear in hospital sewage plants, or in case of out-patient treatment in municipal wastewater treatment plants (WWTP) [32]. Most studies on cytostatics focus on hospital WW, where the amounts measured vary greatly (ng/L – μg/L) within a week, which is indicative of the high variability in emissions [44]. To date, only three studies actually report 5-FU presence in hospital effluents [33, 45, 46] ranging from 27 ng/L to 124 μg/L depending on type of sampling and hospital effluent.

5-FU is also a metabolite generated from another cytostatic drug Capecitabine in a tumour. Capecitabine is available for oral administration and shows smaller gastrointestinal toxicity when compared with intravenously administered 5-FU. These explains a recently increasing consumption of Capecitabine when compared to 5-FU [55].

2.3.6 Environmental Fate

5-FU production and consumption as an anticancer drug may result in its release to the environment through various waste streams.

Reported half-lives for 5-FU [8] in water, soil, air and sediment (15, 30, 140 and 2.4 days, respectively) indicate that 5-FU is not-persistent in water and soil (<60 days), moderately persistent in sediment (>60 days) and highly persistent in air (>2 days). If 5-FU is exposed to a mixture of water, soil, sediment and air, it will be distributed only between water (39%) and soil (61%) according to PBT profiler estimates [8].

An estimated vapour pressure of 2.68×10^{-6} mm Hg at 25°C indicates that if released to air, 5-FU will exist in both the vapour and particulate phases in the atmosphere. Vapour-phase 5-FU will be degraded in the atmosphere by reaction with photochemically-produced $^{\cdot}OH$ radicals (the half-life for this reaction in air is estimated to be 3 days). Particulate-phase 5-FU will be removed from the atmosphere by wet or dry deposition. 5-FU absorbs light at wavelengths 256–266 nm [36] and does not absorb light at wavelengths > 290 nm and therefore is not expected to be susceptible to direct photolysis by sunlight. Based on its K_{oc} of 8 (Table 3), 5-FU is expected to have very high mobility.

Volatilization from moist soil surfaces is not expected to be an important fate process based upon an estimated Henry's Law constant of 1.66×10^{-10} atm $\times$ m^3/mole. 5-FU is not expected to volatilize from dry soil surfaces based upon an estimated vapour pressure of 2.68×10^{-6} mm Hg (Table 3).

100% biodegradation in 5 days using an activated sludge inoculum and the OECD Confirmatory Test suggest that biodegradation is an important environmental fate process [32]. If released into water, 5-FU is not expected to adsorb to suspended solids and sediment based upon the estimated K_{oc} (Table 3), determined from a log K_{ow} of -0.89 (Table 3). Volatilization from water surfaces is not expected to be an important fate process based upon this compound's estimated Henry's Law constant. An estimated BCF of 3 (Table 3) suggests that the potential for bioconcentration in aquatic organisms is low.

Hydrolysis is not expected to be an important environmental fate process since this compound lacks functional groups that hydrolyze under environmental conditions.

3 Conclusions

The aim of this chapter is to describe the environmental fate of selected model HHPh, representatives of three different pharmaceutical groups. They include Ciprofloxacin (antibiotic), Diazepam (sedative / hypnotic, anticonvulsant) and

5-Fluorouracil (anticancer drug). All compounds share a halogenated heterocyclic structure, but differ in their physico-chemical properties. This, together with quantities produced affect their occurrence and fate in the environment.

Ciprofloxacin is a commonly applied antibacterial agent. On the basis of its physico-chemical parameters, it will occur in the aqueous phase (water solubility, K_{ow}). When exposed to soil, suspended solids, sediments and/or biota (K_{oc} of 61,000), it will be adsorbed onto solid particles, but have a low bioconcentration factor in aquatic organisms. In addition, volatilization is not an important fate process and biodegradation is unlikely to be an important elimination route from water. Abiotic degradation, including photolysis and adsorption onto sludge/sediment, is an important option for treatment of CIP-contaminated matrices. Due to its inadequate removal during water treatment and high consumption, CIP has been detected in hospital waste water effluents, municipal waste treatment plants influents and effluents and even in surface waters.

Diazepam is a common anxiolitic-sedative drug. Its physico-chemical parameters suggest, it will be, like CIP, present in the aqueous phase (water solubility, K_{ow}), will not likely adsorb to suspended solids and sediment (K_d), but will bioconcentrate and bioaccumulate (BCF). Volatilization is also not expected to be an important fate process. In addition, in terms of water treatment, biodegradation studies are inconclusive, albeit biomass adaptation and sequential treatment could be a promising option. Advanced oxidation processes have great potential for its removal. Its presence has been detected in WWTP influents and effluents, hospital effluents, surface water and even drinking water.

5-Fluorouracil is one of the most administered anticancer drugs. However, its production and administration is much lower than either CIP or DZ. From its physico-chemical parameters, it is likely that 5-FU will also be mainly distributed in the aqueous phase (K_{ow}). 5-FU will have low bioconcentration potential (BCF) and not adsorb to suspended solids and sediment (K_{oc}). Again, published biodegradation tests are also inconclusive and the compound is not susceptible to direct photolysis; however, indirect photolysis ($^{\bullet}OH$ radicals) is a promising option for its removal. To date, 5-FU has been detected only in hospital effluents.

All three compounds behave differently in the environment with the exception of showing a strong indication of being biologically recalcitrant but susceptible to indirect photolytic degradation. This is due to their halogenated heterocyclic structure. The presented examples are not sufficient to predict the occurrence and fate of other HHPh in the environment and every HHPh with its unique mode of administration and physicochemical characteristics should be addressed individually.

Acknowledgments The research leading to these results has received funding from the Slovenian Research Agency under Research Programme P1-0143 – "Cycling of nutrients and contaminants in the environment, mass balances and modelling of environmental processes and risk analysis" and from the European Community's Seventh Framework Programme (FP7/2007–2013) under grant agreement n. 265264: "CytoThreat": "Fate and effects of cytostatic pharmaceuticals in the environment and the identification of biomarkers for improved risk assessment on environmental exposure".

References

1. Oaks JL, Gilbert M, Virani MZ, Watson RT, Meteyer CU, Rideout BA, Shivaprasad HL, Ahmed S, Chaudhry MJ, Arshad M, Mahmood S, Ali A, Khan AA (2004) Nature 427:630–633
2. Ternes T, Joss A (2006) Human pharmaceuticals, hormones and fragrances: the challenge of micropollutants in urban water management. IWA Publishing, London
3. Foye WO (1995) Parasite chemotherapy. In: Lemke TL, Williams DA, Foye WO (eds) Principles of medicinal chemistry, 4th edn. Williams & Wilkins, Philadephia
4. Johnson AC, Jürgens MD, Williams RJ, Kümmerer K, Kortenkamp A, Sumpter JP (2008) J Hydrol 348:167–175
5. Kosjek T, Heath E (2011) Halogenated heterocycles as pharmaceuticals. In: Iskra J (ed) Halogenated heterocycles: synthesis, application and environment. Springer, Berlin
6. ChemIDplus Advanced (Online) United States National Library of Medicine. http://chem.sis.nlm.nih.gov/chemidplus/. Accessed Apr 2011
7. TOXNET: Hazardous Substances Data Bank (HSDB) (Online) U.S. National Library of Medicine. http://toxnet.nlm.nih.gov/cgi-bin/sis/htmlgen?HSDB. Accessed Apr 2011
8. PBT profiler: Persistent, Bioaccumulative, and Toxic profiles estimated for organic chemicals (Online) U.S. EPA. http://www.pbtprofiler.net/. Accessed Apr 2011
9. Straub JO (2009) Integr Environ Assess Manag 6:540–566
10. Chambers HF, Sande MA (1996) Antimicrobial agents: general considerations. In: Limbird LE, Hardman JG (eds) Goodman & Gilman's the pharmacological basis of therapeutics, 9th edn. McGraw-Hill, New York
11. Nowara A, Burhenne J, Spiteller M (1997) J Agric Food Chem 45:1459–1463
12. Cardoza LA, Knapp CW, Larive CK, Belden JB, Lydy MJ, Graham DW (2005) Water Air Soil Pollut 161:383–398
13. Belden JB, Maul JD, Lydy MJ (2007) Chemosphere 66:1390–1395
14. Lyman WJ (1985) Estimation of physical properties. In: Neely WB, Blau GE (eds) Environmental exposure from chemicals. CRC Press, Boca Raton, FL
15. Kümmerer K, Al-Ahmad A, Mersch-Sundermann V (2000) Chemosphere 40:701–710
16. Al-Ahmad A, Daschner FD, Kümmerer K (1999) Arch Environ Contam Toxicol 37:158–163
17. Meylan WM, Howard PH (1993) Chemosphere 26:2293–2299
18. Coeur-Tourneur C, Cassez A, Wenger JC (2010) J Phys Chem A 114:11645–11650
19. Meylan WM, Howard PH (2005) Chemosphere 61:640–644
20. Torniainen K, Tammilehto S, Ulvi V (1996) Int J Pharm 132:53–61
21. De Bel E, Dewulf J, De Witte B, Van Langenhove H, Janssen C (2009) Chemosphere 77: 291–295
22. Burhenne J, Ludwig M, Spiteller M (1999) Chemosphere 38:1279–1286
23. Vasconcelosa TG, Kummerer K, Henriques DM, Martins AF (2009) J Hazard Mater 169:1154–1158
24. Lam MW, Tantuca K, Mabury SA (2003) Environ Sci Technol 37:899–907
25. Golet EM, Alder AC, Giger W (2002) Environ Sci Technol 36:3645–3651
26. Kosjek T, Perko S, Zupanc M, Zanoški Hren M, Landeka Dragičević T, Kompare B, Heath E (2011) Wat Resl. Sent for publication (July, 2011)
27. Ternes TA, Janex-Habibi ML, Knacker T, Kreuzinger N, Siegrist H (2004) Detailed report related to the overall project POSEIDON (Contract No. EVK1-CT-2000-00047) duration (2001-2004) Assessment of technologies for the removal of pharmaceuticals and personal care products in sewage and drinking water facilities to improve the indirect potable water reuse
28. Johnson DJ, Sanderson H, Brain RA, Wilson CJ, Solomon KR (2007) Ecotoxicol Environ Saf 67:128–139
29. Löffler D, Rombke J, Meller M, Ternes TA (2005) Environ Sci Technol 39:5209–5218
30. Redshaw CH, Cooke MP, Talbot HM, McGrath S, Rowland SJ (2008) J Soils Sediments 8:217–230

31. Calisto V, Esteves VI (2009) Chemosphere 77:1257–1274
32. Kiffmeyer T, Götze HJ, Jursch M, Luders U (1998) Fresenius J Anal Chem 361:185–191
33. Mahnik SN, Lenz K, Weissenbacher N, Mader RM, Fuerhacker M (2007) Chemosphere 66:30–37
34. Kümmerer K, Al-Ahmad A (1997) Acta Hydrochim Hydrobiol 25:166–172
35. Rossi Bautitz I, Pupo Nogueira RF (2010) Catal Today 151:94–99
36. O'Neil MJ (2001) The Merck Index, 13th edn. Merck and Co., Inc., Whitehouse Station, New Jersey
37. Rey RP, Padron AS, Leon LG, Pozo MM, Baluja C (1999) Ozone: Sci Eng 21:69–77
38. Brown KD, Kulis J, Thomson B, Chapman TH, Mawhinney DB (2006) Sci Total Environ 366:772–783
39. Alder AC, Golet E, Ibric S, Giger W (2000) Fate of fluoroquinolone antibiotics during municipal wastewater treatment. In: Keith LH, Needham LL, Jones-Lepp TL (org.) Issues in the analysis of environmental endocrine disruptors. Amer Chem Soc, Div Environ Chem, Preprint Ext Abstr, 219th ACS Nat Meet, 40:103–104
40. Verlicchi P, Galletti A, Petrovic M, Barceló D (2010) J Hydrol 389:416–428
41. Hartmann A, Alder AC, Koller T, Widmer RM (1998) Environ Toxicol Chem 17:377–382
42. Hartmann A, Golet EM, Gartiser S, Alder AC, Koller T, Widmer RM (1999) Arch Environ Contam Toxicol 36:115–119
43. Ternes T, Bonerz M, Schmidt T (2001) J Chromatogr A 938:175–185
44. Kümmerer K (2001) Chemosphere 45:957–969
45. Mahnik SN, Rizovski B, Fuerhacker M, Mader RM (2004) Anal Bioanal Chem 380:31–35
46. Kovalova L (2009) Cytostatics in the aquatic environment: analysis, occurrence, and possibilities for removal. Doctoral Dissertation, RWTH Aachen University, Aachen
47. Kolpin DW, Furlong ET, Meyer MT, Thurman EM, Zaugg SD, Barber LB, Buxton HT (2002) Environ Sci Technol 36:1202–1211
48. Stuer-Lauridsen F, Birkved M, Hansen LP, Holten Lützhøft HC, Halling-Sørensen B (2000) Chemosphere 40:783–793
49. Zuccato E, Calamari D, Natangelo M, Fanelli R (2000) J Lancet 355:1789–1790
50. Tamtam F, Mercier F, Le Bot B, Eurin J, Tuc Dinh Q, Clément M, Chevreuil M (2008) Sci Total Environ 393:84–95
51. Jiang L, Hu X, Yin D, Zhang H, Yu Z (2011) Chemosphere 82:822–828
52. Batt AL, Sungpyo K, Aga DS (2007) Chemosphere 68:428–435
53. Kosjek T, Heath E (2011) TrAC, Trends Anal. Chem. 30:1065–1087
54. Buerge I, Buser HR, Poiger T, Muller MD (2006) Environ Sci Technol 40:7242–7250
55. Tauxe-Wuersch A, de Alencastro LF, Grandjean D, Tarradellas J (2006) J Environ Anal Chem 86:473–485

Top Heterocycl Chem (2012) 27: 269–308
DOI: 10.1007/7081_2011_60

Published online: 3 November 2011

Green Methods in Halogenation of Heterocycles

Jernej Iskra

Abstract This chapter will give an overview on a recent research in direct halogenation of heterocyclic compounds with an aim to present environmentally friendlier methods to the existing ones. First, a brief outline on halogenation strategies that are used in nature will be given to compare the development of man-made halogenation. Major part of the chapter will cover new methods of halogenation of heterocyclic compounds that were reported in recent 5 years and deal with the development of new reactions or transformations of known methods into an environmentally friendlier outfit. Methods were grouped into three sections – electrophilic (molecular halogens, trihalides, N-X reagents, oxidative halogenation, electrochemistry), radical, and nucleophilic halogenation. Each of them was further grouped according to the "green" approach that was studied: solvent-free reaction conditions, alternative solvents (water, ionic liquids), activation (microwaves, ultrasound), or new technologies (microreactors).

Keywords Alternative media · Fluorination · Green chemistry · Halogenation · Oxidative halogenations · Sustainable methods

Contents

J. Iskra
Laboratory of Organic and Bioorganic Chemistry, Department of Physical and Organic Chemistry, "Jožef Stefan" Institute, Jamova 39, 1000 Ljubljana, Slovenia

Centre of Excellence for Integrated Approaches in Chemistry and Biology of Proteins (CIPKeBiP), Jamova 39, 1000 Ljubljana, Slovenia
e-mail: jernej.iskra@ijs.si

Abbreviations

aq.	Aqueous
ar	Aryl
Bn	Benzyl
BOC	Butoxycarbonyl
BPO	Bromoperoxidase
CPO	Chloroperoxidase
DCM	Dichloromethane
DMF	Dimethylformamide
EDC	Ethylene dichloride
E-factor	Environmental factor, defined by mass of waste per mass of product
equiv.	Equivalent
Et	Ethyl
FEC	Fluoroethylene carbonate
HetAr	Heteroaryl
His	Histidine
HPO	Haloperoxidase
IL	Ionic liquid
LDH	Layered double hydroxide
Me	Methyl
MW	Microwaves
NBS	*N*-Bromosuccinimide
NCS	*N*-Chlorosuccinimide
NFSi	*N*-Fluorobenzenesulfonimide
NIS	*N*-Iodosuccinimide
NXS	*N*-Halosuccinimide
POM	Polyoxometalate
PV	Phase-vanishing
rt	Room temperature

SAM	*S*-adenosylmethionine
SDS	Sodium dodecyl sulfate
Selectfluor™ F-TEDA	1-Chloromethyl-4-fluoro-1,4-diazoniabicyclo[2.2.2] octane bis(tetrafluoroborate)
Ser	Serine
TBAB	Tetrabutylammonium bromide
TBAF	Tetrabutylammonium fluoride
TBHP	*tert*-Butylhydroperoxide
Tf	Trifluoromethylsulfonyl
Tyr	Tyrosine
Us	Ultrasound
X	Halogen

1 Introduction

An area of "green" chemistry is developing very intensively in recent years [1–4]. Our society is very dependent on chemicals (pharmaceuticals, agrochemicals, industrial chemicals, plastics, electronics, etc.), which govern high demand. As major industrial processes were developed at the time when environmental concern was not evolved yet, the waste production is in accord with that time – extensive. High demand for chemicals and extensive waste generation during their production leads to an industry that produces a large pressure on the environment. Therefore, it is not surprising that a chemical industry has become a synonym for ecological problems. As a result, tremendous drive was created to design a chemical processes that gives as little waste as possible and so-called green chemistry or sustainable chemistry was born. One of the founders of "green" chemistry movement, Paul Anastas together with John C. Warner defined 12 principles of "green" chemistry and thus laid a foundation for future development of the field.

Twelve Principles of Green Chemistry [1]

1. Prevention of waste
2. Atom Economy
3. Less Hazardous Chemical Syntheses
4. Designing Safer Chemicals
5. Safer Solvents and Auxiliaries
6. Design for Energy Efficiency
7. Use of Renewable Feedstocks
8. Reduce Derivatives
9. Catalysis
10. Design for Degradation
11. Real-time analysis for Pollution Prevention
12. Inherently Safer Chemistry for Accident Prevention

In-depth analysis of environmental parameters of a chemical process should take into account all steps of the synthesis from the initial starting compound to the final product and if possible also its fate after the use. However, that kind of an analysis would not fit into the scope of this book. Rather, the chapter will deal with the development of the environmentally friendlier reaction of halogenation of heterocyclic molecules. The reason to talk about "green" halogenation lies in the properties of halogens that are basic halogenating reagents; they are toxic, hazardous, and corrosive molecules whose reactions are exothermic. Therefore, many points from the list of 12 principles of "green" chemistry should be addressed in reactions of halogenation. The most obvious ones are: "less hazardous chemical syntheses," "safer solvents and auxiliaries," and "inherently safer chemistry for accident prevention." In exothermic reaction, high amount of solvent is needed to absorb the heat, and furthermore, weak selectivity leads to higher amount of auxiliaries due to extensive purification. The most general solution is the use of less reactive halogenating reagents (NBS, Selectfluor, etc.), which brought further problems in the following points: "prevention of waste," "atom economy," and "reduce derivatives." All these issues that are problematic in halogenation reaction show that this reaction holds a great potential for the development of advanced methods that would import benefits in terms of "green" chemistry. Although halogenated heterocyclic molecules could be synthesized during reaction of construction of heterocyclic core, this chapter will deal only with a direct halogenation of heterocyclic compounds due to the importance of this area that was reported in the last 5 years.

In nature, there are many halogenated organic compounds, and if the halogenation reaction is so problematic from the point of "green" chemistry, it would be instructive to see how nature had coped with such a dangerous chemistry. Therefore, we will first discuss on the parallel between natural way of producing halogenated compounds and the ones that were developed in the laboratory.

2 Biological Halogenations

Laboratory or industrial production of halogenated compounds is usually performed by molecular halogens that are toxic and hazardous chemicals. It is difficult to envisage their use in biological organisms. Yet, nature has harnessed halogenation reaction to gain an access to valuable halogenated organic molecules. Nature is always a source of inspiration as it can efficiently produce complex and chiral molecules. Is this true also for halogenation? Could be the natural way of producing halogenated molecules also the best way for their synthesis in the laboratory or industry? Therefore, it is very valuable to learn about the biological halogenation and how this can help us in creating more sustainable chemical processes.

We are gaining very quickly the knowledge on the biological halogenation. Excellent reviews were published on this subject to which reader is directed, while only the most important aspects are presented in this chapter [5–10]. Fifty years ago, there were less than 30 known naturally occurring halogenated molecules,

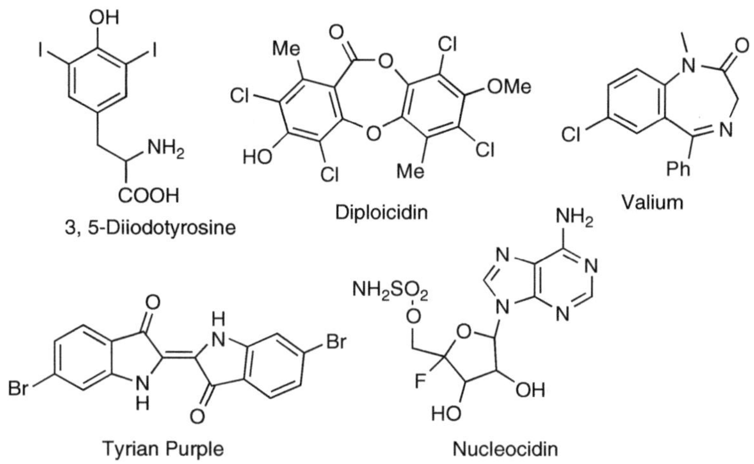

Fig. 1 Some natural halogenated heterocycles [11]

whereas today there are more than 4,500 of them [11]. Majority of them is chlorinated and brominated compounds. In general, chlorinated compounds are produced by terrestrial organisms, while brominated ones are predominately formed by marine organisms. There is surprisingly little fluorinated metabolites (ca. 30) although fluorine is the most abundant halogen in earth's crust. The reason for this small number is inactivation of fluoride in aqueous environment due to strong hydrogen bonds and its high oxidation potential. Although 3,5-diiodotyrosine, found in thyroid glands of mammals, was the first isolated and identified halo-metabolite, only ca. 120 natural iodinated compounds are known. The first halogenated molecule isolated from microorganism, diploicidin, being a halogenated heterocycle, is related directly to this book [11]. Some selected naturally occurring halogenated heterocycles are presented in Fig. 1.

All three possible halogenation strategies – electrophilic, nucleophilic, and radical – occur in the nature. Which of them is operative depends on the metabolite (a product of the biohalogenation), while the presence of halide and its abundance is of course a very important factor. For example, marine environment is rich in bromine metabolites. Fluorine is abundant but difficult to oxidize, and therefore there are only some fluorinated metabolites. On the other hand, iodide is relatively rare and although it is easily oxidized, organo-iodine compounds are not very abundant.

2.1 Electrophilic Biohalogenation

Electrophilic halogenation is the most potent way of producing organo-halogen compounds. Of course organism cannot store such a reactive reagent as it would

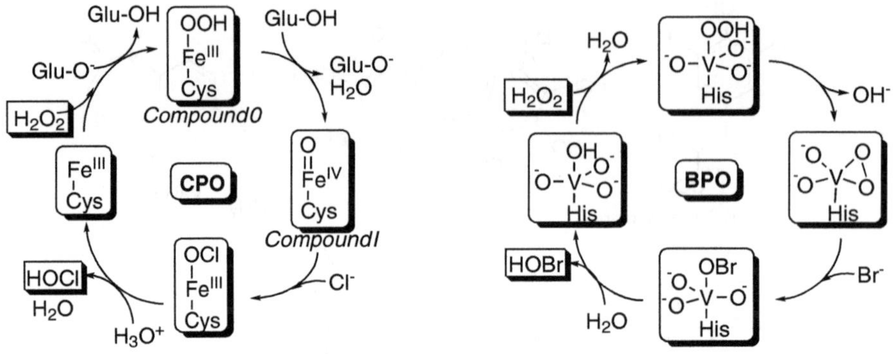

Scheme 1 Catalytic cycle of heme-containing CPO and vanadium-based haloperoxidase

damage neighboring cells. Therefore, the strategy for halogenation involves "on demand" formation of reagent by oxidation of halide. This so-called oxidative halogenation prevails among biohalogenation and it is not surprising that oxidants in this process are hydrogen peroxide and oxygen, the most environmentally acceptable oxidants [9].

The first halogenating enzyme was discovered in a fungus *Caldariomyces fumago* [12]. The enzyme is able to oxidize chloride with hydrogen peroxide in an acidic active site containing heme and hence its name chloroperoxidase (CPO). Heme reacts with hydrogen peroxide to form compound I, which in turn oxidizes chloride into hypochlorite. HOCl is eliminated from the heme upon protonation and acts as a chlorinating reagent (Scheme 1, CPO) [5]. CPO remains the most studied haloperoxidase enzyme (HPO). Later, other heme-dependent HPOs were discovered in different organisms including mammalian enzymes myeloperoxidase, lactoperoxidase, eosinophil peroxidase, and thyroid peroxidase.

At first it looked as if heme-based enzymes are the only possible ones for halogenations. Nevertheless, non-heme HPOs were discovered in marine algae [13]. Their action is bound to hydrogen peroxide as oxidant, but their active site contains vanadium(V) atom incorporated into the protein through the histidine ligand. Vanadium remains in the same oxidation state in the reaction with hydrogen peroxide and forms a peroxo-complex, which oxidizes bromide into hypobromite that further acts as a brominating agent (Scheme 1, BPO). Unlike CPO that is able to catalyze chlorination, bromination, and iodination, V-based HPO generally catalyzes only bromination and iodination.

Another oxidant that is abundant in nature is oxygen. Enzymes capable of catalyzing halogenation using air as oxidant were discovered later [6]. As shown in Scheme 2, their action is based on flavin, where $FADH_2$ is transformed into peroxidic intermediate (FAD-4a-OOH) by oxygen that in turn oxidizes halide into free hypohalous acid. Although data suggest that chloramine of lysine (LysNHCl) could also be formed as reactive specia.

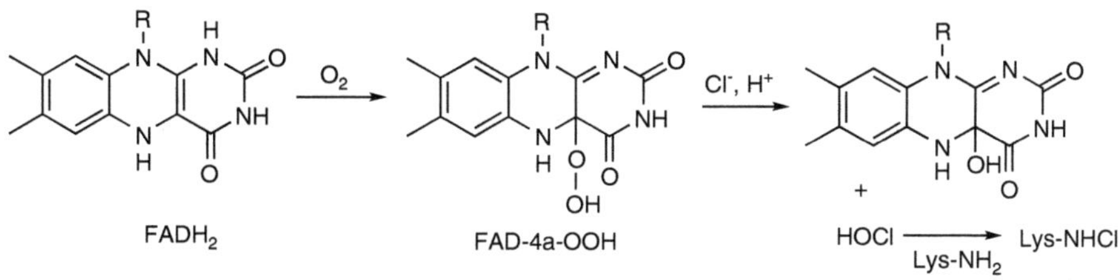

Scheme 2 Flavin-dependent halogenase

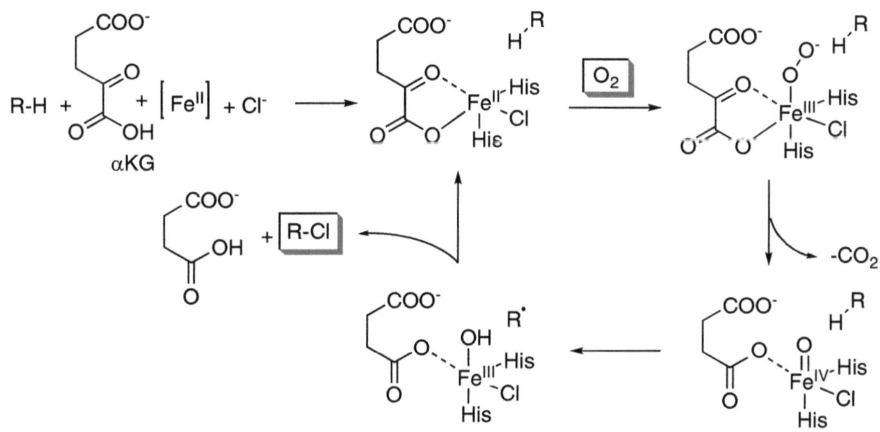

Scheme 3 Non-heme iron and O_2-dependent halogenase

2.2 *Radical Biohalogenation*

Enzymatic radical halogenation was discovered to proceed through oxidative halogenation as well. Non-heme iron and O_2-dependent halogenase are capable of performing halogenations of unactivated carbon sites through radical mechanism, where iron(II), oxygen, chloride, and α-ketoglutarate are needed [14]. A complex of iron(II) with chloride and α-ketoglutarate is oxidized with oxygen into iron(IV) oxo complex, which abstracts hydrogen radical from the substrate and turns into iron(III), while the carbon radical abstracts chlorine radical from the iron (III) complex forming a chlorinated product and iron in the initial two-valent form (Scheme 3). This could also be the pathway to the construction of cyclopropane ring in the biological systems.

2.3 *Nucleophilic Biohalogenation*

Nucleophilic halogenation in biological systems is very rare. One of the reasons is low reactivity of halide ions and the other one is a lack of good electrophilic carbon

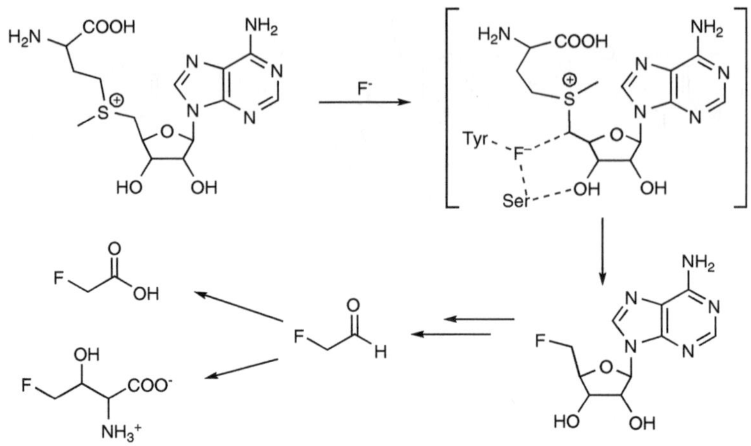

Scheme 4 Mechanism of fluorination by fluorinase

centers. As F^- cannot be oxidized under the biological conditions, it seems that nucleophilic fluorination is the only possible strategy in bio-fluorination. However, an obstacle is strong solvation of fluoride ion in aqueous media. Nevertheless, enzymes are able to invert the reactivity of halides. In the active site of mutants of glycosidase, the observed nucleophilic reactivity ($F^- > Cl^- > Br^-$) is the same as in the organic media or gas phase [15]. This indicates that the enzyme is capable of compensating the strong hydrogen bonds between fluoride and aqueous media within its active site.

The majority of enzymes that synthesize halometabolites by the nucleophilic pathway are *S*-adenosylmethionine (SAM) dependent [16, 17]. Recently, the structure of fluorinase from the *Streptomyces cattleya* has been solved and thus enabled an insight into how enzyme converts fluoride into a good nucleophile in aqucous media (Scheme 4) [18]. It was shown that interaction of fluoride ion in the active site with Ser-158 and Tyr-80 lowers the activation barrier, while electrostatic interactions with positively charged S-atom of SAM also play some part in it. Furthermore, methylthionine in SAM is a very good leaving group being positively charged. Altogether enzymatic acceleration is 10^6-fold [10].

3 Electrophilic Halogenation

Of the different ways that halogen atom can be introduced into an organic molecule, electrophilic halogenation is the most prevalent one with molecular halogens as the most common reagents. Halogenation is in fact oxidation and therefore the power of halogenating reagents can be correlated by their electrode potential (Table 1), although it is only requisite for reactivity, i.e., fluorine has a very high oxidation potential, yet fluorination is unselective.

Table 1 Standard electrode potential of halogens

	Standard electrode potential (V)[a]
$I_{2(s)} + 2e^- \rightarrow 2\,I^-$	0.54
$Br_{2(aq.)} + 2e^- \rightarrow 2\,Br^-$	1.09
$Cl_{2(g)} + 2e^- \rightarrow 2\,Cl^-$	1.36
$F_{2(g)} + 2e^- \rightarrow 2\,F^-$	2.85

[a]Taken from [19]

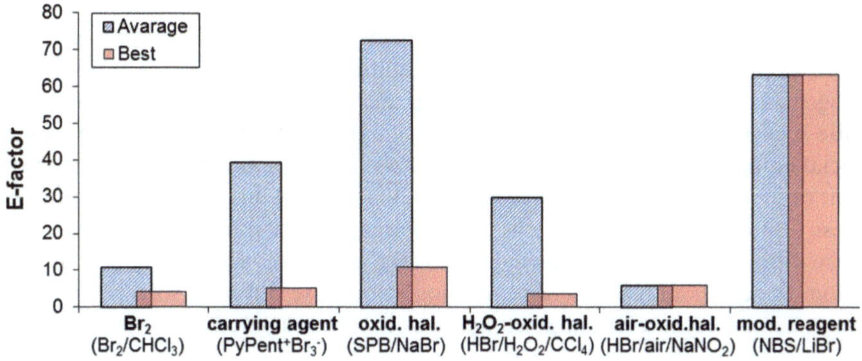

Fig. 2 E-factor for dibromination of alkenes grouped by the method of bromination [20]. The best method written in *parenthesis*

A very instructive case study on "green" halogenation was conducted by Eissen and Lenoir who searched for literature examples of dibromination of alkenes and evaluated their environmental parameters [20]. Three strategies were selected: reactions with bromine, with carrying agents and with in situ generated bromine, i.e., oxidative bromination (Fig. 2; E-factor defines mass of waste in reaction per 1 kg of product). We could see that performing a "green" reaction (carrying agent or oxidative bromination) does not necessary mean that it is "greener" than classical reaction (Br_2 in $CHCl_3$). On the other hand, if care is taken to reduce the amount of substances in the reaction and isolation/purification step, E-factor can be substantially lowered. Judging only on the quantitative measure is not enough, because other parameters are very important, i.e., the quality of waste, its toxicity, safety issues, etc.

Those methods present a general approach to "green" halogenation. An additional one, which is very important in the fluorination, is the use of modified reagents (NFSi, Selectfluor, etc.). Although they decrease atom economy, they also enable selective reactions under mild reaction conditions, which eliminate or simplify isolation/ purification procedure, and increase safety. By proper design recycling of reagent residue might be possible. In oxidative halogenation, the amount of waste is also higher due to the use of oxidant, which is compensated by elimination of the use of molecular halogens and incorporation of both halogen atoms into the molecule, while the type of oxidant is also important.

We will take a look on how the above-mentioned strategies were used in creation of environmentally friendlier halogenation reactions of heterocyclic molecules using solvent-free reaction conditions, alternative solvents (water, ionic liquids), activation (microwaves, ultrasound), or new technologies (microreactors).

3.1 *Molecular Halogens*

Molecular halogens differ quite markedly in their physicochemical properties and reactivity. Therefore, we cannot adopt one unified approach toward "greening" of halogenation. Fluorine is a very reactive gas and oxidizes most organic compounds, while it also reacts with most metals. Chlorine is a corrosive gas that also requires special equipment due to reaction with metals. Bromine is a liquid with a boiling point of 59°C and high vapor pressure. Iodine is a solid that easily sublimes, but it is less reactive than other halogens and requires activation for effective iodination. These differences are also reflected in a plethora of approaches to "green" halogenation. Recent examples in classical halogenation use different solvents and reaction systems to synthesize targeted halogenated heterocycles.

3.1.1 Solvent-Free Reactions

Traditionally, halogenation is performed in chlorinated solvents. Of the different alternative solvents to chlorinated ones, "no solvent" approach is the best. In halogenation, solvent-free reaction is problematic because of high reactivity and exothermic reactions, especially in the case of fluorination.

Fluorine is so reactive that only inert solvents can be used in reactions. Furthermore, F_2 should be diluted by nitrogen to further reduce its reactivity. An example of solvent-free fluorination with F_2 is a preparation of fluoroethylene carbonate (**2**, FEC), which is the main additive in electrolyte of lithium-ion rechargeable batteries. Direct fluorination of **1** with 30% F_2/N_2 in chloroform yields no product, presumably due to reaction of F_2 with $CHCl_3$. Therefore, inert perhaloalkane has to be used. Another possibility is the use of HF as the solvent; however, reaction can only be conducted at low temperature, where fluorination is very slow. In a solvent-free approach, reaction was conducted at higher temperature, higher flux of fluorine was applied (350 ml/min in solvent-free and 50 ml/min in solution), and the yield of FEC remained high (Scheme 5). Further fluorination led to a mixture of difluorinated compounds with **3** as the major one [21, 22].

Instead of using diluted reagent, slow addition to the reaction mixture is also a possibility for harnessing the reactivity of a reagent. An interesting reaction setup is phase-vanishing (PV) reaction, where fluorous solvent acts as a phase screen between a bulk of reagent and the reaction mixture. Reagent is slowly diffused through the fluorous solvent phase screen and enters into the reaction mixture through the whole phase border. If only one equivalent of reagent is used, its

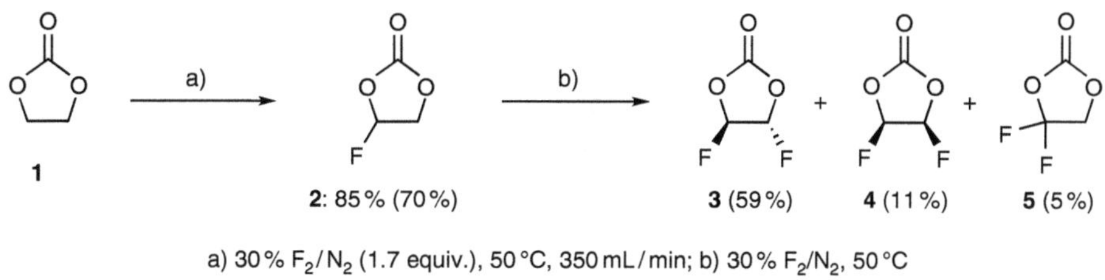

Scheme 5 Solvent-free fluorination of cyclic carbonates with diluted fluorine

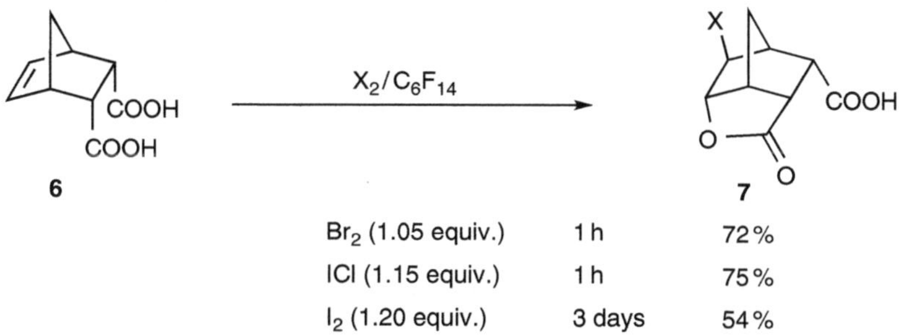

Scheme 6 Phase-vanishing halolactonization

phase vanishes and hence the name "PV reactions" [23, 24]. The reaction system was applied for bromination and chlorination [25, 26]. As the diffusion is so slow and reagent is introduced into the reaction mixture molecule by molecule through the whole phase border, there is no local build-up of heat, and reaction can be performed without any additional solvent in the reaction phase [27]. Halolactonization reaction is an example of using PV method for construction of halogenated heterocycles (Scheme 6) [28].

Iodine needs some form of activation for effective iodination, and in solvent-free systems solid Lewis acid needs to be added in an equimolar amount to promote iodination. Two examples are reported for iodination of heteroaromatics. Silica-supported $Bi(NO_3)_3$ was prepared by co-grinding and used for iodination of thiophene **8** into 2-iodothiophene **9** (58% yield) [29]. The other report used co-grinding of iodine and $AgNO_3$ to iodinate carbazole **10** in 70% yield [30].

3.1.2 Water

Among halogens, iodine is the most challenging for reactions in aqueous media as it is solid and insoluble in water. Iodocyclization to form fused furano-pyrone/coumarin and quinolone structural motifs was studied in water and in aqueous micelles [31]. Initially, reaction in water failed as **11** and I_2 are solids. By addition of surfactant (SDS), reaction is performed within micelles and proceeds in good yields (Scheme 7). Only 5-*exo*-trig cyclization occurred.

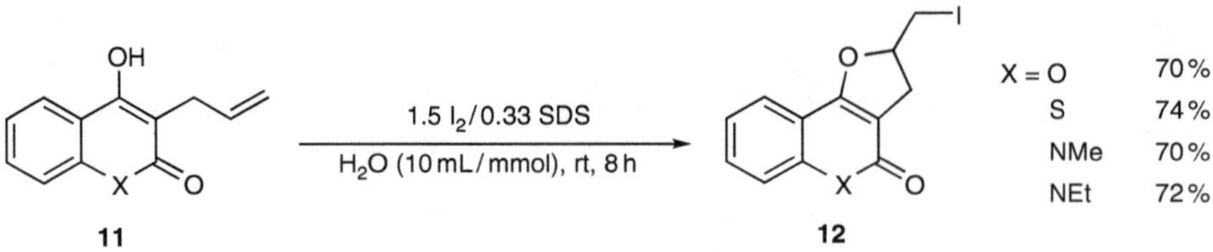

Scheme 7 Halolactonization in aqueous micelles

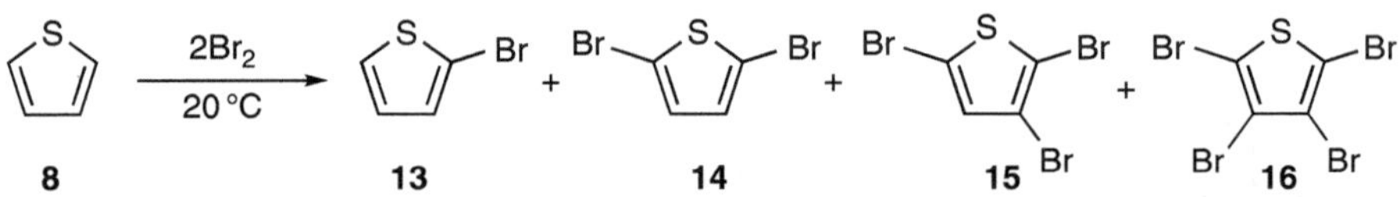

8	13	14	15	16
Batch	8%	79%	13%	/
Interdigital mixer	5%	71%	24%	/
T-junction	11%	71%	17%	1%
Caterpillar mixer	4%	88%	8%	/

Scheme 8 Selectivity of bromination of thiophene **8** in different types of microreactors

On the other hand, there are two reports on a similar iodocyclization of 2-allylphenols in water without addition of surfactants. Reaction proceeded almost quantitatively at 50°C and with 1.1 or 4 equiv. of iodine, respectively [32, 33]. Again, 5-*exo*-trig cyclization proceeded selectively. Furthermore, cyclization into 2,3-dihydrobenzofurans in water at 50°C was more efficient than the same reaction in ethanol or acetonitrile.

3.1.3 Microreactors

Microreactor technology is a very interesting solution for a variety of reactions, and as a scale-up is not problematic, reactions can be directly applied on a larger scale. One of the key fields of microreactor systems is its application for very reactive reagents due to a short contact point between reactants and a high surface/volume ratio. This obviates problems with mass/heat transport that are very difficult in batch reactors, while space-time yields are higher for an order of magnitude. It is not surprising that microreactors were used for halogenations with F_2 and Cl_2 [34–37]. Bromination of thiophene **8** under solvent-free conditions was studied with regard to selectivity of bromination. Micro mixer/tube system was used because both starting compounds are liquids; two types of micro mixer were used – caterpillar micro mixer and triangular interdigital micro mixer – while further comparison was made with T-junction mixing and batch reactors (Scheme 8) [38]. Bromine can easily evaporate and additionally gaseous HBr is formed during the reaction which complicates the design of the microreactor and reaction conditions. Regardless of the temperature, a mixture of monobrominated thiophene

13 and dibrominated thiophene **14** was obtained in reaction with 1 equivalent of Br_2, while at higher temperature tribromo derivative **15** was also formed. **15** became the predominant product when three or four molar equivalents of bromine were used. Microreactor system has orders of magnitude better space-time yield of synthesis of **14** and **15** and no solvent is used during the reaction.

3.1.4 Ultrasound Activation

Ultrasound (Us) irradiation is increasingly used in organic chemistry in recent decades to activate reaction for high yield transformation in short reaction time and mild conditions. Pyrazole-based heterocycles were halogenated by ultrasound activation. A mixture of halogen and **17** in ethyl acetate was sonicated in water bath with ultrasonic cleaner at 40 kHz and 130 W power. Iodine was not a very efficient reagent and reaction yielded only 17% of **18** (Scheme 9). Instead, ICl was used for iodination under Us irradiation. On the other hand, bromine is compatible with this activation method and gave good results [39].

3.2 Trihalides

Halogens have a tendency to form trihalide salts. A well-known example of this principle is a tincture of iodine used as a disinfectant, where solubility of iodine is increased by its transformation into triiodide salt. Similar chemistry is also observed with bromine, although tribromides are not used to increase solubility, but to transform liquid and volatile bromine into solid form and thus increase safety and ease of handling [40]. A disadvantage in using trihalides as reagents is lower atom economy that was addressed by the development of polymer-bound trihalides (polypyridinium and quaternary ammonium and phosphonium polystyrene resins) as recyclable reagents [41]. In iodination, dichloroiodates (I) bound onto tetraalkyl ammonium cation are used for iodination of aromatic and heteroaromatic compounds under basic conditions [42, 43], while polymer-bound analogues were also developed [44, 45].

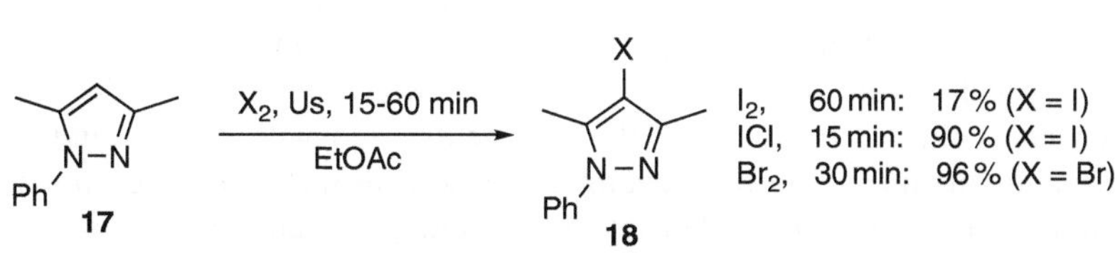

Scheme 9 Ultrasound activation of halogenation of pyrazoles

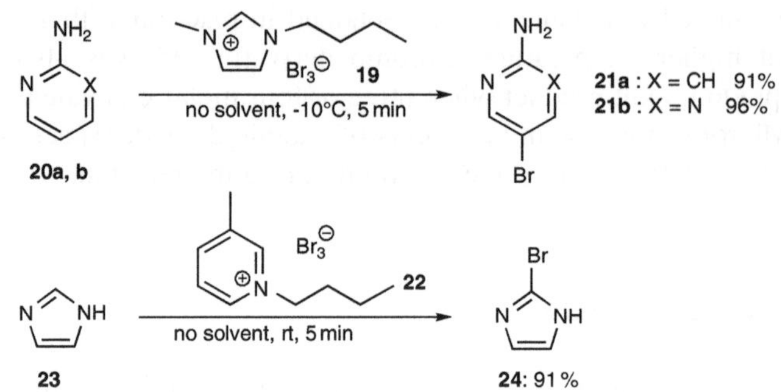

Scheme 10 Tribromide ionic liquids as reagents and solvents

3.2.1 Solvent-Free Reactions with Ionic Liquid Reagents

Tribromides with quaternary ammonium counterparts are usually solids; therefore, they can be used in solvent-free reactions mainly by co-grinding of reagent and substrate to obtain efficient method for bromination of organic substrates [46]. However, when trihalide anion was combined with alkylimidazolium cation, new types of ionic liquids (ILs) were obtained that also act as halogenating agents. Two concepts are joined – solvent-free reactions and ionic liquids [47–49]. [bmim][Br_3] **19** was applied for the bromination of aryl and heteroaryl amines. 2-Aminopyridine **20a** and 2-aminopyrimidine **20b** were efficiently brominated at $-10°C$ (Scheme 10) [50]. Quaternary pyridinium tribromide **22** displayed very similar properties and imidazole **23** was brominated in 5 min at rt [51]. IL **22** was very stable and did not lose any bromine even after prolonged heating under vacuum.

3.2.2 Aqueous Phase

Tetraalkyl ammonium salts are phase-transfer agents that might pose problems during isolation procedure if aqueous phase is used. Using simple alkali metal tribromide aqueous solution could be the reaction media, while organic ammonium cations is avoided leading to a less expensive reaction with a simplified isolation procedure [52]. A report on the bromination in highly concentrated mixture of water and acetonitrile shows how $CaBr_2$–Br_2 system can be used for bromination of various commercially important aromatic molecules [53]. The case of bromination of heteroaromatic 8-hydroxyquinoline **25** is shown in Scheme 11. A solution of heteroarene in a small amount of acetonitrile was mixed with an aqueous solution of $CaBr_2$–Br_2 and stirred for 15 min. Solid product was filtered off, while resulting solution consisting of $CaBr_2$ and HBr was neutralized with $CaCO_3$ and reused. Therefore, no organic waste was generated and no chlorinated organic solvent used presuming that reaction proceeded selectively and quantitatively.

Low solubility of iodine and its low reactivity render iodination in aqueous phase very limited. Using alkali dichloroiodate, iodination in aqueous phase became

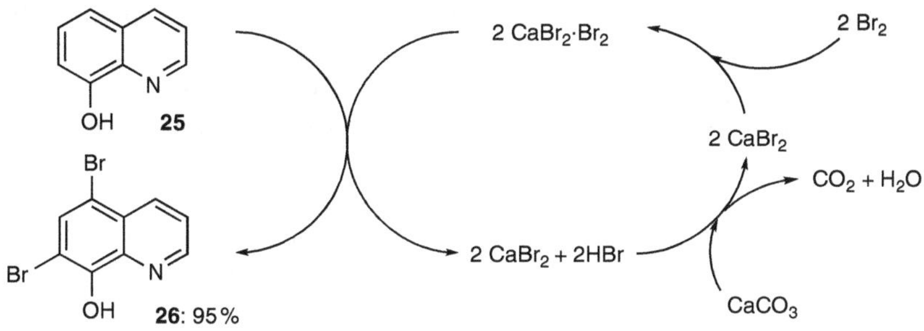

Scheme 11 Bromination with tribromide in aqueous media

Scheme 12 Microwave heating in iodination with dichloroiodates

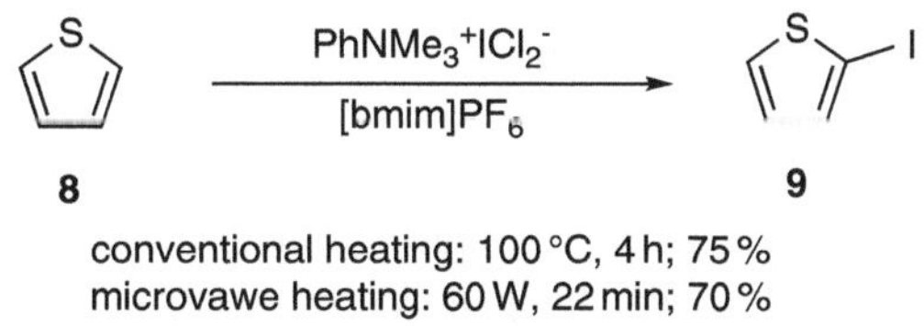

possible due to higher solubility and better reactivity. Various heteroaromatic compounds (8-hydroxyquinoline, imidazoles, pyrazole) were efficiently iodinated with 2 equivalents of $KICl_2$ in aqueous media in less than 6 h at rt [54].

3.2.3 Activation by Microwaves in Ionic Liquids

Dichloroiodate salts, being ionic compounds, are compatible with ILs. $PhMe_3N^+ICl_2^-$ was taken as a reagent to study iodination of arenes and heteroarenes in [bmim] type of ionic liquids under microwave activation. Interestingly, reaction did not proceed in molten salts ($Bu_4N^+Cl^-$), while reaction in [bmim]Cl was also inefficient. On the other hand, iodination proceeded in hydrophobic [bmim]PF_6, which was superior to hydrophilic [bmim]BF_4 in terms of yield and reaction rates (Scheme 12) [55]. Further activation of iodination with microwaves (MW) reduced the reaction time, and by applying microwave heating at 60 W for 22 min, 70% of **9** was isolated.

3.3 N-X Reagents

N-X reagents were implemented into the organic chemistry to substitute toxic, corrosive, and dangerous molecular halogens, especially for fluorination. Therefore, modified reagents increase markedly the safety of reactions, i.e., Selectfluor™, an alternative source of positive fluorine atom, is a stable powder that could be easily handled under standard reaction conditions. A disadvantage of modified halogenating reagents is their low atom economy and disposal of ligand residue. That is why it is of crucial importance to have a green improvement in these reactions. In recent years, research was focused mainly on the development of new reaction media (solvent-free, ionic liquids, etc.) and new activation of halogenation (microwaves, ultrasound).

Modified halogenating reagents are generally solids, and therefore solvent-free reactions usually take place in solid phase (grinding in mortar, ball-milling, etc.) [46]. On the other hand, in the aromatic bromination with NBS in solid phase, it was found that a crystallinity is requisite as in the reactions in melt or in the solution reactivity and product selectivity is lost [56].

3.3.1 Aqueous Phase

Most N-X reagents have a ligand part that is soluble in water. This facilitates separation of halogenated products from the reagent residue, while it also makes recovery of the reagent residue energy consuming. One of the first examples is bromination of olefins with NBS in water published in 1955 [57], while some recent examples include aminohalogenation of alkenes [58, 59] as well as halogenation of arenes and ketones [60, 61].

Fluorination was also studied in aqueous phase and Selectfluor™ was the most frequently used reagent [62]. In heterocyclic chemistry, fluorination of piperidinyl olefins was performed in aqueous acetonitrile ($MeCN/H_2O$ – 1/1) with Selectfluor™ being the most efficient reagent for allylic fluorination, while fluorohydroxylation occurred with the piperidinyl *exo*-olefins [63]. The same solvent mixture was also used in fluorination of tryptamine derivatives **27** (alkaloid based on indole core) where fluorination was coupled with further oxidation to yield 3-fluoro-2-indolinone analogues **28** that are very important structural units in medicinal chemistry (Scheme 13) [64]. Amino function should be protected (**27a**) to have an effective fluorination. Otherwise, oxidation at the amine atom occurs which leads to demethylation and oxidation at the side chain of **27b**. This could be partially avoided using hydrochloride form of alkaloidal tryptamine (**27c**).

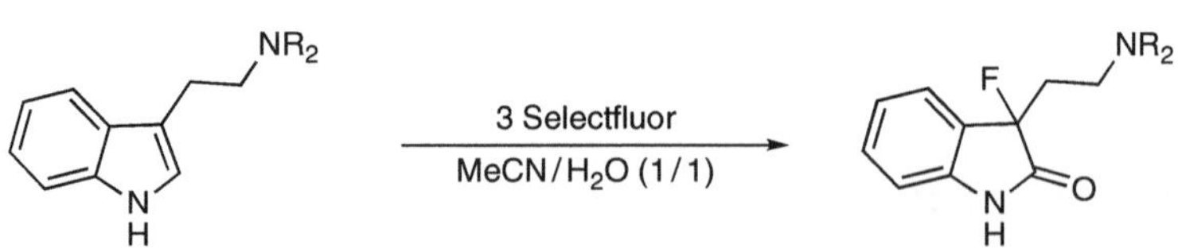

R	React. cond.	Yield (28)
27a : Boc	r.t., overnight	71 %
27b : Me	0°C to r.t., 30 min	13 %
27c : Me·HCl	0°C to r.t., 30 min	35 %

Scheme 13 Fluorination of tryptamines in aqueous media

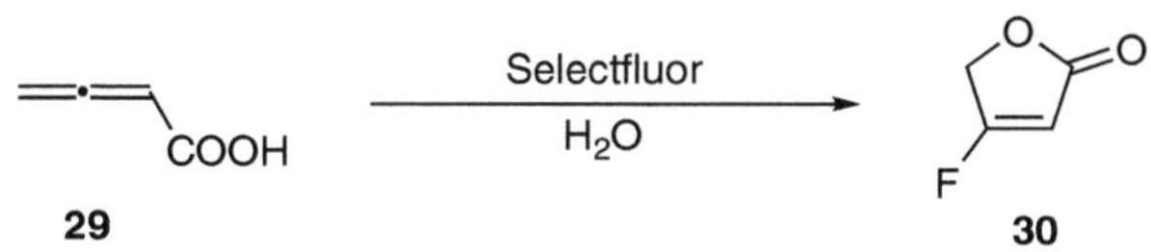

Scheme 14 Fluorolactonization with Selectfluor™ in water

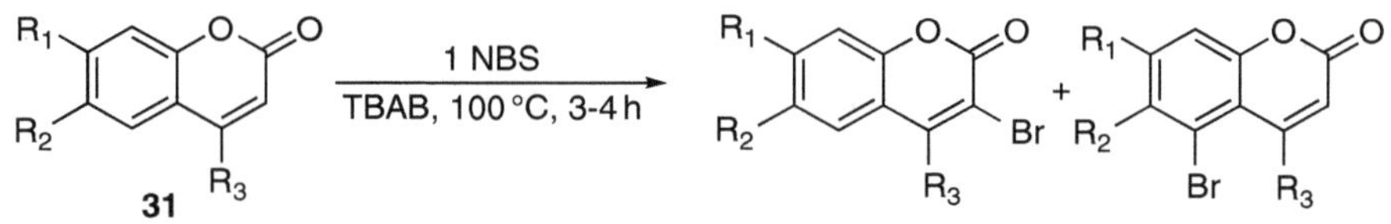

R_1	R_2	R_3	yield (32)	yield (33)
OCH_3	H	CH_3	60%	/
OH	H	H	78%	/
H	H	OH	76%	/
H	OH	H	/	68 %
H	NH_2	H	/	75 %

Scheme 15 Bromination of coumarins in TBAB

In the study on electrophilic fluorocyclization reaction of 2,3-allenoic acids **29** with Selectfluor™, it was reported that reaction in pure water affords β-fluorobutenolides **30** in moderate to high yields (Scheme 14) [65].

3.3.2 Ionic Liquids

An increasing use of room temperature ionic liquids as alternative green solvents is reflected also in halogenation with N-X reagents [66]. Fluorination with Selectfluor™ was studied on aromatic molecules to find that reaction proceeds similarly regardless on the type of IL (hydrophilic [bmim][BF_4] vs. hydrophobic [bmim][PF_6]). More extensive research was performed in halogenation of aromatics, ketones, and olefins with NXS reagents in imidazolium-based ILs with different anions (Br, BF_4, PF_6, HF_2) [66, 67].

Tetrabutylammonium bromide (TBAB, 0.2 g/1 mmol substrate) was used as a media for regioselective monobromination of aromatics and heteroaromatics with 1 equivalent of NBS [68]. Reaction in TBAB was faster than in a classical setup in DMF, although bromination is still selective for monobromination (Scheme 15). It seems that bromide anion acts as H-bond donor, making phenol and aniline derivatives more active toward electrophilic attack of NBS. Results also indicate that formation of tribromide as reagent is less probable. As in classical bromination, addition of Lewis acid (BF_3) activates reaction. Bromination with NBS in TBAB was effective for coumarin derivatives and nucleoside bases uracil and cytosine.

3.3.3 Activation by Microwaves or Ultrasound

Microwaves

Halogenation of activated aromatics can be efficiently made in different reactions systems. On the other hand, deactivated aromatic rings require higher temperatures, longer reaction times, larger amount of reagents, etc. MW are interesting activators as they do not import any additional substances, as is the case with acid catalysts, and the energy consumption is lower due to short reaction times. This was demonstrated in the fluorination of *N*-aryl-3,5-disubstituted pyrazoles with Selectfluor™ F-TEDA in MeCN and 1,450 W MW oven at 10% power [69]. Authors found out that 50-fold reduction of reaction time was observed in comparison with conventional reflux method together with higher product yield. In general, pyrazoles gave 13–71% yield of fluorination.

MW irradiation was used for bromination of deactivated isoxazoles **34b,d,f** and pyrazoles **35a,c,e** with 1.2 equivalents of NBS at 150°C for 10 min [70]. Due to low reactivity of these molecules, acidic solvents had to be used to have a good conversion (Scheme 16). Acetic acid was sufficient for **35a,b**, while more deactivated analogues (**35c–f**) required the use of 5% HNO_3 in AcOH or even CF_3COOH.

In an another approach, MW activation was coupled with solid phase organic synthesis (SPOS). Initial goal of the research was to create iodination method for pyrimidine nucleosides that would use less hazardous reagents. Classical iodination of **36a** with NIS in MeOH in the presence of stoichiometric amount of 2,6-di-*tert*-butylpyridine needed 6 h to yield **37a** in 98%. The same result was obtained only after 3 min of MW irradiation at 200 W, and the addition of base was unnecessary (Scheme 17) [71]. This iodination methodology was applied to SPOS to expand further its synthetic utility. Pyrimidinone bound to Merrifield resin (**36c**) was

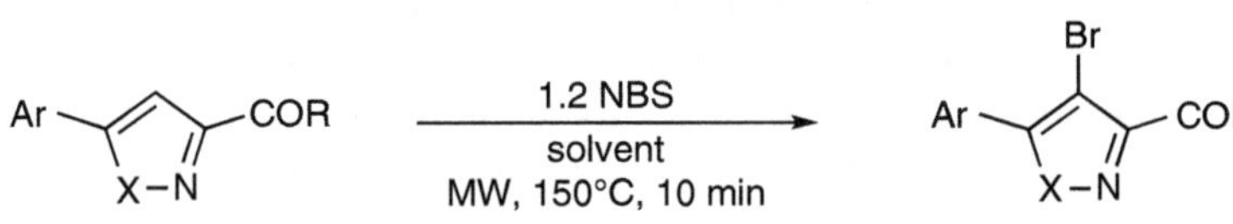

34 **35**

	X	Ar	R	solvent	yield
a	NH	Ph	OMe	AcOH	89%
b	O		NMe_2		87%
c	NH	3, 4-diClPh	OMe	5% HNO_3/AcOH	83%
d	O		OMe		88%
e	NH	3-NO_2Ph	OMe	CF_3COOH	78%
f	O		OMe		90%

Scheme 16 Ring bromination of heterocycles activated by microwaves

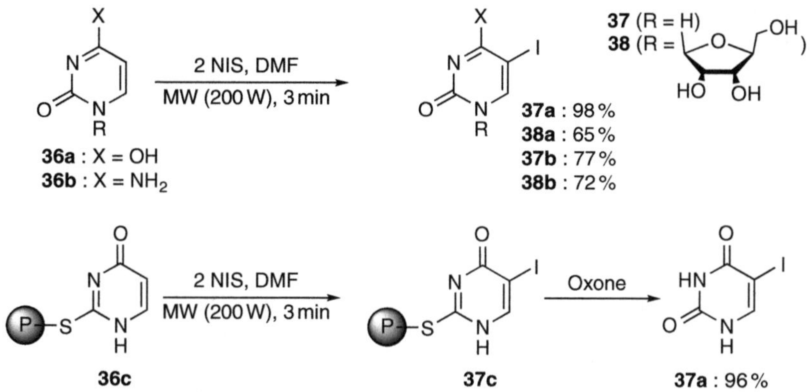

Scheme 17 Solid phase synthesis coupled with MW activation

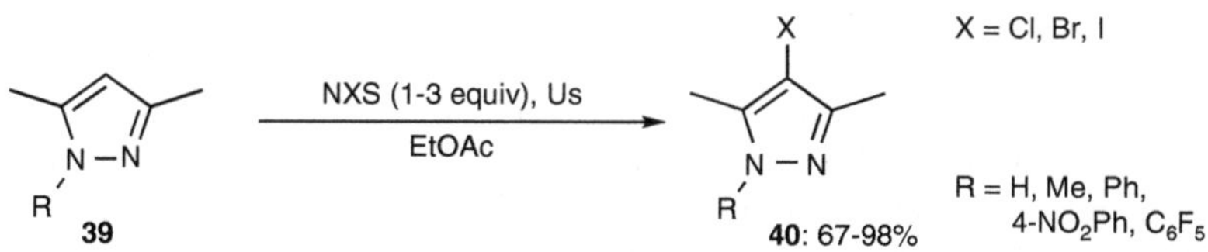

Scheme 18 Ultrasound irradiation in halogenation of pyrazoles

subjected to MW iodination in DMF, resin was washed, and product cleaved from the resin to yield **37a** in 96% yield [71].

36a and **36b** were also brominated with NBS in molten salt ($Bu_4N^+Br^-$, 0.2 g/mmol) in the presence of Montmorillonite K-10 clay (0.5 g/mmol). MW-irradiated reaction gave ca. 10% of better isolated yields, while MW irradiation of 300 W for 2.5 to 4 min was applied in comparison to classical heating at 100°C for 45 min to 2 h [68].

Ultrasound

Microwaves are generally known due to their use in household. Similarly, ultrasound (Us) is known due to its application in medicine and cleaning. As with MW, chemical processes can also be activated by Us without the need for additional substances. Sonochemistry was used for halogenation of highly deactivated pyrazoles **39** with NXS in EtOAc, where Us sonication in water bath with Us cleaner at 40 kHz and 130 W power was applied. All substrates were halogenated in less than 90 min, mainly within 6–20 min (Scheme 18) [39].

Similarly, substituted thiophenes, bi- and terthiophenes, and furan were brominated with NBS at 2 and 5 positions within minutes in EtOAc by applying Us irradiation [72].

Scheme 19 Oxidative halogenation mediated by oxygen or hydrogen peroxide

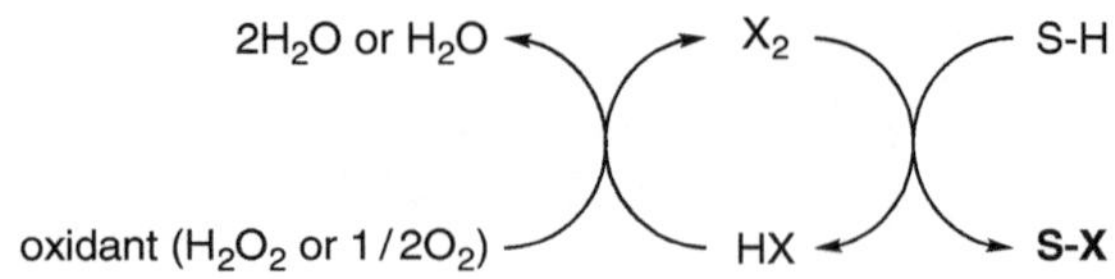

3.4 *Oxidative Halogenation*

Toxicity of molecular halogens forced living organisms to develop a different way of their utilization to gain valuable biological compounds. In Sect. 2, it was shown that a natural way of producing organo-halogen compounds is mainly based on oxidative halogenation. In this method, halogens are formed by oxidation of halides within the enzyme's active site and consumed in situ. A growing ecological awareness among chemists has coincided with an increased understanding of oxidative halogenation in biological systems, which has boosted research in the field of oxidative halogenation. At first, this research was connected with metal-catalyzed reaction as mimics of biological halogenation. In recent years, focus has shifted into the field of "green" chemistry with a development of various reactions of oxidative halogenation, where residual HX is regenerated by various oxidants such as metals, persulfates, and hypervalent iodine oxidants. Among them, hydrogen peroxide and oxygen are the "greenest" oxidants as the only effluent after the reaction is water (Scheme 19) [73]. A synthesis of vinyl chloride is an example of this approach in industry, where waste HCl generated during the production of isocyanate is reused for vinyl chloride production by passing it together with oxygen and ethylene in the gas phase over a Cu^{II} catalyst at a temperature of over 200°C [74].

3.4.1 Aerobic Halogenation

Oxygen is the most abundant and natural oxidant, especially in its diluted form as air. The use of oxygen for organic reactions is both an attractive and challenging research field. Photochemistry is used for harnessing oxygen to oxidize HX into active halogenating agent, where radical addition of bromine or iodine radicals following by incorporation of oxygen leads to conversion of saturated or unsaturated C–C bonds into halogenated ketones [75–77].

It is well known that nitric acid can be used as oxidant in converting HX into active halogenating agent. Furthermore, ILs with nitrate anions were used as oxidants, while reader is directed to the review paper to learn more on the nitrogen species as catalysts in aerobic halogenation with organic molecules [73].

There are two reports on the aerobic halogenation concerning heterocyclic compounds. The first one uses organic nitrite as oxidant for conversion of various heteroaromatic molecules (quinoline, 2-metylfuran, indolizines, coumarin) into brominated derivatives [78]. Although method of aerobic halogenation hold

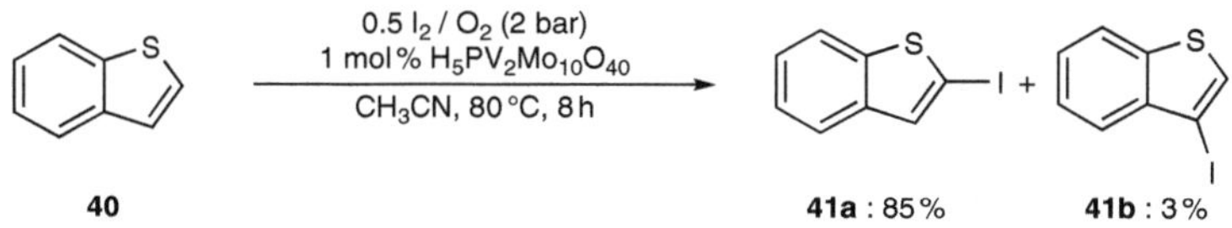

Scheme 20 Polyoxometalate as catalyst for aerobic iodination

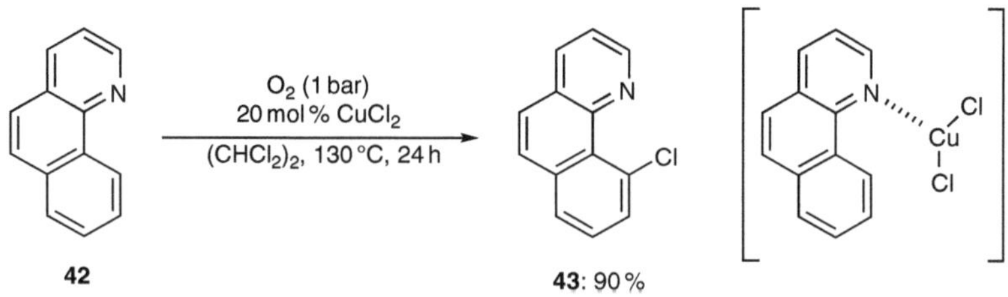

Scheme 21 Cupper-catalyzed aerobic chlorination

"green" potential, the use of equimolar amount of nitrite, higher amount of reagents, and the use of DCM are drawbacks of this reaction. Similarly, the second one reports on the iodocyclization of unsaturated acids and carbamates through 5-*exo*-trig cyclization and also uses organic conditions with $Bu_4N^+I^-$ in DCM, while acid conditions, needed to decompose catalyst into nitrogen oxides, were provided by acetic acid [79].

There are many possibilities for aerobic halogenation catalyzed by metal complexes (V, Cu, Ti, Fe, Mn, Ru) [73]. Polyoxometalate (POM) catalyst with V and Mo in the structure was found to be very active catalyst for aerobic iodination [80]. However, all these processes were studied on the aromatics, and only selected examples in heteroaromatic series were reported. Thiophene **8** was iodinated by POM catalyst ($H_5PV_2Mo_{10}O_{40}$) with 0.5 equiv. of iodine and 1 mol% of POM in acetonitrile under the atmosphere of O_2 (2 bars), and mono- and di-iodinated products were obtained in 2:1 ratio. On the other hand, benzothiphene **40** gave only **41a** with a small amount of **41b** (Scheme 20) [80].

Aerobic chlorination could be achieved by interaction of cupper(II) chloride catalyst with a heterocyclic nitrogen atom followed by oxidation of aromatic ring by Cu(II) and transfer of chloride ion (Scheme 21) [81]. Reaction occurs *ortho* to nitrogen atom and therefore chlorination is governed by the geometry of the molecule, as in **42**.

3.4.2 Hydrogen Peroxide-Mediated Halogenation

Hydrogen peroxide has good environmental character as after the reaction water is the only residue. Because of its unstable nature, diluted form (less than 30%) should be used to increase safety. Oxidative halogenation was first studied in connection with biomimetic halogenation, and metal catalyst in dilute solutions (<0.01 M H_2O_2) was used to mimic the biological conditions. Performing reactions on a laboratory scale generally demands higher concentration of reagents (>0.1 M). In this case, the use of a metal catalyst is superfluous as an acidic solution of hydrogen peroxide oxidizes bromide to bromine (1) and also reduces bromine to bromide (2); the net result of these two compensating reactions is the catalytic decomposition of hydrogen peroxide or *catalase*-type reaction, which is pH dependent and faster at higher pH values (3).

$$H_2O_2 + 2X^- + 2H^+ \rightarrow X_2 + 2H_2O \tag{1}$$

$$H_2O_2 + X_2 \rightarrow O_2 + 2X^- + 2H^+ \tag{2}$$

$$2H_2O_2 \rightarrow 2H_2O + O_2 \tag{3}$$

Metal-Catalyzed

Metal-catalyzed oxidative halogenation evolved out of the biomimetic studies oriented toward the understanding of processes in enzymes, especially vanadium BPO. Therefore, no care was taken to improve the environmental parameters of these reactions, as is the case of oxidative bromination of coumarin, flavone, and aurone derivatives and in halocyclization reactions with large excess of H_2O_2, ammonium bromide, and high amount of V_2O_5 [82, 83].

An example of bromocyclization was presented in the work on the synthesis of marine natural product aplysiapyranoid A **47**, where the first step of the synthesis consisted of bromocyclization of **44** induced by TBHP and vanadium catalyst (Scheme 22). Cyclization product was formed as a mixture of two products, **45a** and **45b**, in a ratio of 80:20 [84].

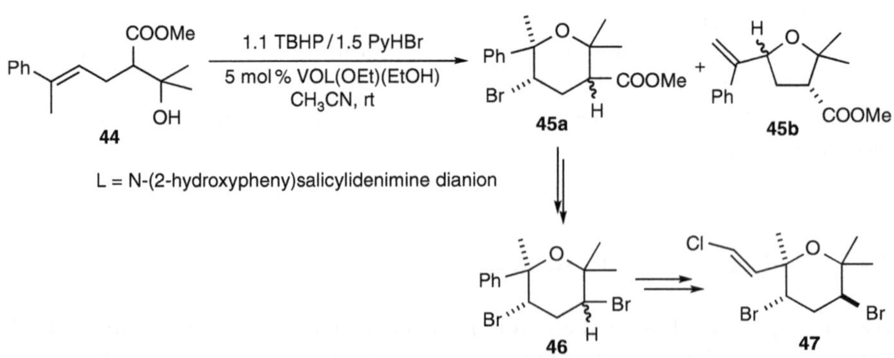

Scheme 22 Vanadium-catalyzed oxidative halogenation

A group of Sels introduced tungstate into the naturally occurring mineral takovite (layered double hydroxide – LDH) to obtain a catalyst for oxidative halogenation with activity close to the enzyme, although very stable toward oxidation [85]. 2,3-Dihydro-4H-pyrane **48** was very efficiently converted into bromo- and iodohydrines **49** [86]. The method was further extended to the halogenation of glycols **50**, although larger excess of halide salt and H_2O_2 was needed (Scheme 23).

Selenium compounds were found as interesting catalysts for oxidative halogenation. Various substituted aryl selenic acids were appropriate catalysts for bromolactonization and iodolactonization with KX and H_2O_2 in the presence of a buffer (pH 6) that provides the protons [87, 88]. A similar reactivity was observed with tellurium analogues [89]. More recently, a reusable selenium catalyst, 4-(hydroxymethyl) phenylbenzyl selenoxide, was reported which is sequestered within halide-permeable xerogels (Fig. 3) [90]. It was observed in the bromolactonization that sequestration not only makes the catalyst 23 times more active than a xerogel-free catalyst in a solution, but it also makes for an easy separation of it from the reaction mixture.

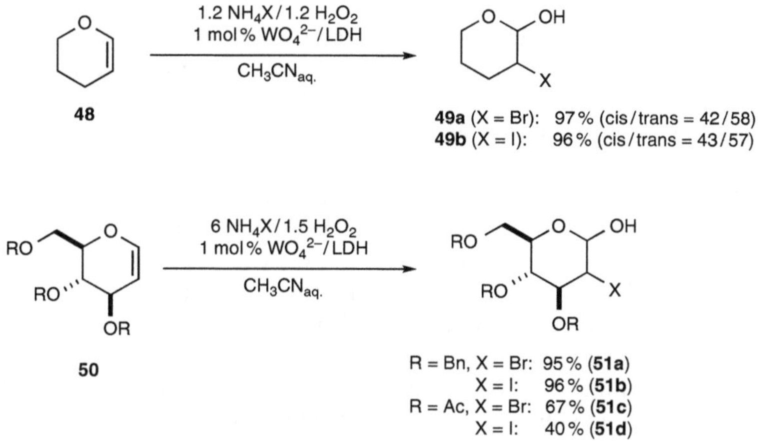

Scheme 23 Layered double hydroxides exchanged with tungstanate as catalysts

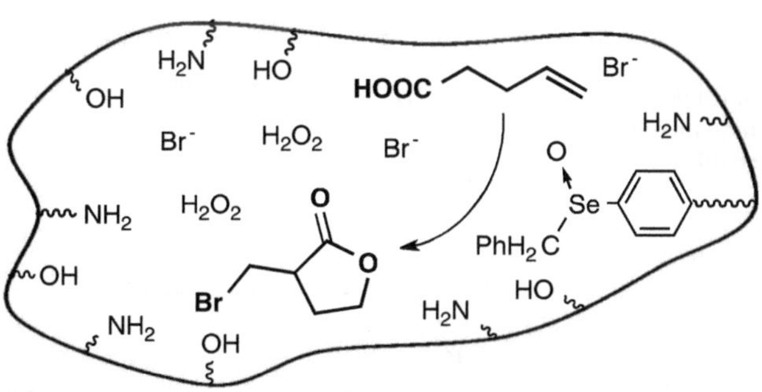

Fig. 3 A reusable selenium catalyst for oxidative bromination in an aqueous environment

Non-catalyzed

After the reaction of oxidative halogenation mediated by hydrogen peroxide, there is no reagent residue apart from water and the reaction occurs with 100% halogen atom economy (4). Due to a higher concentration of reagents, the *catalase*-type decomposition of H_2O_2 (3), that is catalyzed by halogens and halides, becomes more important. This decomposition is pH dependent and inhibited in acidic conditions. If metal halides are used (5), hydroxide would be a side product and H_2O_2 would not survive basic condition thus formed. Therefore, it is necessary to use at least an equivalent of suitable acid to neutralize hydroxide formed during the reaction (6).

$$\mathrm{R-H+HX+H_2O_2 \rightarrow R-X+2H_2O} \quad (4)$$

$$\mathrm{R-H+MX+H_2O_2 \rightarrow R-X+MOH+H_2O} \quad (5)$$

$$\mathrm{R-H+MX+H_2O_2+HA \rightarrow R-X+MA+2H_2O} \quad (6)$$

An example in Scheme 24 shows an oxidative iodination of uracil **36a**, where potassium iodide was used as a source of iodine atom, while sulfuric acid had to be added to prevent decomposition of oxidant [91]. Sulfuric acid is at the same time an activator of the iodination step, and its quantity determines the reactivity of the system.

The H_2O_2-mediated iodination can be achieved using either I_2 or system NaI/acid. Looking only at the source of iodine atom, the use of I_2 appears more economical, since only 0.5 equivalent of I_2 and 0.5 equivalents (or a slight excess) of H_2O_2 is sufficient for good conversion, while there is no residue after the reaction. With this method, 2,3-dihydro-4H-pyrane **48** was iodinated into *trans* addition product **49c** in 61% yield (Scheme 25) [92]. Alternatively, the system NaI/H_2O_2 combined with an acid in a THF/H_2O media gave a higher yield and again only *trans* addition into **49b** occurred [93]. A wide range of acids has been tested,

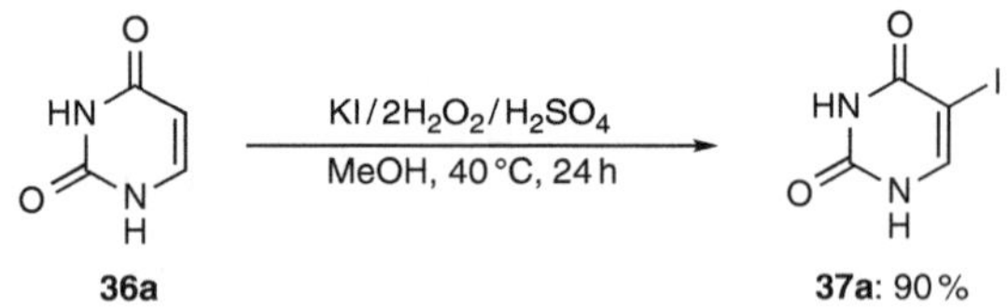

Scheme 24 Non-catalyzed oxidative iodination mediated by hydrogen peroxide

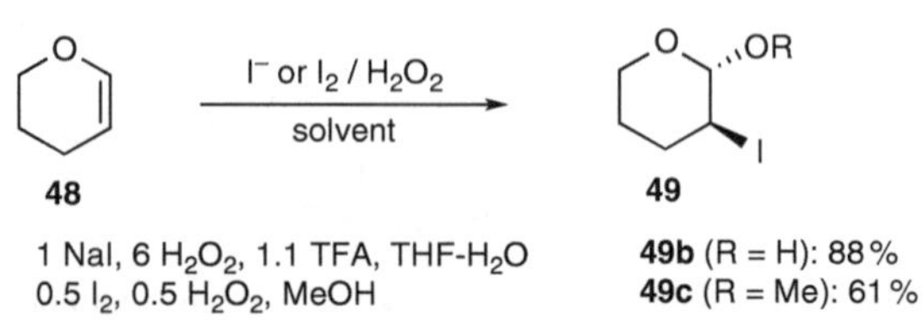

Scheme 25 Oxidative iodination with I_2 or NaI/HA

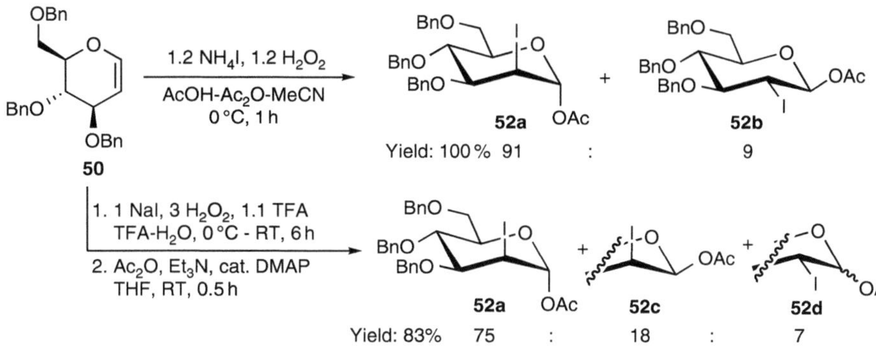

Scheme 26 H_2O_2-mediated oxidative iodination of glucal

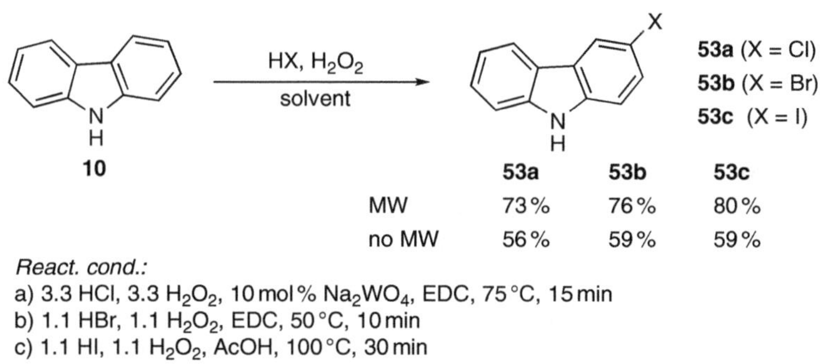

Scheme 27 Microwave-assisted oxidative halogenation of carbazole

such as HBF_4, H_2SO_4, oxalic acid, TFA, H_3PO_4, and Amberlyst 15, with the best result obtained using HBF_4. It was not possible to promote the reaction in such a way as to provide a satisfactory conversion with a weaker acid.

This system was also applied for iodination of protected glycols. Glucal **50** was iodohydroxylated with NaI/H_2O_2 in the presence of TFA and subsequently acetylated (Scheme 26). Reaction was less selective than with **48**, while the main process was again *trans* addition with *manno* isomer **52a** being the predominant product [93]. Slightly different conditions were used with NH_4I in the presence of Ac_2O and AcOH that intercepted the water present in the system and ensured selective conversion into iodoacetate **52** without the intermediate formation of any iodohydrines. Using acetonitrile allowed the reaction temperature to be lowered so as to favor *trans* addition with **52a** as the major product (Scheme 26) [94].

Oxidation of HCl with H_2O_2 is more problematic and needs additional activation in the form of thermodynamic activation through interactions or external heating. Microwave irradiation was used for the activation of oxyhalogenation of carbazole **10** to find that halogenation is faster and yields higher than with conventional protocols, although reaction conditions varies according to the type of halogen atom (Scheme 27) [95].

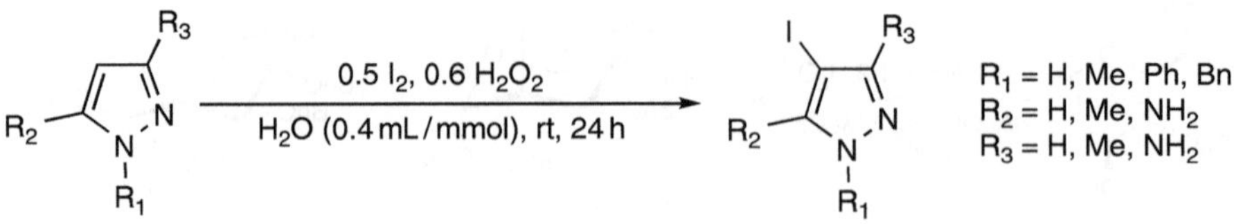

Scheme 28 Iodination of pyrazoles with H_2O_2/I_2 in water

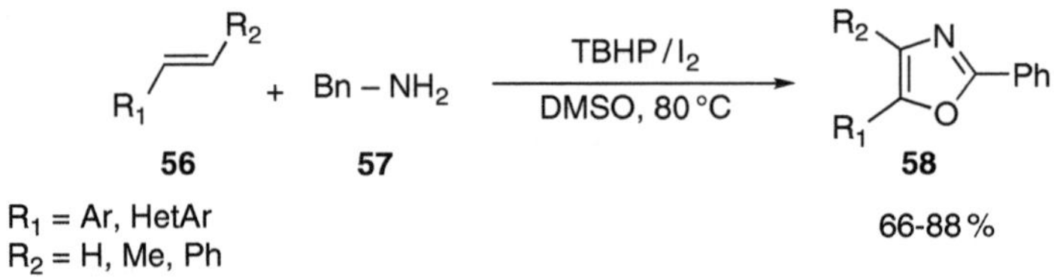

Scheme 29 TBHP/I_2-mediated domino oxidative cyclization

H_2O_2-mediated oxidative halogenation is compatible with aqueous media, and after the reaction all components except halogenated product are dissolved in the aqueous phase [73]. Oxidative iodination was thus applied in aqueous media, although iodine is not soluble in water and results of the iodination of various substituted pyrazoles **54** are presented in Scheme 28. Under these conditions, the reaction proceeds well for unsubstituted as well as *N*-alkyl- and *N*-aryl-substituted pyrazole derivatives. Due to low temperature of reaction, there is no problem with the volatility of iodine [96]. Solid form of H_2O_2–sodium perborate was used as oxidant in iodination of thiophene and imidazole in imidazolium-based ionic liquids [97].

Oxidative iodination was coupled with cyclization reaction to create a domino oxidative cyclization that was mediated by *t*BuOOH (TBHP) and iodine to convert an aryl- and heteroaryl-substituted alkene **56** and benzyl amine **57** into substituted oxazole product **58** (Scheme 29) [98].

3.4.3 Electrochemistry

Electron (or electric current) could be regarded as the greenest reagent as it is waste-free, cheap, energy efficient, and could be easily controlled. Although equipment is simple and not expensive, the setups to deliver electricity in controlled, safe, and effective units are not trivial [99, 100]. There are many challenges that need to be addressed by researchers to bring electrochemistry closer to organic synthesis. In the field of halogenation, electrochemistry is connected with oxidation, either of halide ion into the active halogenating reagent or of organic substrate that further reacts with halide to form a halo-organic compound.

The first option is connected directly with the oxidative halogenation, where electric current is used to transform halide into molecular halogen or positively

42

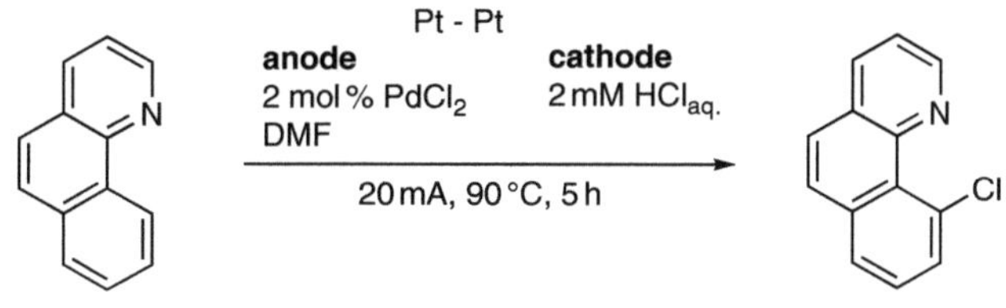

43
quant.

Scheme 30 Electrochemical chlorination catalyzed by Pd^{II}

charged halogen atom that acts further as halogenating reagent. This approach is limited to iodination, bromination, and chlorination. An example of such an approach is selective iodination of aromatic compounds with electrochemically generated I^+ species using micromixing technology [101]. Electrochemical oxida tion of X^- into X^+ was performed under constant current electrolysis at 20 mA, and no additional electrolyte was used in this reaction setup. Chlorination of **42** was quantitative and was completed in 5 h at 90°C (Scheme 30). At the end of the reaction, electrical current is simply switched off and reaction is stopped [102]. In comparison, GC yields after 12 h reactions with NCS or Chloramin-T/$CuCl_2$ were 81% and 85%, respectively.

The second approach, where organic substrate is oxidized first, is characteristic for fluorination. Fluoride is very difficult to oxidize; therefore, the first step of the reaction is oxidation of organic substrate into its cation radical followed by incorporation of fluoride ion into this intermediate. Indole derivatives were difluorinated into 2,3-difluoroindolines by this strategy, in which undivided cell was used for anodic fluorination under constant current with $Et_3N{\cdot}3HF$ in MeCN. Fluorination was more effective when electrolyte was $Et_4NF{\cdot}4HF$ [103]. Acetonitrile could be problematic as it can act as a nucleophile and thus compete with fluoride. Furthermore, it can also play a role in passivation of anode. A solution was found that includes the use of ionic liquids ($Et_4NF{\cdot}nHF$ or imidazolium salts), where IL also serves the role of an electrolyte [104]. The type of salt and solvent markedly influence the course of reaction [105]. Selected examples of anodic fluorination of phthalide **59** in different media are presented on Scheme 31, where the effect of the structure of IL is evident [106]. Fluorodesulfurization of **60** followed very similar trends, and furthermore, recycling of IL was effective when sufficiently high ratio of HF was used (Scheme 31, last line). Anodic fluorination was the most effective when no solvent was used and $Et_4NF{\cdot}4HF$ was both, electrolyte and solvent [107]. Similarly, fluorodesulfurization of phenylthioglycoside under electrochemical conditions gave selectively fluorinated glycoside when $Et_4NF{\cdot}4HF$ was used as reaction media, while in $Et_3N.5HF$ both transformations were operative, α-fluorination and fluorodesulfurization [107].

This new solvent-free electrochemical fluorination was used for the synthesis of FEC **2** and other cyclic ethers and carbonates (Scheme 32). Only a small amount of IL was

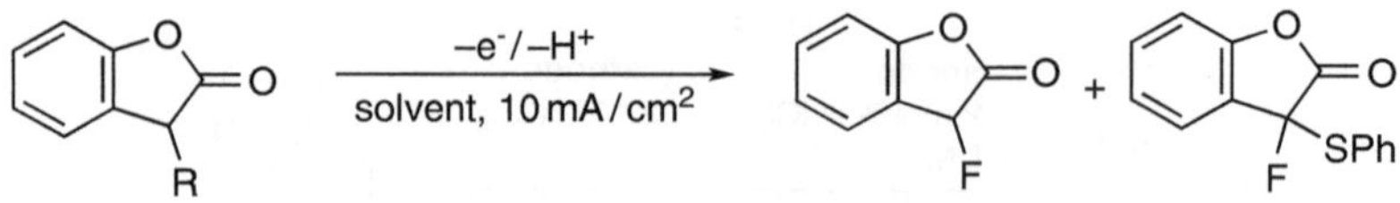

	Solvent	F/mol	Electrolyte	61	62
59 (R = H)	DCM	8	$Et_3N{\cdot}5HF$	23 %	
	MeCN	6	//	16 %	
	[emim][OTf]	4	//	90 %	
	no solvent	4	$Et_3N{\cdot}3HF$	17 %	
60 (R = SPh)	[emim][OTf]	4	$Et_3N{\cdot}5HF$	83 %	
	no solvent	4.2	$Et_3N{\cdot}3HF$	4 %	21%
	//	3.2	$Et_3N{\cdot}5HF$	96 %	
	//	3.3	$Et_4N{\cdot}4HF$	99 %	
	//	4	$Et_4NF{\cdot}5HF$	93 %	
	//	2.5	$Et_4NF{\cdot}4HF$	99 %, 96 %, 93 %, 94 %	

Scheme 31 Ionic liquids as reusable media for anodic fluorination of phthalide

Scheme 32 Anodic monofluorination of cyclic carbonate

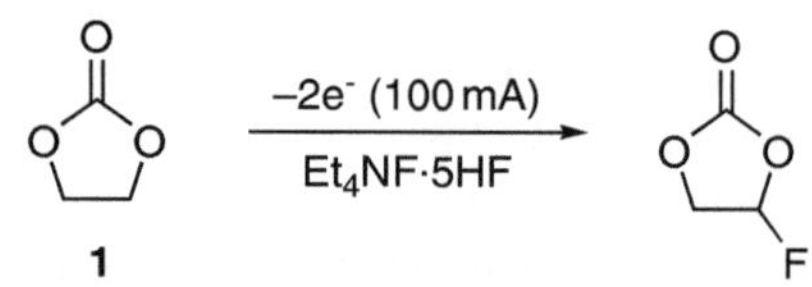

1 **2**: 87 %

used, so that a ratio of F⁻ vs. substrate was only 1.5 to 1. The reaction was more efficient than analogous reactions in the classical IL and in organic solvents, while isolation procedure comprises of extraction or distillation with possible reuse of IL [108].

4 Radical Halogenations

Free radical halogenation is one of the most important methods for the functionalization of sp^3 C-atom. A classical Wohl–Ziegler method uses *N*-bromosuccinimide (NBS) in boiling carbon tetrachloride (CCl_4) in the presence of radical chain initiator (AIBN, etc.) [109]. The reaction is outdated as it requires carbon tetrachloride as solvent that is both ozone-depleting and a greenhouse gas, while its use is strictly regulated. Also, radical chemistry requires a diluted reaction mixture and consequently a larger quantity of solvent must be used. Improved method should also address the issue of low atom economy of NBS. Finally, the use of radical chain initiators and heat further impairs the environmental aspect of reaction. Therefore, there is a plenty of room for the improvement of environmental potential of radical halogenation [110, 111], and this chapter will present recent advances in the field of heterocyclic molecules with halogenation at the heterocycle or at α-position on the side chain of heterocycle.

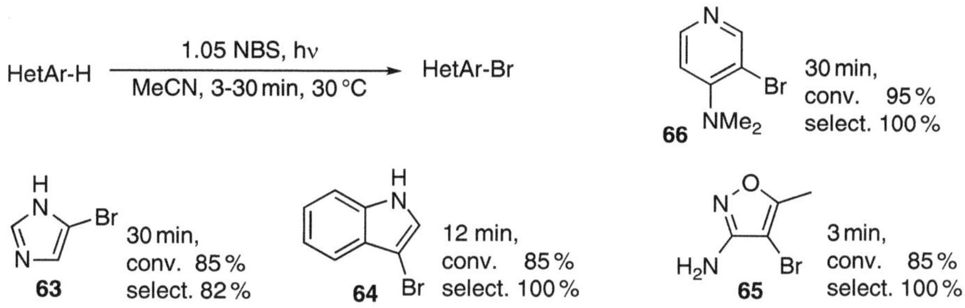

Scheme 33 Photochemical activation of electrophilic bromination of heteroarenes with NBS

4.1 Activation of Ring Halogenation by a Radical Process

NBS is the most general brominating reagent as it does not require any special handling, and the succinimide residue is easily washed by water. When polar solvents are used for reaction instead of chlorinated ones, NBS serves as an electrophilic reagent [112, 113]. As reported by the group of Waghmode, heteroaromatic molecules can be selectively and effectively brominated within minutes by a photochemical activation (mercury vapor 250 W lamp) (Scheme 33) [114]. Regioselectivity was surprisingly high and in the most cases only one regioisomer was formed, i.e., imidazole was brominated mainly at C-5 (**63**) and indole only at C-3 (**64**).

4.2 Radical Bromination at the Side Chain of Heteroarenes

Benzyl bromination is not limited to CCl_4, instead alternative solvents could also be used [110], and furthermore, water is also a very good media for benzyl bromination [61, 115–117]. Polar solvent usually triggers bromination for electrophilic reaction, nevertheless radical reaction can also be operative in this media if there is no sink for "Br^+" in the organic substrate. This was used in a study on the synthesis of halomethyl substituted 1,3-thiazoles **69** as key synthetic intermediates for biologically and pharmaceutically interesting molecules. When side-chain chlorination of 4,5-dimethyl-1,3-thiazoles **68** with NCS was conducted in acetonitrile at 60°C in the presence of radical initiator (AIBN), monochlorinated product was formed with chlorine atom on either of methyl groups. On the other hand, when no initiator was present, 4-methyl group was chlorinated selectively suggesting that radical chain mechanism might not be operative in this example (Scheme 34). Authors showed a very simple synthetic procedure to convert **67** into **69** by first preparing thiazole ring **68** from **67** in *iso*-propanol followed by halogenation in MeCN with simple aqueous work-up (Scheme 34) [118].

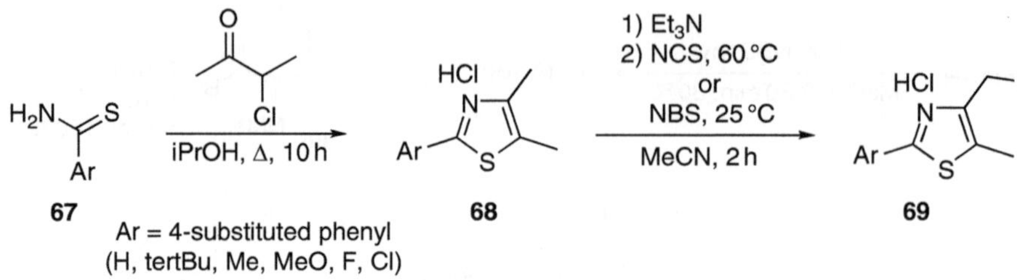

Scheme 34 Radical bromination of sp^3 C-atom in polar solvent

Scheme 35 Solid state regio- and stereo-selective benzyl bromination of diquinoline

A better alternative to various solvents is solvent-free reaction. In this experimental setup, the most important issue is to bring reactants together to react. The group of Rahman substituted CCl_4 with mechanical grinding of a mixture of diquinoline **70** and NBS (Scheme 35). Reactions started only after few hours of grinding after which the bromination finished in an hour. Product was dissolved in a small quantity of acetone and purified by flash chromatography to obtain dibrominated product **71** in good yield [119].

Instead of few hours of mechanical grinding, few minutes of microwave irradiation can be applied. This activation strategy is compatible with solid state side-chain bromination as shown by Goswami et al. [120]. Thoroughly mixed substrate and NBS were subjected to heating with domestic microwave oven (450 W) for 3–9 min (Scheme 36). The method was very effective even for strongly deactivated substrates, although a majority of heterocycles gave a mixture of mono- and dibrominated products with very low selectivity, while protected 2-amino-6-picolines **72b,c** gave also ring-brominated products **74**.

4.3 Radical Bromination in Carbohydrates

Bromination of carbohydrate derivatives is a very useful reaction in functionalization of these molecules. Acetylated β-galactopyranosyl cyanide **75** can be functionalized selectively at C-1 atom by radical bromination, and several protocols were tried to substitute classical Wohl–Ziegler conditions in CCl_4 [121]. Authors

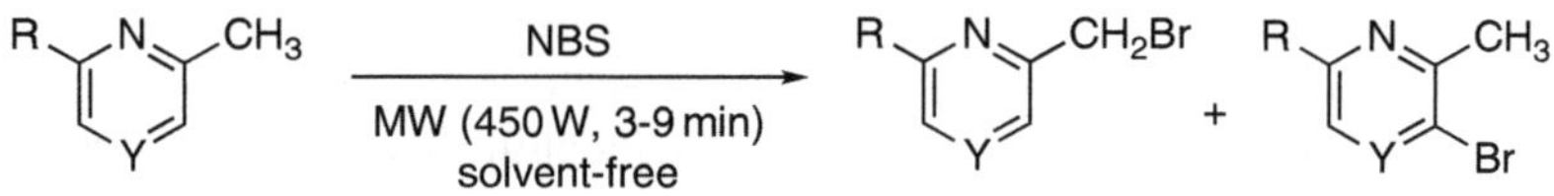

	Y	R	73	74
72a	CH	Me	40%	/
72b	CH	MeCONH	25%	70%
72c	CH	tBuCONH	30%	33%
72d	N	H	30% ($CHBr_2$)	/

Scheme 36 Side-chain bromination of heteroaromatics in solid phase under microwave irradiation

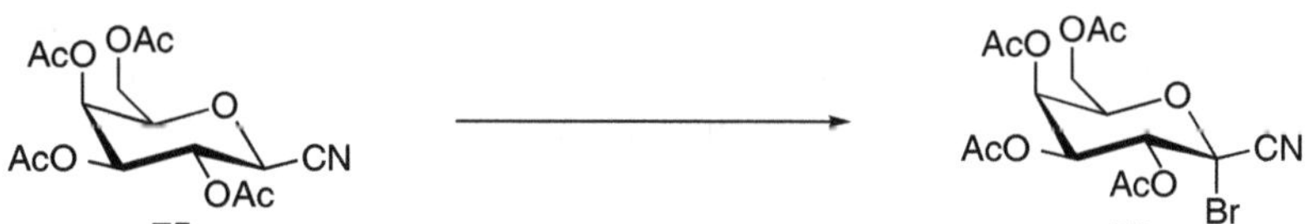

75 → **76**

Br_2, CCl_4, hν, reflux, 0.5h:	88%
Br_2, $PhCF_3$, K_2CO_3, hν, reflux, 1h:	69%
$KBrO_3/Na_2S_2O_4$, H_2O/DCM, rt, 72h:	99%
$KBrO_3/Na_2S_2O_4$, $H_2O/PhCF_3$, rt, 27h:	88%

Scheme 37 Radical bromination in carbohydrate molecules

substituted carbon tetrachloride to benzotrifluoride, and the yield of bromination was somewhat lower (Scheme 37). A second approach was reducing bromate with dithionite in a biphasic system that allows reactions to proceed at room temperature although with longer reaction time. An EtOAc-H_2O system did not yield any brominated product, while DCM-H_2O and $PhCF_3$-H_2O biphasic systems proved to be efficient reaction media for the preparation of **76**. This approach could be regarded as opposite to side-chain oxidative halogenation (see Sect. 3.4, [61]) as active bromine reagent is formed by reducing bromine(V) salt.

5 Nucleophilic Halogenation

An introduction of a halogen atom at the specific site in the heterocyclic compound is possible by a nucleophilic substitution reaction as opposed to electrophilic halogenation, where regioselectivity is more problematic. A necessity is an electron-deficient substrate with a good leaving group. The most problematic reaction among the nucleophilic halogenation from the "green" chemistry point of view is fluorination. Fluoride is a very good nucleophile; however, it interacts very strongly with water and polar solvents, which reduce markedly its nucleophilicity and makes hydroxide a competing nucleophile [122]. Therefore, harsh conditions have to be applied, and further derivatization to activate leaving group is necessary as well as addition of complexes (crown ethers, phase-transfer reagents, etc.) to desolvate fluoride. Important part of the research in nucleophilic fluorination is connected with ^{18}F isotope bound to heterocyclic cores (carbohydrate, pyridines, etc.) as tracers in positron emission tomography [123] that requires fast, efficient, and clean synthesis of fluorinated radiopharmaceuticals.

5.1 *Nucleophilic Halogenation in Protic Solvents*

Protic solvents interact with nucleophilic species making them strongly solvated and thus deactivated. Due to this reason, nucleophilic halogenation is classically performed in aprotic polar solvents such as DMSO, DMF, and MeCN with addition of crown ethers or phase-transfer agents (TBAF) to overcome solubility problems [124]. Interestingly, tertiary alcohols act as if they are not protic solvents. Even more, when used as a solvent, reactions with alkali fluorides are faster and more selective. Different reactivity of alkali fluorides in tertiary alcohols was attributed to the reduction of the strength of ionic bonding due to hydrogen bonding, while fluoride remains "flexible" and thus reactive. Furthermore, a formation of the typical by-products is also reduced [125, 126]. Polyethylene glycols are also very good alternative solvents for nucleophilic fluorination. The use of tri- and tetraethylene glycols as solvents for nucleophilic fluorination joined two effects, promotion of solubility (as with crown ethers) and chelation by two terminal OH groups for activation of fluoride. In polyethers as solvents, KF can be used as a source of fluoride [127]. A recyclable alternative is polymer-supported pentaethylene glycol [128]. Majority of work was performed on the preparation of fluoroalkyl-substituted aromatic and heteroaromatic rings [126], while an example of fluorinated heteroaromatic molecules shows the preparation of 6-[^{18}F]fluoronicotinic acid derivatives in a mixture of *tert*-BuOH and MeCN [129].

5.2 *Nucleophilic Halogenation in Ionic Liquids*

Ionic liquids are a very potent group of "green" solvents for nucleophilic halogenation. Tetrabutylammonium fluoride hydrofluorides were used in oxirane ring opening in carbohydrates with KHF_2 to yield fluorinated derivatives [130, 131]. Imidazolium-based ILs were also used in the preparation of fluorodeoxo sugars. A study on the nucleophilic fluorination of alkyl halides and mesylates with KF showed that IL significantly enhances the reactivity of potassium fluoride and reduces the formation of side products [132]. This finding was further applied in the preparation of 3′-deoxo-3′-[^{18}F]fluorotymidine **78** as an imaging agent of proliferating tumor cells [133]. Aqueous acetonitrile was needed for the extraction of ^{18}F isotope from ion exchange resin, while the reaction was conducted in IL

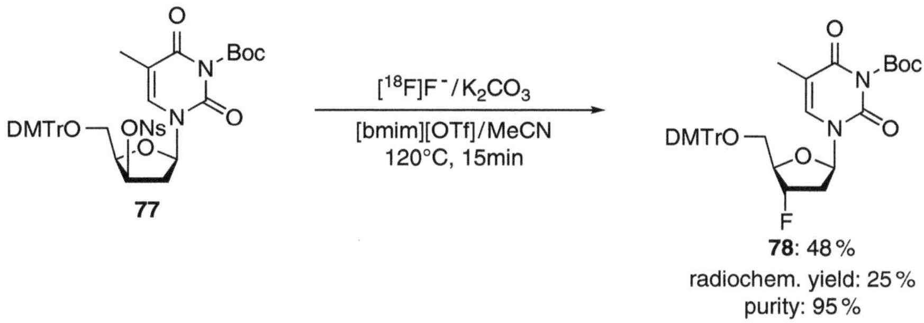

Scheme 38 Preparation of ^{18}F-labeled thymidine derivative in ionic liquid

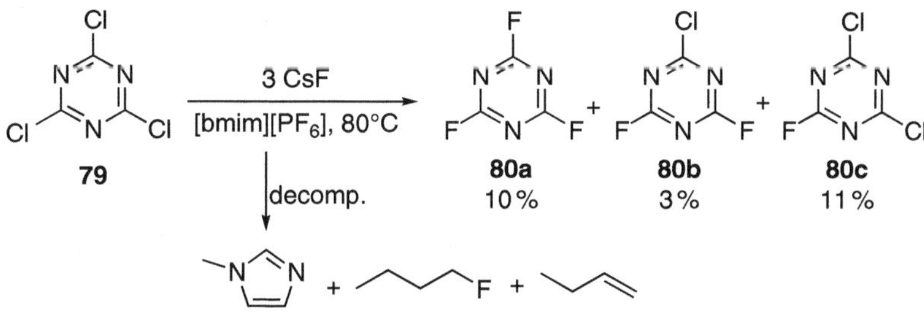

Scheme 39 Halex reaction in ionic liquid

(Scheme 38). Due to the decay of ^{18}F isotope, a short time of reaction is of crucial importance. Using IL protocol, pure **78** was obtained in 70 min including HPLC purification. The reduction of preparation time for these labeled compounds was made possible using IL as a reaction media for the fluorination step, which made time-consuming azeotropic drying step superfluous. This was discovered during the study on the preparation of 2-[^{18}F]fluoro-2-deoxoglucose in ionic liquid without an evaporating step through the help of solid phase synthesis [134].

Halex reaction (halogen exchange) in one of the basic transformation in nucleophilic halogenation and ionic liquids were also used in the preparation of fluorinated heteroaromatic molecules through this method. Activation of nucleophilic substitution reaction in IL makes possible the use of cesium fluoride without the addition of crown ethers or the use of more soluble ammonium salts (Scheme 39). Unfortunately, fluoride ion is so basic to partially decompose IL, although reaction conditions are milder due to activation of reaction by the solvent. This might pose some difficulties in using IL in halex reaction [135].

A very important feature of IL is the possibility of designing solvents for promoting specific chemical reactions. A *tert*-BuOH was found to be a very potent solvent for nucleophilic halogenation, and its structure was incorporated into the IL at the side chain of imidazolium ring to benefit from the synergy between *tert*-BuOH and IL. Thus, modified IL is able to bind alkali metal by its ionic structure and at the same time activated the reaction through interaction of alcohol group with the leaving group. Even less than equimolar amount of *tert*-BuOH-modified IL is sufficient for good conversion [136, 137]. As ionic liquid can be used in a catalytic quantity, the question of its reuse becomes very important. A solution was found in binding an IL onto the polymer beads to obtain a double gain – easier isolation of the product and easier reuse of the IL (Scheme 40). Even more, polymer-bound [hmim][BF_4] shows much higher catalytic activity than the free ionic liquid [138, 139].

5.3 Microwaves and Microreactors

Nucleophilic substitution requires high reaction temperature; therefore, activation with microwave irradiation seems as a good method for promotion of nucleophilic halogenation, especially in the light of radiolabeling with ^{18}F that requires very short reaction times. When MW were used in the synthesis of fluoropyridyl derivatives of [3,2-c]pyrazole-corticosteroids as potential glucocorticoid receptor-mediated imaging agents, through halogen exchange reaction with KF and kryptofix-222 in DMSO, enhancement of reaction rate of nucleophilic fluorination was evident [140]. In the best case, it was 6.2-fold increase in reaction rate of the synthesis of fluoropyridyl glucocorticoids. Even more evident acceleration of fluorination was observed when a hydrate of potassium fluoride was used as a reagent (Scheme 41). Under conventional heating, anhydrous conditions have to be applied to ensure effective fluorination; therefore, spray-dried KF is usually used in anhydrous DMSO. Interestingly, the presence of water does not influence the fluorination when MW heating is used. It seems that MW irradiation promotes

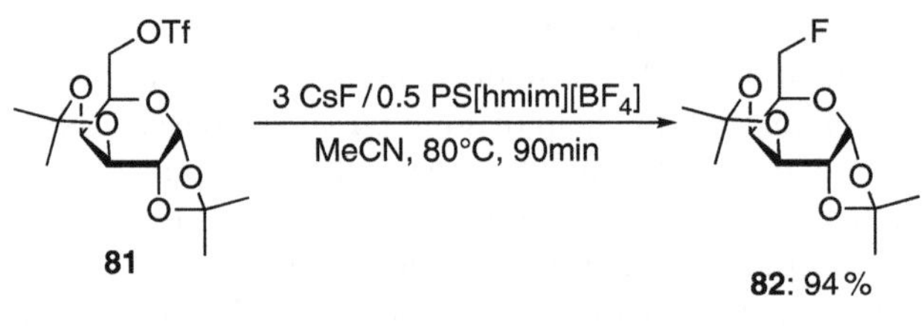

Scheme 40 Polymer-supported ionic liquids as catalysts for nucleophilic substitution

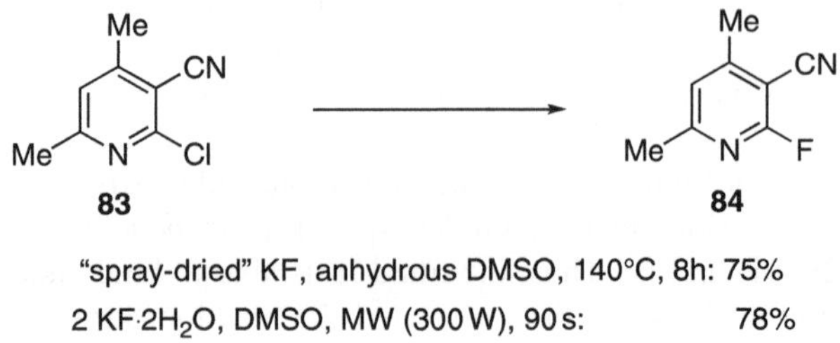

Scheme 41 Microwave-assisted nucleophilic fluorination with aqueous KF

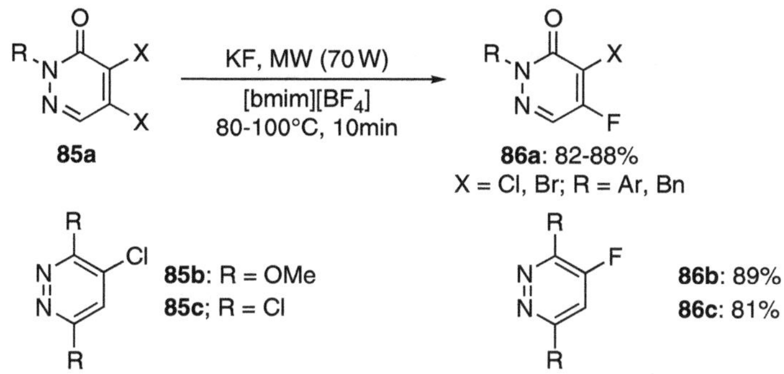

Scheme 42 Microwave activation of nucleophilic fluorination in ionic liquid

Fig. 4 ^{18}F-labeled derivative of WAY-100635

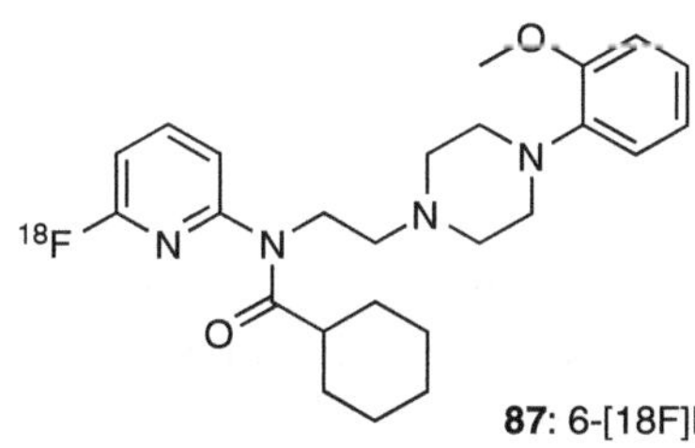

dissociation of KF and desolvation of the fluoride ion and consequently reaction proceeds as in anhydrous environment, although with much greater rate leading to efficient, cheap, and simple nucleophilic fluorination of various heterocycles [141].

A report on nucleophilic fluorination of pyridazine derivatives **85** linked ionic liquids as a reaction media with activation by microwave irradiation [142]. A synergy gave a method that is fast and efficient way to fluoropyridazine derivatives **86** (Scheme 42). Furthermore, anhydrous potassium fluoride was used without the need for its further activation through chelation, neither was the use of more soluble tetraalkylammonium fluorides necessary.

Microwave heating was also applied in the synthesis of radiofluorinated derivatives of 5-HT_{1A} receptor antagonist, WAY-100635 **87** (Fig. 4). In nucleophilic fluorination of bromo-precursor of 6-[^{18}F]fluoro-WAY-100635 **87** under conventional heating (145°C), 10 min was needed for 63% yield [143]. In the same reaction under microwave heating (100 W), 3 min was sufficient for the yield of 61%. In a fluoro-denitration reaction, the yield of **87** was 91% after 1 min of MW irradiation, although additive (K_{222}) was needed for effective fluorination and DMSO was used as a solvent.

Microreactor technology also serves as a tool to fast and efficient reactions with no problems in the scale-up. This technique was used for radiolabeling of PET tracers with ^{18}F isotope through nucleophilic fluorination, where it was applied in reaction with KF/K_{222} in acetonitrile. Another possibility of its application is diaryliodonium salts that were used as precursors of aryl fluorides [144, 145].

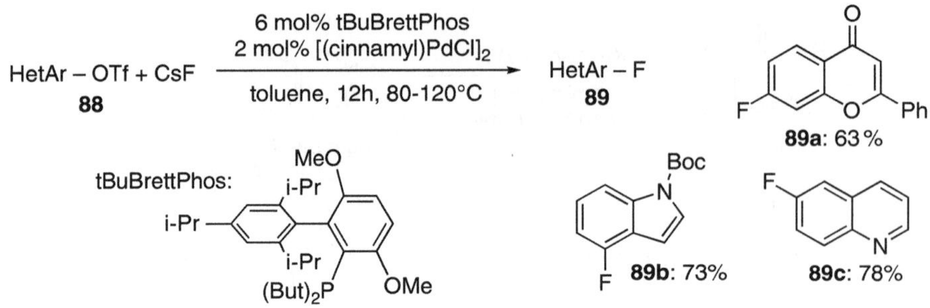

Scheme 43 Metal-catalyzed nucleophilic fluorination

5.4 Metal-Catalyzed Fluorination

Transition metal-promoted fluorination of aromatic rings with "F^{+}" reagents (Selectfluor, etc.) was achieved through the high-valent Pd, Ag, or Au complexes [146–148]. A strategy to incorporate fluoride ion by metal-catalyzed reaction would be even more interesting due to a better atom economy and a higher functional group's compatibility in nucleophilic fluorination. Although it is possible to form $L_nMAr(F)$ complexes (M is Pd or Rh, L is ligand), the problem arises at a reductive elimination step to form an Ar-F compound [149]. Finally, when sterically demanding, electron-rich ligand (tBuBrettPhos) was used for Pd(II) complex, it was possible to claim an effective method for metal-catalyzed fluorination of aryl halides or triflates by reductive elimination using CsF or AgF (Scheme 43). Therefore, this reaction opens a gate for conversion of various triflates **88** into fluorinated analogues **89** that uses CsF, AgF, and even spray-dried KF. This fluorination is regioselective, tolerates wide range of functional groups, and thus can be performed selectively into highly functionalized molecules [150]. Similar process was reported with heteroaryl–aryl complex [151].

6 Concluding Remarks

Halogenation of heterocyclic molecules is an intensively investigated subject because of the importance of the halogenated products per se and as starting compounds for further transformations. The chapter emphasizes only the research in recent 5–10 years that introduced a significant improvement to the halogenation of heterocyclic molecules in the light of creating a more sustainable chemical processes. The aim of this chapter was not to define which strategy is "greener," rather those methods and strategies that hold potential for a sustainable development are reported. Furthermore, the notion of "green" reaction is complex and it is left to the reader to resolve which of the various possibilities is more suited for a specific reaction.

We could expect that halogenation of heterocycles will evolve further, and within this development the existing "green" methods will be reevaluated and compared in the light of new data and findings. Of course, new methods and strategies for sustainable chemistry are expected to emerge that will carry the sustainable halogenation even further.

References

1. Anastas PT, Warner JC (1998) Green chemistry: theory and practice. Oxford University Press, Oxford
2. Anastas P, Horvath IT (2007) Chem Rev 107:2169–2173
3. Clark JH, Macquarrie DJ (2002) Handbook of green chemistry & technology. Blackwell Science, Oxford
4. Uozumi Y (2010) Synlett:1988–1989
5. Butler A, Sandy M (2009) Nature 460:848–854
6. Blasiak LC, Drennan CL (2009) Acc Chem Res 42:147–155
7. Fujimori DG, Walsh CT (2007) Curr Opin Chem Biol 11:553–560
8. van Pee KH, Dong CJ, Flecks S, Naismith J, Patallo EP, Wage T (2006) Adv Appl Microbiol:127–157
9. Vaillancourt FH, Yeh E, Vosburg DA, Garneau-Tsodikova S, Walsh CT (2006) Chem Rev 106:3364–3378
10. O'Hagan D (2006) J Fluorine Chem 127:1479–1483
11. Gribble GW (1998) Acc Chem Res 31:141–152
12. Shaw PD, Hager LP (1959) J Am Chem Soc 81:6527–6528
13. Winter JM, Moore BS (2009) J Biol Chem 284:18577–18581
14. Vaillancourt FH, Yeh E, Vosburg DA, O'Connor SE, Walsh CT (2005) Nature 436: 1191–1194
15. Zechel DL, Reid SP, Nashiru O, Mayer C, Stoll D, Jakeman DL, Warren RAJ, Withers SG (2001) J Am Chem Soc 123:4350–4351
16. O'Hagan D, Schaffrath C, Cobb SL, Hamilton JTG, Murphy CD (2002) Nature 416:279–279
17. O'Hagan D, Harper DB (1999) J Fluorine Chem 100:127–133
18. Dong C, Huang F, Deng H, Schaffrath C, Spencer JB, O'Hagan D, Naismith JH (2004) Nature 427:561–565
19. Weast RC (1974) Handbook of chemistry and physics. CRC, Cleveland
20. Eissen M, Lenoir D (2008) Chem Eur J 14:9830–9841
21. Takehara M, Watanabe S, Nanbu N, Ue M, Sasaki Y (2004) Synth Commun 34:1367–1375
22. Kobayashi M, Inoguchi T, Iida T, Tanioka T, Kumase H, Fukai Y (2003) J Fluorine Chem 120:105–110
23. Iskra J (2006) Lett Org Chem 3:170–175
24. Van Zee N, Dragojlovic V (2010) Chem Eur J 16:7950–7958
25. Ryu I, Matsubara H, Yasuda S, Nakamura H, Curran DP (2002) J Am Chem Soc 124: 12946–12947
26. Iskra J, Stavber S, Zupan M (2003) Chem Commun:2496–2497
27. Podgoršek A, Stavber S, Zupan M, Iskra J (2006) Eur J Org Chem:483–488
28. Windmon N, Dragojlovic V (2008) Beilstein J Org Chem 4:29
29. Alexander VM, Khandekar AC, Samant SD (2003) Synlett:1895–1897
30. Yusubov MS, Tveryakova EN, Krasnokutskaya EA, Perederyna IA, Zhdankin VV (2007) Synth Commun 37:1259–1265
31. Majumdar KC, Ghosh D, Mondal S (2011) Synthesis:599–602
32. Yadav AK, Singh BK, Singh N, Tripathi RP (2007) Tetrahedron Lett 48:6628–6632

33. Fousteris M, Chevrin C, Le Bras J, Muzart J (2006) Green Chem 8:522–523
34. Loeb P, Loewe H, Hessel V (2004) J Fluorine Chem 125:1677–1694
35. Chambers RD, Fox MA, Holling D, Nakano T, Okazoe T, Sandford G (2005) Lab Chip 5:191–198
36. Chambers RD, Holling D, Rees AJ, Sandford G (2003) J Fluorine Chem 119:81–82
37. Chambers RD, Spink RCH (1999) Chem Commun:883–884
38. Löb P, Hessel V, Klefenz H, Löwe H, Mazanek K (2005) Lett Org Chem 2:767–779
39. Stefani HA, Pereira CMP, Almeida RB, Braga RC, Guzen KP, Cella R (2005) Tetrahedron Lett 46:6833–6837
40. Jain SL, Sain B (2010) Russ Chem Rev 79:683–692
41. Cristiano R, Walls AD, Weiss RG (2010) J Phys Org Chem 23:904–909
42. Kosynkin DV, Tour JM (2001) Org Lett 3:991–992
43. Quintin M, Lewin G (2004) Tetrahedron Lett 45:3635–3638
44. Sumi Mitra S, Sreekumar K (1997) Indian J Chem B 36:133–136
45. Šket B, Zupet P, Zupan M (1989) J Chem Soc Perkin Trans 1:2279–2281
46. Pravst I, Zupan M, Stavber S (2009) Curr Org Chem 13:47–70
47. Hosseinzadeh R, Tajbakhsh M, Mohadjerani M, Lasemi Z (2009) Monatsh Chem 140:57–60
48. Chiappe C, Leandri E, Pieraccini D (2004) Chem Commun:2536–2537
49. Bortolini O, Bottai M, Chiappe C, Conte V, Pieraccini D (2002) Green Chem 4:621–627
50. Le Z-G, Chen Z-C, Hu Y, Zheng Q-G (2004) Synthesis:2809–2812
51. Borikar SP, Daniel T, Paul V (2009) Tetrahedron Lett 50:1007–1009
52. Kumar L, Sharma V, Mahajan T, Agarwal DD (2009) Org Process Res Dev 14:174–179
53. Kumar L, Mahajan T, Sharma V, Agarwal DD (2011) Ind Eng Chem Res 50:705–712
54. Garden SJ, Torres JC, Melo SCD, Lima AS, Pinto AC, Lima ELS (2001) Tetrahedron Lett 42:2089–2092
55. Tilve RD, Kanetkar VR (2005) Synth Commun 35:1313–1318
56. Sarma JARP, Nagaraju A, Majumdar KK, Samuel PM, Das I, Roy S, McGhie AJ (2000) J Chem Soc Perkin Trans 2(6):1119–1123
57. Guss CO, Rosenthal R (1955) J Am Chem Soc 77:2549–2549
58. Wu XL, Wang GW (2009) Tetrahedron 65:8802–8807
59. Wu XL, Wang GW (2007) J Org Chem 72:9398–9401
60. Podgoršek A, Stavber S, Zupan M, Iskra J (2007) Green Chem 9:1212–1218
61. Podgoršek A, Stavber S, Zupan M, Iskra J (2009) Tetrahedron 65:4429–4439
62. Stavber G, Zupan M, Jereb M, Stavber S (2004) Org Lett 6:4973–4976
63. Chang M-Y, Lee N-C, Lee M-F, Huang Y-P, Lin C-H (2010) Tetrahedron Lett 51:5900–5903
64. Seki T, Fujiwara T, Takeuchi Y (2011) J Fluorine Chem 132:181–185
65. Zhou C, Ma ZC, Gu ZH, Fu CL, Ma SM (2008) J Org Chem 73:772–774
66. Pavlinac J, Zupan M, Laali KK, Stavber S (2009) Tetrahedron 65:5625–5662
67. Pingali SRK, Madhav M, Jursic BS (2010) Tetrahedron Lett 51:1383–1385
68. Ganguly NC, De P, Dutta S (2005) Synthesis:1103–1108
69. Sloop JC, Jackson JL, Schmidt RD (2009) Heteroat Chem 20:341–345
70. Li G, Kakarla R, Gerritz SW (2007) Tetrahedron Lett 48:4595–4599
71. Paolini L, Petricci E, Corelli F, Botta M (2003) Synthesis:1039–1042
72. Arsenyan P, Paegle E, Belyakov S (2010) Tetrahedron Lett 51:205–208
73. Podgoršek A, Zupan M, Iskra J (2009) Angew Chem Int Ed 48:8424–8450
74. Wittcoff HA, Reuben BG, Plotkin JS (2004) Industrial organic chemicals. Wiley, Hoboken
75. Tada N, Ban K, Hirashima S-I, Miura T, Itoh A (2010) Org Biomol Chem 8:4701–4704
76. Nobuta T, Hirashima S-I, Tada N, Miura T, Itoh A (2010) Tetrahedron Lett 51:4576–4578
77. Nobuta T, Hirashima S-I, Tada N, Miura T, Itoh A (2010) Synlett:2335–2339
78. Gavara L, Boisse T, Rigo B, Hénichart JP (2008) Tetrahedron 64:4999–5004
79. Liu H, Pan Y, Tan CH (2008) Tetrahedron Lett 49:4424–4426
80. Branytska OV, Neumann R (2003) J Org Chem 68:9510–9512
81. Chen X, Hao XS, Goodhue CE, Yu JQ (2006) J Am Chem Soc 128:6790–6791

82. Khan AT, Goswami P (2005) Tetrahedron Lett 46:4937–4940
83. Bora U, Bose G, Chaudhuri MK, Dhar SS, Gopinath R, Khan AT, Patel BK (2000) Org Lett 2:247–249
84. Greb M, Hartung J, Kohler F, Spehar K, Kluge R, Csuk R (2004) Eur J Org Chem:3799–3812
85. Sels B, De Vos D, Buntinx M, Pierard F, Kirsch-De Mesmaeker A, Jacobs P (1999) Nature 400:855–857
86. Sels B, Levecque P, Brosius R, De Vos D, Jacobs P, Gammon DW, Kinfe HH (2005) Adv Synth Catal 347:93–104
87. Drake MD, Bateman MA, Detty MR (2003) Organometallics 22:4158–4162
88. Goodman MA, Detty MR (2004) Organometallics 23:3016–3020
89. Abe M, You Y, Detty MR (2002) Organometallics 21:4546–4551
90. Bennett SM, Tang Y, McMaster D, Bright FV, Detty MR (2008) J Org Chem 73:6849–6852
91. Iskra J, Stavber S, Zupan M (2004) Synthesis:1869–1873
92. Jereb M, Zupan M, Stavber S (2005) Green Chem 7:100–104
93. Barluenga J, Marco-Arias M, Gonzalez-Bobes F, Ballesteros A, Gonzalez JM (2004) Chem Eur J 10:1677–1682
94. Gammon DW, Kinfe HH, De Vos DE, Jacobs PA, Sels BF (2004) Tetrahedron Lett 45:9533–9536
95. Bogdal D, Lukasiewicz M, Pielichowski J (2004) Green Chem 6:110–113
96. Kim MM, Ruck RT, Zhao D, Huffman MA (2008) Tetrahedron Lett 49:4026–4028
97. Bhilare SV, Deorukhkar AR, Darvatkar NB, Salunkhe MM (2008) Synth Commun 38:2881–2888
98. Jiang H, Huang H, Cao H, Qi C (2010) Org Lett 12:5561–5563
99. Matthews MA (2001) Pure Appl Chem 73:1305–1308
100. Kreysa G, Jüttner K (2007) Chem Eng 114:50–55
101. Midorikawa K, Suga S, Yoshida JI (2006) Chem Commun:3794–3796
102. Kakiuchi F, Kochi T, Mutsutani H, Kobayashi N, Urano S, Sato M, Nishiyama S, Tanabe T (2009) J Am Chem Soc 131:11310–11311
103. Yin B, Wang L, Inagi S, Fuchigami T (2010) Tetrahedron 66:6820–6825
104. Fuchigami T, Tajima T (2005) J Fluorine Chem 126:181–187
105. Fuchigami T (2007) J Fluorine Chem 128:311–316
106. Hasegawa M, Ishii H, Fuchigami T (2003) Green Chem 5:512–515
107. Hasegawa M, Fuchigami T (2004) Electrochim Acta 49:3367–3372
108. Hasegawa M, Ishii H, Cao Y, Fuchigami T (2006) J Electrochem Soc 153:D162–D166
109. Smith MB, March J (2001) March's advanced organic chemistry: reactions, mechanisms and structure, 5th edn. Wiley, New York, pp 911–914
110. Zupan M, Iskra J (2009) "Greening" of free radical benzylic bromination. In: Pearlman JT (ed) Green chemistry research trends. Nova Publishers, New York, pp 243–264
111. Tanko JM, Blackert JF (1994) Science 263:203–205
112. Mitchell RH, Lai YH, Williams RV (1979) J Org Chem 44:4733–4735
113. Carreno MC, Ruano JLG, Sanz G, Toledo MA, Urbano A (1995) J Org Chem 60:5328–5331
114. Chhattise PK, Ramaswamy AV, Waghmode SB (2008) Tetrahedron Lett 49:189–194
115. Podgoršek A, Stavber S, Zupan M, Iskra J (2006) Tetrahedron Lett 47:7245–7247
116. Podgoršek A, Stavber S, Zupan M, Iskra J (2006) Tetrahedron Lett 47:1097–1099
117. Shaw H, Perlmutter HD, Gu C, Arco SD, Quibuyen TO (1997) J Org Chem 62:236–237
118. Yamane T, Mitsudera H, Shundoh T (2004) Tetrahedron Lett 45:69–73
119. Rahman ANMM, Bishop R, Tan R, Shan N (2005) Green Chem 7:207–209
120. Goswami S, Dey S, Jana S, Adak AK (2004) Chem Lett 33:916–917
121. Czifrak K, Somsak L (2002) Tetrahedron Lett 43:8849–8852
122. Furuya T, Klein JEMN, Ritter T (2010) Synthesis:1804–1821
123. Le Bars D (2006) J Fluorine Chem 127:1488–1493
124. Kuhnast B, de Bruin B, Hinnen F, Tavitian B, Dollé F (2004) Bioconjugate Chem 15: 617–627

125. Kim DW, Jeong HJ, Lim ST, Sohn M-H, Katzenellenbogen JA, Chi DY (2008) J Org Chem 73:957–962
126. Sachin K, Jeong H-J, Lim ST, Sohn M-H, Chi DY, Kim DW (2011) Tetrahedron 67: 1763–1767
127. Lee JW, Yan H, Jang HB, Kim HK, Park S-W, Lee S, Chi DY, Song CE (2009) Angew Chem Int Ed 48:7683–7686
128. Jadhav VH, Jang SH, Jeong H-J, Lim ST, Sohn M-H, Chi DY, Kim DW (2010) Org Lett 12:3740–3743
129. Olberg DE, Arukwe JM, Grace D, Hjelstuen OK, Solbakken M, Kindberg GM, Cuthbertson A (2010) J Med Chem 53:1732–1740
130. Mastihubová M, Biely P (2001) Tetrahedron Lett 42:9065–9067
131. Honda T, Masuda T, Yoshida S, Arai M, Kobayashi Y, Yamashita M (2002) Bioorg Med Chem Lett 12:1921–1924
132. Kim DW, Song CE, Chi DY (2002) J Am Chem Soc 124:10278–10279
133. Moon BS, Lee KC, An GI, Chi DY, Yang SD, Choi CW, Lim SM, Chun KS (2006) J Labelled Compd Radiopharm 49:287–293
134. Kim HW, Jeong JM, Lee Y-S, Chi DY, Chung K-H, Lee DS, Chung J-K, Lee MC (2004) Appl Radiat Isot 61:1241–1246
135. Murray CB, Sandford G, Korn SR (2003) J Fluorine Chem 123:81–84
136. Shinde SS, Chi HM, Lee BS, Chi DY (2009) Tetrahedron Lett 50:6654–6657
137. Oh Y-H, Jang HB, Im S, Song MJ, Kim S-Y, Park S-W, Chi DY, Song CE, Lee S (2011) Org Biomol Chem 9:418–422
138. Kim DW, Chi DY (2004) Angew Chem Int Ed 43:483–485
139. Kim DW, Hong DJ, Jang KS, Chi DY (2006) Adv Synth Catal 348:1719–1727
140. Kahn MGC, Konde E, Dossou F, Labaree DC, Hochberg RB, Hoyte RM (2006) Bioorg Med Chem Lett 16:3454–3458
141. Shestopalov AM, Fedorov AE, Rodinovskaya LA, Shestopalov AA, Gakh AA (2009) Tetrahedron Lett 50:5257–5259
142. Sun L, Pei W (2007) Chin J Chem 25:1005–1007
143. Karramkam M, Hinnen F, Berrehouma M, Hlavacek C, Vaufrey F, Halldin C, McCarron JA, Pike VW, Dollé F (2003) Biorg Med Chem 11:2769–2782
144. Lu S, Chun JH, Pike VW (2010) J Labelled Compd Radiopharm 53:234–238
145. Chun J-H, Lu S, Lee Y-S, Pike VW (2010) J Org Chem 75:3332–3338
146. Gouverneur V (2009) Science 325:1630–1631
147. Vigalok A, Kaspi AW (2010) In: Vigalok A (ed) C-X bond formation (Topics in Organometallic Chemistry), vol 31. Springer, Berlin, pp 19–38
148. Engle KM, Mei T-S, Wang X, Yu J-Q (2011) Angew Chem Int Ed 50:1478–1491
149. Grushin VV (2002) Chem Eur J 8:1007–1014
150. Watson DA, Su M, Teverovskiy G, Zhang Y, García-Fortanet J, Kinzel T, Buchwald SL (2009) Science 325:1661–1664
151. Withbroe GJ, Singer RA, Sieser JE (2008) Org Process Res Dev 12:480–489

Index

C

MIX
Papier aus verantwortungsvollen Quellen
Paper from responsible sources
FSC® C105338

If you have any concerns about our products,
you can contact us on
ProductSafety@springernature.com

In case Publisher is established outside the EU,
the EU authorized representative is:
Springer Nature Customer Service Center GmbH
Europaplatz 3, 69115 Heidelberg, Germany

Printed by Libri Plureos GmbH
in Hamburg, Germany